SUZUKI GS & GSX 250, 400 & 450 Twins Owners Workshop Manual

by Chris Rogers
with an additional chapter on the UK GSX250/400 EZ and US GS450 models
by Pete Shoemark

Models covered

GS250 T. 249cc. Introduced UK and US 1980
GSX250 E. 249cc. Introduced UK only January 1980
GSX400 E. 399cc. Introduced UK only August 1980
GSX400 T. 399cc. Introduced UK only June 1981
GS450 E. 448cc. Introduced US only September 1979
GS450 L. 448cc. Introduced US only February 1980
GS450 S. 448cc. Introduced US only November 1979
GS450 T. 448cc. Introduced US only September 1980
GS450 TX. 448cc. Introduced US only September 1981

ISBN 978 1 85010 253 3

© J H Haynes & Co. Ltd. 1990

All rights reserved. No part of this book may be reproduced or transmitted in any form or by any means, electronic or mechanical, including photocopying, recording or by any information storage or retrieval system, without permission in writing from the copyright holder.

(736-7P6)

J H Haynes & Co. Ltd.
Haynes North America, Inc

www.haynes.com

British Library Cataloguing in Publication Data
Rogers, Chris Suzuki GS/GSX 250, 400 & 450 twins.— (Owners workshop manual) 1. Suzuki motorcycle I. Title   II. Shoemark, Pete   III. Series 629.28'775   TL448.S8 ISBN 1-85010-253-8
Library of Congress Catalog Card Number
86-80033

Acknowledgements

Our thanks are due to Fran Ridewood & Co. Wells, Somerset, who supplied the machine featured in the photographs throughout this Manual and who was also kind enough to allow the Author full access to the various model types present on his showroom floor whilst supplying any necessary technical information and advice.

Our thanks are also due to Heron Suzuki (GB) Ltd who gave permission to use their line drawings and to the technical department of Heron Suzuki (GB) Ltd., who supplied service information for the machines covered in this Manual.

The GSX250 EZ model featured on the front cover was supplied by Jim Bartlet of Castle Cary, Somerset.

Finally, we would like to thank the Avon Rubber Company, who kindly supplied information and technical assistance on tyre fitting; NGK Spark Plugs (UK) Ltd for information on sparking plug maintenance and electrode conditions, and Renold Ltd for advice on chain care and renewal.

About this manual

The purpose of this manual is to present the owner with a concise and graphic guide which will enable him to tackle any operation from basic routine maintenance to a major overhaul. It has been assumed that any work would be undertaken without the luxury of a well-equipped workshop and a range of manufacturer's service tools.

To this end, the machine featured in the manual was stripped and rebuilt in our own workshop, by a team comprising a mechanic, a photographer and the author. The resulting photographic sequence depicts events as they took place, the hands shown being those of the author and the mechanic.

The use of specialised, and expensive, service tools was avoided unless their use was considered to be essential due to risk of breakage or injury. There is usually some way of improvising a method of removing a stubborn component, provided that a suitable degree of care is exercised.

The author learnt his motorcycle mechanics over a number of years, faced with the same difficulties and using similar facilities to those encountered by most owners. It is hoped that this practical experience can be passed on through the pages of this manual.

Where possible, a well-used example of the machine is chosen for the workshop project, as this highlights any areas which might be particularly prone to giving rise to problems. In this way, any such difficulties are encountered and resolved before the text is written, and the techniques used to deal with them can be incorporated in the relevant sections. Armed with a working knowledge of the machine, the author undertakes a considerable amount of research in order that the maximum amount of data can be included in this manual.

Each Chapter is divided into numbered sections. Within these sections are numbered paragraphs. Cross reference throughout the manual is quite straightforward and logical. When reference is made 'See Section 6.10' it means Section 6, paragraph 10 in the same Chapter. If another Chapter were intended the reference would read, for example, 'See Chapter 2, Section 6.10'. All the photographs are captioned with a section/paragraph number to which they refer and are relevant to the Chapter text adjacent.

Figures (usually line illustrations) appear in a logical but numerical order, within a given Chapter. Fig. 1.1 therefore refers to the first figure in Chapter 1.

Left-hand and right-hand descriptions of the machines and their components refer to the left and right of a given machine when the rider is seated normally.

Whilst every care is taken to ensure that the information in this manual is correct no liability can be accepted by the author or publishers for loss, damage or injury caused by any errors in or omissions from the information given.

Motorcycle manufacturers continually make changes to specifications and recommendations, and these, when notified, are incorporated into our manuals at the earliest opportunity.

Contents

The 1981 Suzuki GS250 T model

The 1981 Suzuki GSX400 E model

Engine unit of the 1981 Suzuki GS250 T

Introduction to the Suzuki GS and GSX 250 and 400 models

Although the Suzuki Motor Company Limited commenced manufacturing motorcycles as early as 1936, it was not until 1963 that their machines were first imported into the UK. The first of the twin cylinder models, the T10, became available during 1964, and it was immediately obvious that this model would be well-received by holders of a provisional driving licence, who are restricted to an engine capacity of 250 cc. Its popularity was mostly due to its impressive performance, in terms of both speed and fuel economy, and its advanced specification, including an electric starter and a hydraulic rear brake.

The number of models available from Suzuki has increased steadily; in 1971, one could choose a model ranging in capacity from 50 cc to 750 cc; in the last two or three years, in common with the other large Japanese companies, their range of models has increased quite dramatically. There is now an even larger choice of machines available, with engine capacities still starting at 50 cc but now continuing on up to 1100 cc, and encompassing machines designed for such varying roles as Grand Prix road racing, all forms of off-road competition, and long-distance road travel. With such a variety of machine designations to cater for, heavy demands were placed upon the Suzuki engineers to create the different engine types and configurations necessary for the machines' specific intended purpose. This they have done with typical Japanese alacrity to produce a range of engines including water-cooled, square-four cylinder two-strokes, single-cylinder two-strokes, and four-stroke one, two and four-cylinder types.

The 250 cc and 400 cc model types covered in this Manual are all based around a similar engine design, a sophisticated dohc, four valves per cylinder, parallel twin, and all have similar cycle components with only one exception, that of frame front downtube design.

Suzuki supply these machines in two basic styles, that of a sporting roadster and of a US custom design, the only differences being in the styling of the fuel tank, seat and sidepanel units with variations in construction of the exhaust systems and the fitting of either cast alloy or wire-spoked wheels.

It will be found when using this Manual, that wherever necessary, reference is made to the various differences in equipment between each model type. Each model type is identified by letters which appear each side of the machine capacity rating. The first of the two suffix letters refers to the model type, that is E for the standard roadster version and T for the Custom model. The second suffix letter refers to the model year, that is T for the 1980 models and X for the 1981 models. With the 250 cc machines, the prefix letters differ with the model type, that is a GS model is of the Custom design whereas a GSX model is of the standard roadster design. This ruling does not however apply to the 400 cc models, where the GSX prefix covers both model designs thereby avoiding confusion with the earlier two valves per cylinder 400 cc models.

All versions of the 250 cc and 400 cc models are available in the UK, the first GSX250 being introduced in January of 1980 and being followed in August of the same year by both the GS250 and GSX400 models. At the time of writing this Manual, only the GS250 model had been introduced to the US market, making its appearance in March of 1980.

Dimensions and weights

	GSX250 ET	GSX250 EX	GS250 TT,TX	GSX400 ET,EX	GS400 TT,TX
Overall length	2075 mm (81.7 in)	2060 mm (81.1 in)	2075 mm (81.7 in)	2075 mm (81.7 in)	2100 mm (82.7 in)
Overall width	885 mm (34.8 in)	755 mm (29.7 in)	885 mm (34.8 in)	755 mm (29.7 in)	845 mm (33.3 in)
Overall height	1155 mm (45.5 in)	1090 mm (42.9 in)	1155 mm (45.5 in)	1075 mm (42.3 in)	1145 mm (45.1 in)
Wheelbase	1370 mm (53.9 in)	1360 mm (53.3 in)	1370 mm (53.9 in)	1385 mm (54.5 in)	1375 mm (54.1 in)
Ground clearance	140 mm (5.5 in)	155 mm (6.1 in)	140 mm (5.5 in)	155 mm (6.1 in)	140 mm (5.5 in)
Dry weight	160 kg (353 lb)	160 kg (353 lb)	158 kg (347 lb)	175 kg (385 lb)	174 kg (382 lb)

Note: Slight variations in some figures may be found, depending on the country or state of original delivery and year of manufacture.

Ordering spare parts

When ordering spare parts for any Suzuki, it is advisable to deal direct with an official Suzuki agent who should be able to supply most of the parts ex-stock. Parts cannot be obtained from Suzuki direct and all orders must be routed via an approved Agent even if the parts required are not held in stock. Always quote the engine and frame numbers in full, especially if parts are required for earlier models.

The frame and engine numbers are stamped on a Manufacturer's Plate which is attached to the steering head on the left-hand side. The frame number is stamped on the frame itself on the right-hand side of the steering head. The engine number is stamped on the upper crankcase.

Use only genuine Suzuki spares. Some pattern parts are available that are made in Japan and may be packed in similar looking packages. They should only be used if genuine parts are hard to obtain or in an emergency, for they do not normally last as long as genuine parts, even though there may be a price advantage.

Some of the more expendable parts such as sparking plugs, bulbs, tyres, oils and greases etc., can be obtained from accessory shops and motor factors, who have convenient opening hours and can often be found not far from home. It is also possible to obtain parts on a Mail Order basis from a number of specialists who advertise regularly in the motorcycle magazines.

Location of frame number

Location of engine number

Safety first!

Professional motor mechanics are trained in safe working procedures. However enthusiastic you may be about getting on with the job in hand, do take the time to ensure that your safety is not put at risk. A moment's lack of attention can result in an accident, as can failure to observe certain elementary precautions.

There will always be new ways of having accidents, and the following points do not pretend to be a comprehensive list of all dangers; they are intended rather to make you aware of the risks and to encourage a safety-conscious approach to all work you carry out on your vehicle.

Essential DOs and DON'Ts

DON'T start the engine without first ascertaining that the transmission is in neutral.

DON'T suddenly remove the filler cap from a hot cooling system – cover it with a cloth and release the pressure gradually first, or you may get scalded by escaping coolant.

DON'T attempt to drain oil until you are sure it has cooled sufficiently to avoid scalding you.

DON'T grasp any part of the engine, exhaust or silencer without first ascertaining that it is sufficiently cool to avoid burning you.

DON'T allow brake fluid or antifreeze to contact the machine's paintwork or plastic components.

DON'T syphon toxic liquids such as fuel, brake fluid or antifreeze by mouth, or allow them to remain on your skin.

DON'T inhale dust – it may be injurious to health (see *Asbestos* heading).

DON'T allow any spilt oil or grease to remain on the floor – wipe it up straight away, before someone slips on it.

DON'T use ill-fitting spanners or other tools which may slip and cause injury.

DON'T attempt to lift a heavy component which may be beyond your capability – get assistance.

DON'T rush to finish a job, or take unverified short cuts.

DON'T allow children or animals in or around an unattended vehicle.

DON'T inflate a tyre to a pressure above the recommended maximum. Apart from overstressing the carcase and wheel rim, in extreme cases the tyre may blow off forcibly.

DO ensure that the machine is supported securely at all times. This is especially important when the machine is blocked up to aid wheel or fork removal.

DO take care when attempting to slacken a stubborn nut or bolt. It is generally better to pull on a spanner, rather than push, so that if slippage occurs you fall away from the machine rather than on to it.

DO wear eye protection when using power tools such as drill, sander, bench grinder etc.

DO use a barrier cream on your hands prior to undertaking dirty jobs – it will protect your skin from infection as well as making the dirt easier to remove afterwards; but make sure your hands aren't left slippery. Note that long-term contact with used engine oil can be a health hazard.

DO keep loose clothing (cuffs, tie etc) and long hair well out of the way of moving mechanical parts.

DO remove rings, wristwatch etc, before working on the vehicle – especially the electrical system.

DO keep your work area tidy – it is only too easy to fall over articles left lying around.

DO exercise caution when compressing springs for removal or installation. Ensure that the tension is applied and released in a controlled manner, using suitable tools which preclude the possibility of the spring escaping violently.

DO ensure that any lifting tackle used has a safe working load rating adequate for the job.

DO get someone to check periodically that all is well, when working alone on the vehicle.

DO carry out work in a logical sequence and check that everything is correctly assembled and tightened afterwards.

DO remember that your vehicle's safety affects that of yourself and others. If in doubt on any point, get specialist advice.

IF, in spite of following these precautions, you are unfortunate enough to injure yourself, seek medical attention as soon as possible.

Asbestos

Certain friction, insulating, sealing, and other products – such as brake linings, clutch linings, gaskets, etc – contain asbestos. *Extreme care must be taken to avoid inhalation of dust from such products since it is hazardous to health.* If in doubt, assume that they *do* contain asbestos.

Fire

Remember at all times that petrol (gasoline) is highly flammable. Never smoke, or have any kind of naked flame around, when working on the vehicle. But the risk does not end there – a spark caused by an electrical short-circuit, by two metal surfaces contacting each other, by careless use of tools, or even by static electricity built up in your body under certain conditions, can ignite petrol vapour, which in a confined space is highly explosive.

Always disconnect the battery earth (ground) terminal before working on any part of the fuel or electrical system, and never risk spilling fuel on to a hot engine or exhaust.

It is recommended that a fire extinguisher of a type suitable for fuel and electrical fires is kept handy in the garage or workplace at all times. Never try to extinguish a fuel or electrical fire with water.

Note: *Any reference to a 'torch' appearing in this manual should always be taken to mean a hand-held battery-operated electric lamp or flashlight. It does* **not** *mean a welding/gas torch or blowlamp.*

Fumes

Certain fumes are highly toxic and can quickly cause unconsciousness and even death if inhaled to any extent. Petrol (gasoline) vapour comes into this category, as do the vapours from certain solvents such as trichloroethylene. Any draining or pouring of such volatile fluids should be done in a well ventilated area.

When using cleaning fluids and solvents, read the instructions carefully. Never use materials from unmarked containers – they may give off poisonous vapours.

Never run the engine of a motor vehicle in an enclosed space such as a garage. Exhaust fumes contain carbon monoxide which is extremely poisonous; if you need to run the engine, always do so in the open air or at least have the rear of the vehicle outside the workplace.

The battery

Never cause a spark, or allow a naked light, near the vehicle's battery. It will normally be giving off a certain amount of hydrogen gas, which is highly explosive.

Always disconnect the battery earth (ground) terminal before working on the fuel or electrical systems.

If possible, loosen the filler plugs or cover when charging the battery from an external source. Do not charge at an excessive rate or the battery may burst.

Take care when topping up and when carrying the battery. The acid electrolyte, even when diluted, is very corrosive and should not be allowed to contact the eyes or skin.

If you ever need to prepare electrolyte yourself, always add the acid slowly to the water, and never the other way round. Protect against splashes by wearing rubber gloves and goggles.

Mains electricity and electrical equipment

When using an electric power tool, inspection light etc, always ensure that the appliance is correctly connected to its plug and that, where necessary, it is properly earthed (grounded). Do not use such appliances in damp conditions and, again, beware of creating a spark or applying excessive heat in the vicinity of fuel or fuel vapour. Also ensure that the appliances meet the relevant national safety standards.

Ignition HT voltage

A severe electric shock can result from touching certain parts of the ignition system, such as the HT leads, when the engine is running or being cranked, particularly if components are damp or the insulation is defective. Where an electronic ignition system is fitted, the HT voltage is much higher and could prove fatal.

Working conditions and tools

When a major overhaul is contemplated, it is important that a clean, well-lit working space is available, equipped with a workbench and vice, and with space for laying out or storing the dismantled assemblies in an orderly manner where they are unlikely to be disturbed. The use of a good workshop will give the satisfaction of work done in comfort and without haste, where there is little chance of the machine being dismantled and reassembled in anything other than clean surroundings. Unfortunately, these ideal working conditions are not always practicable and under these latter circumstances when improvisation is called for, extra care and time will be needed.

The other essential requirement is a comprehensive set of good quality tools. Quality is of prime importance since cheap tools will prove expensive in the long run if they slip or break when in use, causing personal injury or expensive damage to the component being worked on. A good quality tool will last a long time, and more than justify the cost.

For practically all tools, a tool factor is the best source since he will have a very comprehensive range compared with the average garage or accessory shop. Having said that, accessory shops often offer excellent quality tools at discount prices, so it pays to shop around. There are plenty of tools around at reasonable prices, but always aim to purchase items which meet the relevant national safety standards. If in doubt, seek the advice of the shop proprietor or manager before making a purchase.

The basis of any tool kit is a set of open-ended spanners, which can be used on almost any part of the machine to which there is reasonable access. A set of ring spanners makes a useful addition, since they can be used on nuts that are very tight or where access is restricted. Where the cost has to be kept within reasonable bounds, a compromise can be effected with a set of combination spanners – open-ended at one end and having a ring of the same size on the other end. Socket spanners may also be considered a good investment, a basic $3/8$ in or $1/2$ in drive kit comprising a ratchet handle and a small number of socket heads, if money is limited. Additional sockets can be purchased, as and when they are required. Provided they are slim in profile, sockets will reach nuts or bolts that are deeply recessed. When purchasing spanners of any kind, make sure the correct size standard is purchased. Almost all machines manufactured outside the UK and the USA have metric nuts and bolts, whilst those produced in Britain have BSF or BSW sizes. The standard used in USA is AF, which is also found on some of the later British machines. Others tools that should be included in the kit are a range of crosshead screwdrivers, a pair of pliers and a hammer.

When considering the purchase of tools, it should be remembered that by carrying out the work oneself, a large proportion of the normal repair cost, made up by labour charges, will be saved. The economy made on even a minor overhaul will go a long way towards the improvement of a toolkit.

In addition to the basic tool kit, certain additional tools can prove invaluable when they are close to hand, to help speed up a multitude of repetitive jobs. For example, an impact screwdriver will ease the removal of screws that have been tightened by a similar tool, during assembly, without a risk of damaging the screw heads. And, of course, it can be used again to retighten the screws, to ensure an oil or airtight seal results. Circlip pliers have their uses too, since gear pinions, shafts and similar components are frequently retained by circlips that are not too easily displaced by a screwdriver. There are two types of circlip pliers, one for internal and one for external circlips. They may also have straight or right-angled jaws.

One of the most useful of all tools is the torque wrench, a form of spanner that can be adjusted to slip when a measured amount of force is applied to any bolt or nut. Torque wrench settings are given in almost every modern workshop or service manual, where the extent to which a complex component, such as a cylinder head, can be tightened without fear of distortion or leakage. The tightening of bearing caps is yet another example. Overtightening will stretch or even break bolts, necessitating extra work to extract the broken portions.

As may be expected, the more sophisticated the machine, the greater is the number of tools likely to be required if it is to be kept in first class condition by the home mechanic. Unfortunately there are certain jobs which cannot be accomplished successfully without the correct equipment and although there is invariably a specialist who will undertake the work for a fee, the home mechanic will have to dig more deeply in his pocket for the purchase of similar equipment if he does not wish to employ the services of others. Here a word of caution is necessary, since some of these jobs are best left to the expert. Although an electrical multimeter of the AVO type will prove helpful in tracing electrical faults, in inexperienced hands it may irrevocably damage some of the electrical components if a test current is passed through them in the wrong direction. This can apply to the synchronisation of twin or multiple carburettors too, where a certain amount of expertise is needed when setting them up with vacuum gauges. These are, however, exceptions. Some instruments, such as a strobe lamp, are virtually essential when checking the timing of a machine powered by CDI ignition system. In short, do not purchase any of these special items unless you have the experience to use them correctly.

Although this manual shows how components can be removed and replaced without the use of special service tools (unless absolutely essential), it is worthwhile giving consideration to the purchase of the more commonly used tools if the machine is regarded as a long term purchase Whilst the alternative methods suggested will remove and replace parts without risk of damage, the use of the special tools recommended and sold by the manufacturer will invariably save time.

SUZUKI GS/GSX 250 & 400

Check list

Weekly or every 300 miles (500 km)

1. Check tyre pressures and tyre condition
2. Safety check
3. Legal check
4. Check the engine/transmission oil level
5. Lubricate the final drive chain
6. Check the level of electrolyte in the battery
7. Check the prop stand spring and pivot for correct operation
8. Inspect the centre stand pivots and spring for security
9. Check the level of hydraulic fluid in the front brake master cylinder
10. Check the operation of the throttle twistgrip and cable

Monthly or every 600 miles (1000 km)

1. Check the tension of the wheel spokes
2. Inspect the cast alloy wheels for cracks or chipping
3. Adjust the final drive chain
4. Check the hydraulic brake components for signs of leakage

Four monthly or every 2000 miles (3000 km)

1. Clean, lubricate and adjust the final drive chain
2. Lubricate control cables
3. Lubricate pivots and any exposed linkages
4. Remove and clean the air filter

Six monthly or every 3000 miles (5000 km)

1. Clean and adjust the sparking plug
2. Check and if necessary adjust the ignition timing
3. Check and adjust the valve clearances
4. Change the engine/transmission oil and renew the oil filter
5. Check the engine/transmission oil pressure
6. Synchronise and adjust the carburettor
7. Adjust the throttle cable
8. Check and adjust the clutch
9. Lubricate the rear brake pedal operating mechanism
10. Check and adjust steering head bearing play

Annually or every 6000 miles (10 000 km)

1. Remove and clean the oil strainer
2. Renew the sparking plug
3. Lubricate and examine the speedometer and tachometer drive cables
4. Change the front brake hydraulic fluid
5. Change the front fork oil

Two yearly or every 12 000 miles (20 000 km)

1. Renew the fuel lines
2. Renew the hydraulic brake hose
3. Grease the swinging arm bearings
4. Inspect and grease the steering head bearings

General adjustments and examination

1. Examine the condition of the front brake pads and renew if necessary
2. Examine the condition of the rear brake shoes and adjust or renew if necessary

Adjustment data

Tyre pressures	GS models	GSX400T models	All other models
Front			
Normal riding:			
Solo	21 psi (1.50 kg/cm²)	21 psi (1.50 kg/cm²)	25 psi (1.75 kg/cm²)
Dual	25 psi (1.75 kg/cm²)	25 psi (1.75 kg/cm²)	25 psi (1.75 kg/cm²)
Continuous high speed riding:			
Solo	25 psi (1.75 kg/cm²)	25 psi (1.75 kg/cm²)	28 psi (2.00 kg/cm²)
Dual	28 psi (2.00 kg/cm²)	28 psi (2.00 kg/cm²)	32 psi (2.25 kg/cm²)
Rear			
Normal riding:			
Solo	25 psi (1.75 kg/cm²)	25 psi (1.75 kg/cm²)	28 psi (2.00 kg/cm²)
Dual	32 psi (2.25 kg/cm²)	32 psi (2.25 kg/cm²)	32 psi (2.25 kg/cm²)
Continuous high speed riding:			
Solo	28 psi (2.00 kg/cm²)	25 psi (1.75 kg/cm²)	32 psi (2.25 kg/cm²)
Dual	32 psi (2.25 kg/cm²)	36 psi (2.50 kg/cm²)	36 psi (2.50 kg/cm²)

Sparking plug type
250 UK models — NGK DR8ES or ND X27ESR-U
250 US models — NGK D9EA or ND X27ES-U
400 models — NGK DR8ES-L or ND X24ESR-U

Sparking plug gap 0.6 – 0.7 mm (0.024 – 0.028 in)

Valve clearances (cold)
Inlet and Exhaust 0.08 – 0.13 mm (0.003 – 0.005 in)

Ignition timing
Below 1650 ± 100 rpm 20° BTDC
Above 3500 ± 100 rpm 40° BTDC

Tick over (idle speed)
250 models 1250 ± 100 rpm
400 models 1100 ± 50 rpm

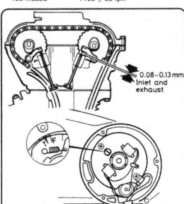

0.08 – 0.13 mm
Inlet and exhaust

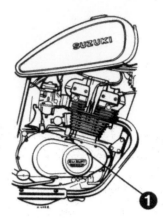

Component	Quantity	Grade
① Engine/transmission		SAE 10W/40
250 models:		
At oil change	2.0 lit (3.6/4.2 Imp/US pint)	
At oil and filter change	2.6 lit (4.6/5.4 Imp/US pint)	
400 models:		
At oil change	2.6 lit (4.6/5.4 Imp/US pint)	
At oil and filter change	2.9 lit (5.1/6.0 Imp/US pint)	
② Front forks (per leg)		
GS250 (US)	150cc (5.28/5.07 Imp/US fl oz)	Fork oil
Other models	150cc (5.28/5.07 Imp/US fl oz)	50/50 mixture of SAE 10W/30 motor oil and ATF
③ Steering head bearings	As required	High melting point grease
④ Final drive chain	As required	Aerosol chain lubricant
⑤ Wheel bearings	As required	High melting point grease
⑥ Swinging arm	As required	High melting point grease
⑦ Control cables	As required	Light machine oil
⑧ Pivot points	As required	High melting point grease

ROUTINE MAINTENANCE GUIDE

Refer to Chapter 7 for information relating to GSX250/400 EZ and all 450 models

Routine maintenance

Periodic routine maintenance is a continuous process that commences immediately the machine is used and continues until the machine is no longer fit for service. It must be carried out at specified mileage recordings or on a calendar basis if the machine is not used regularly, whichever is the sooner. Maintenance should be regarded as an insurance policy, to help keep the machine in the peak of condition and to ensure long, trouble-free service. It has the additional benefit of giving early warning of any faults that may develop and will act as a safety check, to the obvious advantage of both rider and machine alike.

The various maintenance tasks are described under their respective mileage and calendar headings. Accompanying photos or diagrams are provided, where necessary. It should be remembered that the interval between the various maintenance tasks serves only as a guide. As the machine gets older, is driven hard, or is used under particularly adverse conditions, it is advisable to reduce the period between each check.

For ease of reference each service operation is described in detail under the relevant heading. However, if further general information is required it can be found within the manual in the relevant Chapter.

Although no special tools are required for routine maintenance, a good selection of general workshop tools are essential. Included in the tools must be a range of metric ring or combination spanners, and a set of Allen keys (socket wrenches).

Check the tyre pressures with a gauge that is known to be accurate

---

Weekly or every 300 miles (500 km)

1 Tyre pressures

Refer to the pressure table on page 10 and check the tyre pressures with a pressure gauge that is known to be accurate. Always check the pressure when the tyres are cold. If the machine has travelled a number of miles, the tyres will have become hot and consequently the pressure will have increased. A false reading will therefore result.

It is recommended that a small pocket gauge is purchased and carried on the machine, as the readings on garage forecourt gauges can vary and may often be inaccurate.

It is essential that the tyres are kept inflated to the correct pressure at all times. Under- or over-inflated tyres can lead to accelerated rates of wear, and more importantly, can render the machine inherently unsafe. Whilst this may not be obvious during normal riding, it can become painfully and expensively so in an emergency situation, as the tyres' adhesion limits will be greatly reduced.

The pressures given are those recommended for the tyres fitted as original equipment. If replacement tyres are purchased, the pressure settings may vary. Any reputable tyre distributor will be able to give this information.

2 Safety check

Give the machine a close visual inspection, checking for loose nuts and fittings, frayed control cables etc. Check the tyres for damage, especially splitting on the sidewalls. Remove any stones or other objects caught between the treads. This is particularly important on the front tyre, where rapid deflation due to penetration of the inner tube will almost certainly cause total loss of control.

The tyres should also be checked for wear, bearing in mind that tyres worn beyond the legal limit will seriously limit the roadholding capabilities of the machine, thus presenting a potential danger to both the rider and other road users. The manufacturer's recommended limit for tyre tread depth is 1.6 mm (0.06 in) on the front tyre and 2.0 mm (0.08 in) on the rear tyre.

3 Legal check

Ensure that the lights, horn and trafficators function correctly, also the speedometer. Note that if bulbs have to be renewed, then they must have the same rating as the original. Remember that if different wattage bulbs are used in the traffic indicators, the flashing rate will be altered.

Remember also that it is a legal offence to have excessively worn or damaged tyres fitted to the machine.

4   Engine/gearbox oil level

Unscrew the filler plug which is situated mid-way along the upper surface of the right-hand crankcase cover. It will be noted that the plug incorporates a dipstick which should be wiped off using a clean, lint-free rag. Place the plug back in position, but do not screw it home; allow it to rest in position on the edge of the orifice. Remove the plug and note the level of the oil on the dipstick, which should be between the two level marks. Note that if the machine has been ridden recently, it should be allowed to stand for a few minutes to allow the oil clinging to the internal surfaces to drain down into the sump. The machine should be positioned upright on level ground before carrying out this check.

If required, replenish the oil reservoir with SAE 10W/40 engine oil. Do not run the engine with the oil level lower than the minimum level line and do not overfill the reservoir.

Check the engine/gearbox oil level with the dipstick provided

If required, replenish the oil reservoir with SAE 10W/40 engine oil

5   Final drive chain lubrication

In order that final drive chain life be extended as much as possible, regular lubrication and adjustment is essential. This is particularly so when the chain is not enclosed or is fitted to a machine transmitting high power to the rear wheel.

Intermediate lubrication should take place at the weekly or 300 mile (500 km) service with the chain in position on the machine. The chain may be lubricated whilst it is on the machine by the application of one of the proprietary chain greases contained in an aerosol can. Ordinary engine oil can be used, though owing to the speed with which it is flung off the rotating chain, its effective life is limited.

6   Battery electrolyte level

The transparent plastic case of the battery permits the upper and lower levels of the electrolyte to be observed when the right-hand side panel has been removed. Maintenance is normally limited to keeping the electrolyte level between the prescribed upper and lower limits and by making sure that the vent pipe is not blocked and remains correctly routed. The lead plates and their separators can be seen through the transparent case, a further guide to the general condition of the battery.

Unless acid is spilt, as may occur if the machine falls over, the electrolyte should always be topped up with distilled water, to restore the correct level. If acid is spilt on any part of the machine, it should be neutralised with an alkali such as washing soda and washed away with plenty of water, otherwise serious corrosion will occur. Top up with sulphuric acid of the correct specific gravity (1.280) only when spillage has occurred. Check that the vent pipe is well clear of the frame tubes or any of the other cycle parts, for obvious reasons.

Note the instructions attached to the battery cover for the correct routing of the vent pipe.

Regular final drive chain lubrication is essential

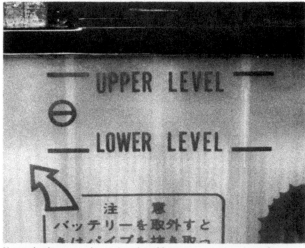

Keep the battery electrolyte level between the level marks

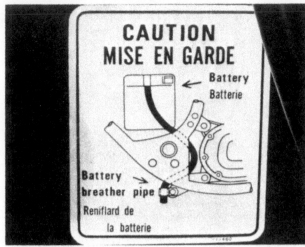

Note the instructions for routing the battery vent pipe

7 Prop stand examination

Check that the prop stand pivot bolt is secure and well lubricated and that the spring is in good condition and not overstretched. If the stand does not retract properly or extends whilst the machine is in motion, a serious accident is more or less inevitable.

8 Centre stand examination

Inspect each of the two centre stand pivot bolts for security and check that the return spring is in good condition. A missing pivot bolt or a broken or weak spring may cause the stand to fall whilst the machine is in motion, with the resulting danger that once it catches in some obstacle the balance of the machine will be drastically affected.

9 Front brake hydraulic fluid level

Check the level of the hydraulic fluid in the master cylinder reservoir mounted on the handlebars. The level can be seen through the transparent reservoir and should be between the upper and lower level marks. Ensure that the handlebars are in the central position when a level reading is taken and also when the cap and diaphragm are removed. Replenish the reservoir with an hydraulic fluid of the following specifications:

| UK | SAE J1703 |
| USA | DOT 3 or DOT 4 |

Care should be taken that a specified fluid is used. An incorrect fluid may perish the piston seals and cause brake failure.

10 Throttle operation

Check that the throttle twistgrip turns smoothly over the full operating range whilst opening the throttle and that it returns to the closed position automatically. This check should be carried out with the handlebars placed at full right-lock, full left-lock and in the central position.

Inspect the throttle cable for signs of damage, kinking or deterioration and check that there is no danger of it becoming trapped between moving components.

Monthly or every 600 miles (1000 km)

Complete the tasks listed under the weekly/300 mile (500 km) heading and then carry out the following:

1 Spoke tension

On machines fitted with the conventional type of wire-spoked wheel, check the spokes for tension, by gently tapping each one with a metal object. A loose spoke is identifiable by the low pitch noise emitted when struck. If any one spoke needs considerable tightening it will be necessary to remove the tyre and inner tube in order to file down the protruding spoke end. This will prevent it from chafing through the rim band and piercing the inner tube.

2 Cast alloy wheel inspection

On machines fitted with cast alloy wheels, carry out a careful check of the complete wheel for cracks and chipping, particularly at the spoke roots and at the edge of the rim. Check also for damage to the lacquer coating of the wheel. If this finish has been penetrated, the wheel will have started to corrode. Corrosion will be indicated by the presence of a whitish-grey powder deposit which will be caused by oxidisation of the bared metal.

If any one of these faults are found, the machine should not be ridden again until reference is made to Section 2 of Chapter 5 and the offending wheel either reconditioned or replaced with a serviceable item.

3 Final drive chain adjustment

Check the slack in the final drive chain. The correct up and down movement, as measured at the mid-point of the chain lower run, should be 20 – 30 mm (0.8 – 1.2 in) with the machine positioned on its centre stand. Adjustment should be carried out as follows. Check that the rear wheel is clear of the ground and free to rotate. Remove the split-pin from the wheel spindle and slacken the wheel nut a few turns. Loosen the locknuts on the two chain adjuster bolts, and slacken off the brake torque rod nuts.

Rotation of the adjuster bolts in a clockwise direction will tighten the chain. Tighten each bolt a similar number of turns so that wheel alignment is maintained. This can be verified by checking that the mark on the outer face of each chain adjuster is aligned with the same aligning mark on each fork end. With the adjustment correct, tighten the wheel nut and fit a new split pin. Finally, retighten the adjuster bolt locknuts and tighten the torque rod nuts, securing them by means of the split pins.

4 Front brake inspection

Check the front brake master cylinder, hoses and caliper units for signs of leakage. Pay particular attention to the condition of the hoses, which should be renewed without question if there are signs of cracking, splitting or other exterior damage.

Note the alignment marks provided when adjusting the final drive chain

Secure each adjuster bolt with its locknut

Four monthly or every 2000 miles (3000 km)

Complete the tasks listed under the weekly and monthly service headings and then carry out the following:

1   Final drive chain lubrication

Lubrication of the rear chain should be carried out at short intervals as described in Section 5 of the Weekly or 300 mile (500 km) maintenance schedule. On machines fitted with a non-standard chain using a spring joining link, more thorough lubrication should be carried out with the chain off the machine. The original chain has no spring link and therefore swinging arm removal must take place before the chain can be removed. In this case it is suggested that the chain be cleaned thoroughly whilst still in position, using a proprietary degreaser such as 'Gunk' or 'Jizer' and then lubricated using graphited grease from an aerosol can. Where a chain having a spring link is used, separate the chain by removing the spring link and run it off the sprockets. If an old chain is available, interconnect the old and new chain, before the new chain is run off the sprockets. In this way the old chain can be pulled into place on the sprockets and then used to pull the regreased chain into place with ease.

Clean the chain thoroughly in a paraffin bath and then finally with petrol. The petrol will wash the paraffin out of the links and rollers which will then dry more quickly. Remember that petrol is highly flammable, and care should be taken to avoid accidents.

Allow the chain to dry and then immerse it in a molten lubricant such as Linklyfe or Chainguard. These lubricants must be used hot and will achieve better penetration of the links and rollers. They are less likely to be thrown off by centrifugal force when the chain is in motion.

Refit the newly greased chain onto the sprocket, fitting the spring link. This is accomplished most easily when the free ends of the chain are pushed into mesh on the rear wheel sprocket. The spring link must be fitted so that the closed end faces the normal direction of chain travel.

2   Control cable lubrication

Lubricate the control cables thoroughly with motor oil or an all-purpose oil. A good method of lubricating the cables is shown in the accompanying illustration, using a plasticine funnel. This method has the disadvantage that the cables usually need removing from the machine. An hydraulic cable oiler which pressurises the lubricant overcomes this problem. Do not lubricate nylon lined cables (which may have been fitted as replacements), as the oil may cause the nylon to swell, thereby causing total cable seizure.

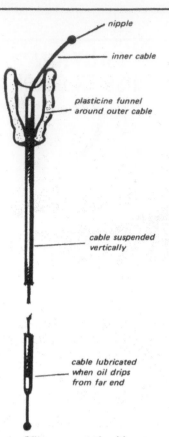

Oiling a control cable

3   General lubrication

Work around the machine, applying grease or oil to any exposed pivot points such as those of the mainstand, propstand, handlebar levers, rear brake rod linkage, etc.

4   Air filter cleaning

To gain access to the air filter element, detach the seat unit and release the element housing lid by sliding its two retaining clips forward off their locations. It was found in practice that these clips were very difficult to dislodge, the best method of removal being to push one of the clips completely off its location with the flat of a large screwdriver, thus taking the tension off the remaining clip which may then be removed without much effort.

The element is retained within its housing by a plastic grid. This grid may be bent in two and lifted from its location within the housing to allow the element to be pulled clear. Inspect the element for tears; on no account should a torn element be refitted.

The element is oil impregnated and should be cleaned thoroughly in a non-flammable solvent such as white spirit (available as Stoddard solvent in the US) to remove all the old oil and dust. After cleaning, squeeze out the sponge to remove the solvent and then allow a short time for any remaining solvent to evaporate. Do not wring out the sponge as this will cause damage and will lead to the need for early renewal. Reimpregnate the sponge with clean engine oil and gently squeeze out the excess. If the element has become very badly clogged or hardened with age, then it should be replaced with a new item. If the machine has been run in a dusty atmosphere, it is advisable to increase the frequency of cleaning and re-impregnating the element. Never run the engine without the element connected to the carburettor because the carburettor is specially jetted to compensate for the addition of this component. The resulting weak mixture will cause overheating of the engine with the probable risk of severe engine damage.

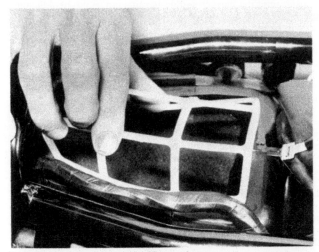

Refit the air filter element and its retainer

Locate the element retainer beneath its locating spigots

Secure the filter housing lid by sliding each clip into position

Six monthly or every 3000 miles (5000 km)

Complete the tasks listed under the weekly, monthly and four monthly headings and then carry out the following:

1 Sparking plug cleaning and checking

Detach the sparking plug cap, and using the correct spanner, remove the sparking plug. Clean the electrodes using a wire brush followed by a strip of fine emery cloth or paper. Check the plug gap with a feeler gauge, adjusting it if necessary to within the range 0.6 – 0.7 mm (0.024 – 0.028 in). Make adjustments by bending the outer electrode, never the inner (central) electrode.

Before fitting the sparking plug, smear the threads with a graphited grease; this will aid subsequent removal.

2 Ignition timing checking and adjustment

In order to check the accuracy of the ignition timing, it is first necessary to remove the small circular cover from the right-hand crankcase cover.

The ignition timing can only be checked using a stroboscopic timing lamp. In this way an additional task, that of checking the correct function of the ATU, may be accomplished simultaneously with checking the ignition timing.

Connect the strobe to the left-hand cylinder high tension lead, following the makers instructions. If an external 12 volt power source is required it is best not to use the machine's battery as spurious impulses can be picked up from the electrical system. A separate 12 volt car or motorcycle battery is preferable.

Start the engine and illuminate the ATU through the inspection aperture in the signal generating unit pick-up plate. With the engine running below 1500 rpm, the 'F' mark on the ATU should be in exact alignment with the index mark.

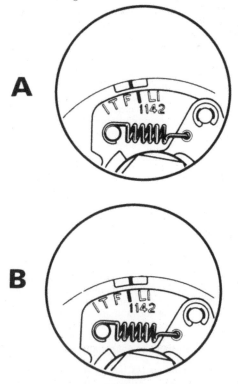

Ignition timing alignment mark on ATU assembly

A   Timing is retarded, move plate anti-clockwise
B   Timing is advanced, move plate clockwise

If the ignition timing is found to be slightly incorrect, the two screws which hold the signal generator mounting plate should be slackened, to facilitate movement of the plate. Move the plate a slight amount in the required direction, retighten the screws and recheck the timing. Repeat this process, if necessary, until the timing is correct.

To check the ATU for correct operation, increase the engine speed up to 3500 rpm whilst observing the timing marks with the aid of the strobe. If, when increasing the engine speed from the commencement of advance at 1500 rpm, the timing marks are seen to move erratically, or if the advance range has altered appreciably, the ATU should be inspected for wear or malfunctioning as described in Section 8 of Chapter 3.

3 Valve clearance checking and adjustment

It is important that the correct valve clearance is maintained to ensure the proper operation of the valve assemblies. A small amount of free play is designed into each valve train to allow for expansion of the various engine components. If the setting deviates greatly from that specified, a marked drop in performance will be evident. In the case of the clearance becoming too great, it will be found that valve operation will be noisy, and performance will drop off as a result of the valves not opening fully. If on the other hand, the clearance is too small, the valves may not close completely. This will not only cause loss of compression, but will also cause the exhaust valve to burn out very quickly. In extreme cases, the valve head may strike the piston crown, causing extensive damage to the engine. The clearances should be checked and adjusted with the engine cold, by the following method.

Depending on the model type being serviced, it may be found necessary to remove the fuel tank in order to gain enough access to the top of the engine to allow removal of the cylinder head cover. Full details of seat and tank removal are given in Section 15 of Chapter 4 and Section 2 of Chapter 2 respectively.

Detach the tachometer drive cable from the cylinder head cover by unscrewing its knurled retaining ring. Detach the breather pipe from the breather cover and pull the HT lead suppressor caps from the sparking plugs. Tuck all of these items clear of the cylinder head cover. Loosen and partially unscrew both sparking plugs, leaving them in situ until the cylinder head cover is removed and the engine is ready to be rotated. Free the tachometer drive gear assembly from the cylinder head cover by unscrewing the single crosshead screw and removing the retaining plate to allow the assembly to be pulled from position. Remove the four cylinder head cover end caps, each of which is retained by two countersunk screws.

Remove the cylinder head cover by unscrewing its retaining bolts and easing it clear of its two locating dowels. Unscrew the retaining bolts a little at a time at first and in a diagonal sequence to avoid any risk of distortion to the cover. It should be noted for reference when refitting that the bolts are of differing lengths, the two longer bolts passing through the locating dowels. Where necessary, tap around the gasket face with a soft-faced mallet to aid separation. Finally, remove the small circular cover from the right-hand crankcase cover in order to expose the ignition components located beneath it.

Place a spanner on the engine turning hexagon, located beneath the ATU central retaining bolt head, and rotate the crankshaft until the 'T' mark on the '2' side of the ATU is in alignment with the index mark on the static plate and the notches in the camshaft ends are pointing outwards as shown in the table accompanying this text. With the engine set in this position, the left-hand pair of exhaust valves may be checked for clearance as follows.

Using a feeler gauge, check the clearance between each rocker arm adjuster and the end of the valve stem with which it makes contact. The clearance between each adjuster and stem should be identical to that listed in the Specifications at the beginning of Chapter 1 for the particular model type.

If the clearances are found to be incorrect, adjustment can be made by slackening each locknut concerned and turning the

adjuster until the correct setting is obtained. Hold the adjuster still and tighten the locknut. Then recheck the setting before passing to the next valve.

The feeler gauge should be a light sliding fit between the adjuster and the valve stem.

Note that valve clearances should always be adjusted with the engine COLD, otherwise a false reading will be obtained. Badly adjusted tappets normally give a pronounced clicking noise from the vicinity of the cylinder head.

Once the clearances of the left-hand pair of exhaust valves are found to be satisfactory, refer again to the table accompanying this text and rotate the crankshaft through 180° until the camshaft notches are in the positions indicated. The right-hand pair of exhaust valves may now be checked for clearance by following the previously noted procedure.

Continue to check the clearances of the two pairs of inlet valves by turning the crankshaft in further increments of 180° whilst referring to the table for the respective positions of the camshaft notches and the pair of valves to check.

On completion of checking and adjusting the valve clearances, refit or reconnect all the disturbed engine components using a reverse sequence to that given for removal whilst noting the following points. Inspect and, if necessary, renew both the cylinder head cover and ignition pick-up assembly cover gaskets. Tighten the cylinder head cover retaining bolts evenly and in a diagonal sequence until the specified torque loading of 6.5 – 7.0 lbf ft (0.9 – 1.0 kgf m) is reached. Clean and degrease the threads of each of the chromed end cap retaining screws and coat them with a thread locking compound before inserting the screws in position. Finally, inspect and if necessary, renew the oil seals on the tachometer drive gear assembly before lightly lubricating them with clean engine oil prior to inserting the assembly into position.

ATU alignment mark

Cam Position	Notch 1 position	
	Intake Camshaft	Exhaust Camshaft
Ⓐ L. EX.	◧	◧
Ⓑ R. EX.	◨	◨
Ⓒ L. IN.	⊟	⊟
Ⓓ R. IN.	⊡	⊡

Camshaft notch positions

Valve clearance alignment marks

Adjust each valve clearance by rotating the adjuster screw

Hold the adjuster in position and tighten its locknut

to renew the filter cover O-ring at the same time. This will obviate the possibility of any oil leaks.

The by-pass valve, which allows a continued flow of lubrication if the element becomes clogged, is an integral part of the filter. For this reason routine cleaning of the valve is not required since it is renewed regularly.

With the new filter element located in its housing, refit the retaining spring and push the cover into position over its retaining studs. Locate the retaining nuts and washers onto the stud threads tightening them evenly and in small increments to prevent the spring slipping from its location in the cover and filter end.

Finally, fit a new sealing washer beneath the head of the drain plug and refit and tighten the plug. Pour the correct amount of SAE 10W/40 engine oil through the crankcase filler point and check its level with the filler cap dipstick before refitting the filler cap and starting the engine.

Never run the engine without the filter element or increase the period between the recommended oil changes or oil filter changes.

Remove the sump plug to drain the engine/gearbox oil

4 Oil filter renewal and engine/gearbox oil change

It is recommended that the oil filter element be renewed at every oil change and that the oil strainer be cleaned at every second oil change.

The oil filter element is contained within a semi-isolated chamber in the base of the lower crankcase, closed by a finned cover retained by three domed nuts and plain washers. The oil strainer is contained within the rearmost of the two compartments covered by the finned sump cover and is in the form of a circular-framed metal gauze plate which is retained to the crankcase by three screws. To effect efficient draining of the oil, start and run the engine until it reaches full operating temperature; this will heat and therefore thin the oil. Position a suitable receptacle beneath both the drain plug in the centre of the finned sump cover and the filter cover. Remove the drain plug and allow the oil to drain into the receptacle. A coil spring is fitted between the cover and the filter element to keep the latter seated firmly in position. Be prepared for the cover to fly off after removal of the retaining nuts and their washers. Allow any oil contained within the filter housing to drain into the receptavle and withdraw the element from the housing.

No attempt should be made to clean the oil filter element; it must be renewed. When renewing the filter element it is wise

Remove the filter housing cover to expose the element

5 Engine/gearbox oil pressure check

Prior to carrying out a check to determine the engine/gearbox oil pressure, refer to the weekly check list in this Chapter and check the oil level. Carry out an inspection of the unit for any signs of serious oil leakage that may affect the oil pressure.

A blanking plug is fitted to the right-hand crankcase cover, directly below the ignition pick-up assembly housing. This plug should be substituted by a suitable adaptor piece to which the pressure gauge can be attached, via a flexible hose. The gauge should have a scale reading of 0 – 10 kg/sq cm (0 – 140 psi).

With the pressure gauge connected and the crankcase oil level correct, start the engine and allow it to run for 10 minutes at 2000 rpm (summer conditions) or 20 minutes at 2000 rpm (winter conditions). On completion of this warming-up period raise the engine speed to 3000 rpm, when the pressure gauge should show a reading of 3.0 – 5.5 kg/sq cm (42.7 – 78.2 psi). A low oil pressure reading may be due to worn main and big-end bearings. It may also be caused by a worn oil pump, a blocked oil strainer or oil filter element.

Do not omit to check the condition of, and if necessary renew, the sealing washer located beneath the head of the blanking plug before refitting the plug to the crankcase cover.

6 Carburettor synchronisation and adjustment

In order that the engine maintains the best possible performance at all times, the carburettors must always remain correctly adjusted and synchronised. This check is essential and is made in two stages. The first stage is adjustment of the tickover (idle) speed and mixture strength by means of the shared throttle stop screw and the pilot screws respectively. Note that on US models the pilot mixture settings (controlled by the pilot screws) are preset at the factory so that the exhaust emissions comply with EPA limits. On these models adjustment should be confined to the idle speed and synchronisation. See Chapter 2 Section 9 for further details. The second stage, which is the synchronisation of the two carburettors, requires the use of two vacuum gauges together with the necessary adaptors and connection tubes. Vacuum gauges are somewhat expensive. For this reason and because both stages of adjustment require some expertise it is strongly recommended that the machine is returned to a Suzuki Service Agent who will be able to carry out the work quickly and efficiently and for a reasonable charge.

Before carrying out adjustments of the carburettors, it is important to check that the following items are adjusted correctly; valve clearances, ignition timing and sparking plug gaps. Many engine faults which at first are thought to be due to carburettor maladjustment can often be traced to those components listed above. Both stages of adjustment must be carried out with the engine at normal working temperature, preferably after the machine has been taken for a short run.

Start the engine and by means of the throttle stop screw located between the two instruments, set the engine tick-over speed to within the range 1150 – 1350 rpm (250 models) or 1050 – 1150 rpm (400 models). The mixture strength must now be adjusted on each carburettor in turn. Select one carburettor and screw in the pilot screw fully until it can be felt to seat lightly. **Do not overtighten** because the screw may break. Now unscrew the pilot screw until the engine speed is at its highest. This should occur between $1\frac{1}{4}$ and $1\frac{3}{4}$ turns out. With the engine running at the highest speed within the specified range, the pilot adjustment for the carburettor in question is now correct. Repeat the procedure on the second carburettor and then readjust the tick-over to the specified speed. Refer to the following Section in this Chapter and check and adjust the amount of free play in the throttle cable.

Synchronisation of the two carburettors, as mentioned above, requires the use of a pair of vacuum gauges and the correct adaptors for interconnection with the inlet tracts. If the necessary equipment is available, it should be connected up and calibrated following the manufacturer's recommendations. A blanking plug in the form of a crosshead screw is fitted to each side of the cylinder just forward of the inlet stud flange. After

removal of these screws and their sealing washers, the adaptors may be fitted.

Start the engine and set the speed to 1750 rpm with the throttle stop screw. If the gauge readings differ, rotate the throttle valve link adjuster screw either anti-clockwise or clockwise until the two readings are equal. Once the readings are equal, lock the adjuster screw in position with its locknut, reset the engine speed to 1750 rpm with the throttle stop screw and recheck the gauge readings. Finally, reset the engine speed to idle and remove the gauge assembly from the machine. Check that the sealing washer fitted to each blanking plug is in good condition and renewed if necessary before refitting and tightening the plugs.

Remove each blanking screw to facilitate fitting of the vacuum gauge adaptors

Rotate the throttle valve link adjuster screw to equalise the vacuum gauges

7 Throttle cable adjustment

To check the throttle cable for correct adjustment, locate the adjuster at the carburettor assembly retaining clip and pull upwards on the cable to check that there is 0.5 mm (0.02 in) of free play between the cable end and the adjuster. If necessary, carry out adjustment by loosening the adjuster locknut and turning the adjuster. Once the desired amount of free play is achieved, hold the adjuster steady and retighten the locknut.

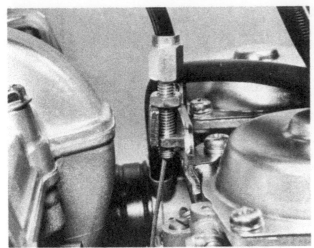

Carry out adjustment of the throttle cable by rotating the adjuster in its retaining clip

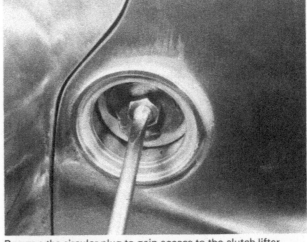

Remove the circular plug to gain access to the clutch lifter mechanism adjuster screw

8 Clutch adjustment

Carry out adjustment of the clutch cable and lifting mechanism as follows. Gain access to the clutch lifter mechanism adjuster screw and locknut by removing the circular plug from the rearmost section of the left-hand crankcase cover. Loosen the locknut on the adjuster screw and rotate the screw inwards until it can be felt to abut against the end of the pushrod. To gain the necessary running clearance unscrew the screw by $\frac{1}{4}$ – $\frac{1}{2}$ a turn and then tighten the locknut. The cable should be adjusted so that there is 4 mm (0.16 in) play measured between the handlebar stock and lever before the clutch commences lifting. Cable adjustment can be carried out by loosening the locknut of the adjuster located at the gearbox sprocket cover end of the cable and rotating the adjuster until the correct amount of play is achieved. Lock the adjuster in position by tightening the locknut whilst holding the adjuster to prevent it from turning and carry out any fine adjustment on the cable by rotating the knurled adjuster ring at the handlebar stock.

9 Rear brake pedal shaft and operating cam spindle lubrication

In order to maintain smooth and efficient operation of the rear brake operating assembly, it is essential that the brake pedal shaft and the cam spindle are both cleaned and re-lubricated in accordance with the instructions given in Section 14 of Chapter 4 and Section 13 of Chapter 5 respectively.

10 Steering head bearing check and adjustment

Place the machine on the centre stand so that the front wheel is clear of the ground. If necessary, place blocks below the crankcase to prevent the motorcycle from tipping forwards.

Grasp the front fork legs near the wheel spindle and push and pull firmly in a fore and aft direction. If play is evident between the upper and lower steering yokes and the head lug casting, the steering head bearings are in need of adjustment. Imprecise handling or a tendency for the front forks to judder may be caused by this fault.

Bearing adjustment is correct when the adjuster ring is tightened until resistance to movement is felt and then loosened $\frac{1}{8}$ to $\frac{1}{4}$ of a turn. The adjuster ring should be rotated by means of a C-spanner.

Take great care not to overtighten the nut. It is possible to place a pressure of several tons on the head bearings by overtightening even through the handlebars may seem to turn quite freely. Overtight bearings will cause the machine to roll at low speeds and give imprecise steering. Adjustment is correct if

Carry out clutch cable adjustment by rotating the coarse adjuster...

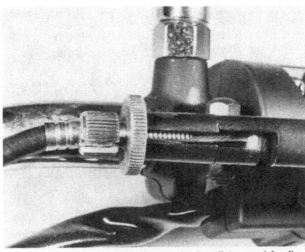

... before using the adjuster ring at the handlebar end for fine adjustment

there is no play in the bearings and the handlebars swing to full lock either side when the machine is on the centre stand with the front wheel clear of the ground. Only a light tap on each end should cause the handlebars to swing.

Annually or every 6000 miles (10 000 km)

Complete the tasks listed under all the previous service interval headings and then carry out the following:

1   Oil strainer cleaning

The oil strainer should be removed for cleaning at every second oil filter element change. To gain access to the strainer, remove the bolts that retain the finned sump cover in position. These bolts should be loosened a little at a time at first and in a diagonal sequence to prevent any risk of the cover becoming distorted. Before removing the bolts, note the position of the electrical cable retaining clips located beneath two of the bolt heads for reference when refitting. It will be found that even with all the retaining bolts removed, the cover will be stuck fast in position and will need to be tapped quite sharply around to mating face with a soft-faced mallet in order to free it.

With the strainer removed from the crankcase, wash it and the sump cover in fuel whilst using a soft bristle brush to remove any stubborn traces of contamination. Be sure to observe the necessary fire precautions when carrying out this cleaning procedure.

Carefully inspect the strainer gauze for any signs of damage. If the gauze is split or holed, it must be renewed as it no longer forms an effective barrier between the sump and oil pump.

Fit the oil strainer back into the crankcase housing, taking care to tighten fully its retaining screws. Ensure that the mating surfaces of the sump cover and the lower crankcase half have been cleaned of all old gasket material and jointing compound and are properly degreased. Smear an even coating of Suzuki Bond No 4 (or equivalent) over the area of sump cover mating surface indicated in the accompanying photograph and over the corresponding area of crankcase mating surface. Fit the new gasket into position on the crankcase and place the sump cover in position on top of it. Refit the sump cover retaining bolts into their respective locations, remembering to position the electrical cable retaining clips on their previously noted positions beneath the bolt heads. Tighten the retaining bolts evenly and in a diagonal sequence, noting the final torque loading figures given in the Specifications Section of Chapter 1.

Smear an even coating of Suzuki Bond No 4 over the indicated crankcase and sump cover mating surfaces before fitting a new gasket

2   Sparking plug renewal

Remove and discard the existing sparking plugs regardless of their condition. Although the plugs may give acceptable performance after they have reached this mileage, it is unlikely that they are still working at peak efficiency.

The correct plug types are listed in the Specifications Section of Chapter 3. Before fitting each new plug, adjust its gap to 0.6 – 0.7 mm (0.024 – 0.028 in).

3   Instrument drive cable examination and lubrication

It is advisable to detach the speedometer and tachometer drive cables from time to time in order to check whether they are adequately lubricated and whether the outer cables are compressed or damaged at any point along their run. A jerky or sluggish movement at the instrument head can often be attributed to a cable fault.

To grease the cable, uncouple both ends and withdraw the inner cable. (On some model types this may not be possible, in which case a badly seized cable will have to be renewed as a complete assembly).

After removing any old grease, clean the inner cable with a petrol soaked rag and examine the cable for broken strands or other damage. Do not check the cable for broken strands by passing it through the fingers and palm of the hand, this may well cause a painful injury if a broken strand snags the skin. It is best to wrap a piece of rag around the cable and pull the cable through it, any broken strands will snag on the rag.

Regrease the cable with high melting point grease, taking care not to grease the last six inches closest to the instrument head. If this precaution is not observed, grease will work into the instrument and immobilise the sensitive movement.

If the cable breaks, it is usually possible to renew the inner cable alone, provided the outer cable is not damaged or compressed at any point along its run. Before inserting the new inner cable, it should be greased in accordance with the instructions given in the preceding paragraph. Try to avoid tight bends in the run of the cable because this will accelerate wear and make the instrument movement sluggish.

4   Front brake hydraulic fluid renewal

Because of the possible deterioration of brake fluid with constant use and the possible build-up of contamination in the fluid, either from external sources or from materials within the system itself, Suzuki recommend that the brake fluid be changed once every year or at the mileage interval stated above.

The oil strainer is secured in position with three screws

To drain the brake system of used fluid, position a clean container below the caliper unit and attach a plastic tube from the bleed screw on top of the caliper unit to the container. Open the bleed screw one complete turn and drain the system by operating the brake lever until the master cylinder reservoir is empty. Close the bleed screw and remove the pipe.

Care must be taken during both draining and the following replenishing procedure, to ensure that no brake fluid is allowed to come into contact with the component parts of the motorcycle. Hydraulic brake fluid is a very effective paint remover and will have an adverse effect on any plastic cycle parts with which it comes into contact.

Position the handlebars in the central position so that the hydraulic fluid reservoir is in a suitable position for carrying out a fluid level check. Remove the reservoir top and diaphragm and fill the unit with an hydraulic fluid of SAE J1703 (UK) or DOT 3/DOT 4 (USA) specification to the upper level mark. Do not vary from these specifications as an incorrect fluid may perish the piston seals and cause brake failure.

The brake system should now be bled of air by referring to the information given in Section 9 of Chapter 5. Examination of the used fluid drained from the system will give a good indication as to the condition of such component parts as the piston seals, hose bore, piston bores, etc., by reference to the type of contamination found.

5   Front fork oil renewal

With the machine placed on its centre stand so that it is resting securely in a level position, remove the two U-clamps that retain the handlebar assembly to the upper yoke. This will allow full access to the fork leg cap bolts once the handlebar assembly has been eased back onto a padded area of the fuel tank. Loosen the pinch bolt at either side of the upper yoke. Attend to each leg separately so that the leg which remains undisturbed will support the machine.

Position a container below the fork leg drain screw located above and to the rear of the wheel spindle and remove the drain screw with its sealing washer. Unscrew the fork cap bolt.

Allow the damping fluid to drain and then pump the forks up and down to drain any residual oil. Refit and tighten the drain plug after checking that the sealing washer is in good condition. Replenish the fork leg with 150 cc (5.28/5.07 Imp/US fl oz) of the recommended fluid.

Various mixtures of oil types are recommended by Suzuki for the model types covered in this Manual; these are all listed in the Specifications Section of Chapter 4 along with the correct torque loading figures for the upper yoke pinch bolts, the fork leg cap bolts and the handlebar clamp retaining bolts. Repeat the draining sequence and refilling for the second fork leg.

Remove the drain screw from each fork leg

Before refitting each cap bolt, check the condition of the O-ring located beneath its head. These O-rings should be renewed if they are found to be flattened, perished, or in any way damaged. Note that the handlebar assembly must be refitted in the upper yoke clamps with the punch marks in the correct positions; that is with the punch mark on the handlebar in line with the rear mating face of the lower clamp.

Ensure that the clamp retaining bolts are tightened evenly so that the gaps between the front and rear mating surfaces of each clamp are equal.

Two yearly or every 12 000 miles (20 000 km)

Complete the tasks listed under all the previous service interval headings and then carry out the following:

1   Fuel line renewal

Because of the effect both time and use will have on the type of fuel line fitted to the machines covered in this Manual, Suzuki recommend that both lines fitted to the fuel tap be renewed at the above stated interval. Fuel lines that have been on a machine that has been in constant use for this period of time will have begun to harden and possibly crack or split. It is also possible that the surface of the bore of the fuel line will have begun to break down, thus causing contamination of both carburettors.

Remember to take adequate fire precautions when renewing the fuel lines and, if necessary, renew any fatigued securing clips.

2   Brake hose renewal

Because of the exposed fitted position of the brake hose on the machine and of the effect the elements will have on its material structure, Suzuki recommend that it is replaced with a new item at the above stated service interval.

Full details of brake hose examination, removal and fitting, as well as brake system bleeding, may be found in the relevant Sections of Chapter 5 of this Manual.

3   Swinging arm bearing lubrication

As no grease nipple is fitted to the tubular crossmember of the swinging arm fork it is necessary to remove the complete fork assembly from the machine in order to carry out efficient inspection and lubrication of the two bushes and caged needle roller bearings upon which the fork assembly pivots.

Full details of fork removal, examination, renovation and refitting are given in Section 9 of Chapter 4.

4   Steering head bearing lubrication

As with the swinging arm assembly, some considerable amount of dismantling is necessary in order to gain access to the steering head bearings for the purpose of inspection and lubrication. Full details of steering head assembly removal and filling as well as bearing examination and renovation are included in Chapter 4 of this Manual.

General adjustments and examination

1   Front brake pad wear

Brake pad wear depends largely on the conditions in which the machine is ridden and at what speed. It is difficult therefore to give precise inspection intervals, but it follows that pad wear should be checked more frequently on a hard ridden machine.

The condition of each pad can be checked easily whilst still on the machine. The pads have a red groove around their outer periphery which can be seen through an inspection window provided in the caliper casing. This window is closed by a small cover which must be prised from position in order to inspect the pads. If wear has reduced either or both pads in the caliper down to the red line the pads should be renewed as a pair.

To gain access to the pads for renewal, detach the brake caliper from its support bracket by removing the two securing bolts and pull the caliper rearwards to expose both pads. Withdraw the pads together with the shim plate.

Fit the new pads together with the shim plate and refit the caliper by reversing the removal procedure. The caliper piston should be pushed fully into the caliper so that there is sufficient clearance between the brake pads to allow the caliper to fit over the disc. Note the torque figure given in the Specifications Section of this Chapter before refitting and tightening the two caliper support bracket securing bolts.

Ensure that during the pad removal and refitting operation, the brake lever is not operated at any time. This will operate and displace the caliper piston; reassembly is then considerably more difficult.

In the interests of safety, always check the function of the brakes; pump the brake lever several times to restore full braking power, before taking the machine on the road.

2 Rear brake shoe wear

Brake shoe wear depends largely on the conditions in which the machine is ridden and at what speed. Because of this, it is difficult to give precise inspection intervals but it follows that the brake shoes should be checked for wear more frequently on a hard ridden machine.

The condition of the brake shoes can be checked quite easily with the wheel still in situ. With the rear brake assembly properly adjusted, as described in Section 14 of Chapter 5, apply the brake and observe the position of the line inscribed on the end of the brake cam spindle in relation to the indicator line cast in the brake backplate. The brake shoes are worn beyond limits if the line on the cam spindle is seen to have moved outside the area covered by the indicator line.

Details of brake shoe removal and fitting are given in Section 13 of Chapter 5.

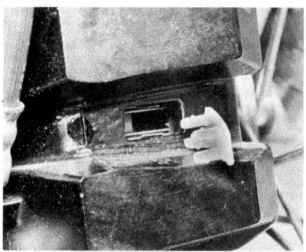

Inspect the brake pads for wear through the caliper window

Detach the brake caliper from its support bracket...

...and withdraw each brake pad from position

An indication of rear brake shoe lining wear is provided

Castrol Lubricants

Castrol Engine Oils
Castrol Grand Prix

Castrol Grand Prix 10W/40 four stroke motorcycle oil is a superior quality lubricant designed for air or water cooled four stroke motorcycle engines, operating under all conditions.

Castrol Super TT Two Stroke Oil

Castrol Super TT Two Stroke Oil is a superior quality lubricant specially formulated for high powered Two Stroke engines. It is readily miscible with fuel and contains selective modern additives to provide excellent protection against deposit induced pre-ignition, high temperature ring sticking and scuffing, wear and corrosion.
Castrol Super TT Two Stroke Oil is recommended for use at petrol mixture ratios of up to 50:1.

Castrol R40

Castrol R40 is a castor-based lubricant specially designed for racing and high speed rallying, providing the ultimate in lubrication. Castrol R40 should never be mixed with mineral-based oils, and further additives are unnecessary and undesirable. A specialist oil for limited applications.

Castrol Gear Oils
Castrol Hypoy EP90

An SAE 90 mineral-based extreme pressure multi-purpose gear oil, primarily recommended for the lubrication of conventional hypoid differential units operating under moderate service conditions. Suitable also for some gearbox applications.

Castrol Hypoy Light EP 80W

A mineral-based extreme pressure multi-purpose gear oil with similar applications to Castrol Hypoy but an SAE rating of 80W and suitable where the average ambient temperatures are between 32°F and 10°F. Also recommended for manual transmissions where manufacturers specify an extreme pressure SAE 80 gear oil.

Castrol Hypoy B EP80 and B EP90

Are mineral-based extreme pressure multi-purpose gear oils with similar applications to Castrol Hypoy, operating in average ambient temperatures between 90°F and 32°F. The Castrol Hypoy B range provides added protection for gears operating under very stringent service conditions.

Castrol Greases

Castrol LM Grease

A multi-purpose high melting point lithium-based grease suitable for most automotive applications, including chassis and wheel bearing lubrication.

Castrol MS3 Grease

A high melting point lithium-based grease containing molybdenum disulphide. Suitable for heavy duty chassis application and some CV joints where a lithium-based grease is specified.

Castrol BNS Grease

A bentone-based non melting high temperature grease for ultra severe applications such as race and rally car front wheel bearings.

Other Castrol Products

Castrol Girling Universal Brake and Clutch Fluid

A special high performance brake and clutch fluid with an advanced vapour lock performance. It is the only fluid recommended by Girling Limited and surpasses the performance requirements of the current SAE J1703 Specification and the United States Federal Motor Vehicle Safety Standard No. 116 DOT 3 Specification.
In addition, Castrol Girling Universal Brake and Clutch fluid fully meets the requirements of the major vehicle manufacturers.

Castrol Fork Oil

A specially formulated fluid for the front forks of motorcycles, providing excellent damping and load carrying properties.

Castrol Chain Lubricant

A specially developed motorcycle chain lubricant containing non-drip, anti corrosion and water resistant additives which afford excellent penetration, lubrication and protection of exposed chains.

Castrol Everyman Oil

A light-bodied machine oil containing anti-corrosion additives for both household use and cycle lubrication.

Castrol DWF

A de-watering fluid which displaces moisture, lubricates and protects against corrosion of all metals. Innumerable uses in both car and home. Available in 400gm and 200gm aerosol cans.

Castrol Easing Fluid

A rust releasing fluid for corroded nuts, locks, hinges and all mechanical joints. Also available in 250ml tins.

Castrol Antifreeze

Contains anti-corrosion additives with ethylene glycol. Recommended for the cooling system of all petrol and diesel engines.

Chapter 1 Engine, clutch and gearbox

Refer to Chapter 7 for information relating to GS450 and GSX250/400 EZ models

Contents

Specifications

Note: Specifications for 400 models are shown only where they differ from that of 250 models

Engine

	250 models	400 models
Type	Twin cylinder, air cooled, DOHC, 4-stroke	
Bore	60.0 – 60.015 mm	67.0 – 67.015 mm
	(2.362 – 2.363 in)	(2.637 – 2.638 in)
Stroke	44.2 mm	56.6 mm
	(1.740 in)	(2.228 in)
Capacity	249 cc	399 cc
	(15.2 cu in)	(24.34 cu in)
Compression ratio	10.5 : 1	10.0 : 1

Cylinder barrel

Standard bore	60.0 – 60.015 mm (2.362 – 2.363 in)	67.0 – 67.015 mm (2.637 – 2.638 in)
Service limit	60.095 mm (2.366 in)	67.080 mm (2.641 in)
Barrel to head face distortion limit	0.10 mm (0.004 in)	
Cylinder bore to piston clearance	0.035 – 0.045 mm (0.0014 – 0.0018 in)	0.050 – 0.060 mm (0.0020 – 0.0024 in)
Service limit	0.120 mm (0.0047 in)	

Piston

Outside diameter	59.960 – 59.975 mm (2.360 – 2.361 in)	66.945 – 66.960 mm (2.635 – 2.636 in)
Service limit	59.880 mm (2.357 in)	66.880 mm (2.633 in)
Gudgeon pin OD	15.995 – 16.0 mm (0.629 – 0.630 in)	17.996 – 18.0 mm (0.708 – 0.709 in)
Service limit	15.980 mm (0.629 in)	17.980 mm (0.708 in)
Gudgeon pin hole bore	16.002 – 16.008 mm (0.630 – 0.632 in)	18.002 – 18.008 mm (0.708 – 0.709 in)
Service limit	16.030 mm (0.631 in)	18.030 mm (0.710 in)
Ring groove width:		
Top and second	1.21 – 1.23 mm (0.047 – 0.048 in)	
Oil	2.51 – 2.53 mm (0.099 – 0.10 in)	

Piston rings

Ring to groove clearance service limit:		
Top	0.180 mm (0.0071 in)	
Second	0.150 mm (0.006 in)	
Ring thickness:		
Top	1.175 – 1.190 mm (0.046 – 0.047 in)	
Second	1.170 – 1.190 mm (0.046 – 0.047 in)	
End gap (fitted):		
Top	0.10 – 0.25 mm (0.004 – 0.010 in)	0.10 – 0.030 mm (0.004 – 0.012 in)
Service limit	0.70 mm (0.028 in)	0.70 mm (0.028 in)
Second	0.10 – 0.30 mm (0.004 – 0.012 in)	
Service limit	0.7 mm (0.028 in)	
End gap (free):		
Top	6.5 mm (0.26 in)	9.5 mm (0.37 in)
Service limit	5.2 mm (0.20 in)	7.6 mm (0.30 in)
Second	8.5 mm (0.33 in)	10.0 mm (0.39 in)
Service limit	6.8 mm (0.27 in)	8.0 mm (0.31 in)

Valves

Valve stem OD:		
Inlet	5.460 – 5.475 mm (0.215 – 0.216 in)	
Exhaust	5.445 – 5.460 mm (0.214 – 0.215 in)	
Stem runout service limit	0.05 mm (0.002 in)	
Stem end length service limit	3.6 mm (0.14 in)	
Valve guide ID	5.50 – 5.512 mm (0.216 – 0.217 in)	
Stem to guide clearance:		
Inlet	0.025 – 0.052 mm (0.001 – 0.002 in)	
Service limit	0.090 mm (0.0035 in)	
Exhaust	0.040 – 0.067 mm (0.002 – 0.003 in)	
Service limit	0.100 mm (0.0039 in)	
Valve head radial runout service limit	0.03 mm (0.001 in)	
Valve head diameter:		
Inlet	20.90 – 21.10 mm (0.823 – 0.831 in)	22.90 – 23.10 mm (0.90 – 0.91 in)
Exhaust	17.90 – 18.10 mm (0.705 – 0.713 in)	19.90 – 20.10 mm (0.78 – 0.79 in)
Valve seat width	0.9 – 1.1 mm (0.035 – 0.043 in)	
Valve lift	7.0 mm (0.28 in)	

Valve springs

Free length service limit:	
Inner	31.9 mm (1.26 in)
Outer	35.5 mm (1.40 in)

Valve clearances (cold)

Inlet and exhaust .... 0.08 – 0.13 mm (0.003 – 0.005 in)

Camshafts

Cam height .... 34.650 – 34.690 mm (1.364 – 1.365 in)
Service limit .... 34.350 mm (1.352 in)
Journal OD .... 21.959 – 21.980 mm (0.864 – 0.865 in)
Bearing surface ID .... 22.0 – 22.013 mm (0.866 – 0.867 in)
Journal oil clearance:
    GSX 400T, and EX models .... 0.032 – 0.066 mm (0.0013 – 0.0026 in)
    Service limit .... 0.150 mm (0.006 in)
    All others .... 0.020 – 0.054 mm (0.001 – 0.002 in)
    Service limit .... 0.150 mm (0.006 in)
Camshaft runout service limit .... 0.10 mm (0.004 in)

Camshaft chain

Maximum length (measured over 20 pitch length) .... 157.80 mm (6.213 in)

Rockers

Rocker arm spindle OD .... 11.973 – 11.984 mm (0.471 – 0.472 in)
Rocker arm bore ID .... 12.0 – 12.018 mm (0.472 – 0.473 in)

Cylinder head

Distortion limit .... 0.10 mm (0.004 in)

Crankshaft

Runout service limit	0.05 mm (0.002 in)	
Crankpin OD	31.976 – 32.0 mm (1.259 – 1.260 in)	33.976 – 34.0 mm (1.337 – 1.338 in)
Crankpin width	20.10 – 20.15 mm (0.791 – 0.793 in)	23.10 – 23.15 mm (0.909 – 0.911 in)
Journal OD	31.976 – 32.0 mm (1.259 – 1.260 in)	
Journal oil clearance	0.020 – 0.044 mm (0.001 – 0.002 in)	
Service limit	0.080 mm (0.003 in)	
Thrust clearance	0.05 – 0.25 mm (0.002 – 0.010 in)	–
Service limit	0.35 mm (0.014 in)	–
Thrust bearing thickness	–	2.95 – 2.98 mm (0.116 – 0.117 in)
Service limit	–	2.85 mm (0.112 in)

Connecting rods

Small-end ID	16.006 – 16.014 mm (0.630 in)	18.006 – 18.014 mm (0.709 in)
Service limit	16.040 mm (0.631 in)	18.040 mm (0.710 in)
Big-end width	19.95 – 20.0 mm (0.785 – 0.787 in)	22.95 – 23.0 mm (0.904 – 0.906 in)
Big-end oil clearance	0.024 – 0.048 mm (0.001 – 0.002 in)	
Service limit	0.080 mm (0.003 in)	
Big-end side clearance	0.10 – 0.20 mm (0.004 – 0.008 in)	
Service limit	0.30 mm (0.012 in)	

Balancer shaft

Journal OD .... 31.984 – 32.0 mm (1.259 – 1.260 in)
Journal oil clearance .... 0.020 – 0.044 mm (0.001 – 0.002 in)
Service limit .... 0.080 mm (0.003 in)

Clutch

Spring free length service limit	33.6 mm (1.32 in)	38.4 mm (1.51 in)
Friction plate thickness	2.9 – 3.1 mm (0.11 – 0.12 in)	
Service limit	2.6 mm (0.10 in)	
Friction plate maximum warpage	0.2 mm (0.008 in)	
Friction plate tongue width	11.8 – 12.0 mm (0.46 – 0.47 in)	15.8 – 16.0 mm (0.62 – 0.63 in)
Service limit	11.0 mm (0.43 in)	15.0 mm (0.59 in)
Plain plate thickness	1.54 – 1.66 mm (0.061 – 0.065 in)	
Plain plate maximum warpage	0.10 mm (0.004 in)	

Gearbox

Type	6-speed constant mesh	
Primary reduction ratio	3.125 : 1 (75/24)	2.714 : 1 (76/28)
Gear ratios:		
1st	2.500 : 1 (30/12)	2.461 : 1 (32/13)
2nd	1.625 : 1 (26/16)	1.777 : 1 (32/18)
3rd	1.210 : 1 (23/19)	1.380 : 1 (29/21)
4th	1.000 : 1 (21/21)	1.125 : 1 (27/24)
5th	0.863 : 1 (19/22)	0.961 : 1 (25/26)
6th	0.782 : 1 (18/23)	0.851 : 1 (23/27)
Final reduction ratio:		
GSX250 models	3.285 : 1 (46/14)	
GX250 models	3.133 : 1 (47/15)	
GSX400 ET	2.933 : 1 (44/15)	
GSX400 EX	2.812 : 1 (45/16)	
GSX400T models	2.687 : 1 (43/16)	
Selector fork claw end thickness	5.3 – 5.4 mm (0.209 – 0.213 in)	
Selector fork claw end to gear pinion groove clearance	0.10 – 0.30 mm (0.004 – 0.012 in)	
Service limit	0.50 mm (0.020 in)	
Gear pinion groove width	5.5 – 5.6 mm (0.217 – 0.220 in)	

Torque wrench settings

	lbf ft	kgf m
Cylinder head bolt	5.0 – 8.0	0.7 – 1.1
Cylinder head nuts:		
250 models	16.0 – 20.0	2.2 – 2.8
400 models	25.5 – 30.0	3.5 – 4.0
Cylinder head cover bolts	6.5 – 7.0	0.9 – 1.0
Cam chain tensioner bolts	4.5 – 6.0	0.6 – 0.8
Cam chain tensioner locknut	6.5 – 10.0	0.9 – 1.4
Cam chain tensioner shaft locknut	6.0 – 7.0	0.8 – 1.0
Cam chain tensioner shaft assembly	22.5 – 25.5	3.1 – 3.5
Camshaft journal retaining bolts	6.0 – 8.5	0.8 – 1.2
Camshaft sprocket retaining bolts	6.5 – 8.5	0.9 – 1.2
Valve clearance adjuster locknuts	6.5 – 8.0	0.9 – 1.1
Rocker arm shaft stopper bolts	6.0 – 7.0	0.8 – 1.0
Counterbalance shaft centre bolt	25.5 – 32.5	3.5 – 4.5
Connecting rod cap retaining nuts	21.5 – 24.5	3.0 – 3.4
Primary drive gear pinion retaining nut:		
250 models	36.0 – 50.5	5.0 – 7.0
400 models	65.1 – 79.5	9.0 – 11.0
Starter clutch Allen bolts	11.0 – 14.5	1.5 – 2.0
Starter motor retaining bolts	3.0 – 5.0	0.4 – 0.7
Clutch hub retaining nut:		
250 models	21.5 – 36.0	3.0 – 5.0
400 models	30.0 – 43.5	4.0 – 6.0
Clutch spring retaining bolts	3.0 – 4.5	0.4 – 0.6
Gearchange stopper arm pivot bolt	11.0 – 16.5	1.5 – 2.3
Alternator rotor retaining bolt:		
250 models	43.5 – 50.5	6.0 – 7.0
400 models	65.1 – 70.0	9.0 – 10.0
Automatic timing unit retaining bolt	9.5 – 16.5	1.3 – 2.3
Crankcase bolts:		
Nos 1 – 8	14.5 – 17.5	2.0 – 2.4
Nos 9 – 12	6.5 – 9.5	0.9 – 1.3
6 mm	7.0	1.0
8 mm	14.5	2.0
Gearbox sprocket retaining nut	36.0 – 50.5	5.0 – 7.0
Sump cover retaining bolts	7.0	1.0
Oil pressure regulator	12.5 – 14.5	1.7 – 2.0
Oil pressure switch	9.5 – 12.5	1.3 – 1.7
Neutral stopper housing	13.0 – 20.0	1.8 – 2.8
Engine mounting bolts:		
8 mm	14.5 – 21.5	2.0 – 3.0
10 mm	21.5 – 26.76	3.0 – 3.7
Exhaust pipe clamp bolts	6.5 – 10.0	0.9 – 1.4

1 General description

The engine/gearbox unit fitted to the Suzuki GS/GSX 250/400 models is a four valve per cylinder, double overhead camshaft, vertical parallel twin, built in unit with the primary transmission and gearbox and of conventional light alloy construction.

The horizontally split crankcase incorporates a one-piece forged crankshaft which has its two crankpins arranged at 180° to one another. The crankshaft is supported by four renewable plain main bearings. The connecting rods have split big-end eyes, also with renewable bearing shells.

In order to reduce engine vibration, a single balance shaft is fitted within the crankcase, mounted forward of the crankshaft and supported on a renewable plain bearing at either end. The shaft is driven by a pinion mounted on the crankshaft, inboard of the primary drive pinion.

The multi-plate clutch is fitted to the right-hand side of the engine, primary drive being through two spur gears. A six speed constant mesh gearbox then transmits power through a roller chain to the rear wheel.

The camshaft drive chain, which is driven from a sprocket mounted centrally on the crankshaft, passes upwards through a tunnel in the cylinder block to the inlet and exhaust camshafts mounted on the cylinder head. The chain is maintained in correct tension by an automatic tensioner unit bolted to the rear of the cylinder block.

The valves are operated in pairs by the cam lobes via forked rocker arms which pivot on spindles retained within the cylinder head. Valve adjustment is achieved through adjuster screws threaded through the fork ends of each rocker arm.

Wet sump lubrication is supplied by a gear driven trochoidal pump, the oil passing through a wire gauze trap and paper filter element under pressure to all working surfaces of the engine.

2 Operations with engine/gearbox unit in frame

It is not necessary to remove the engine/gearbox unit from the frame in order to remove and refit the following component parts:

1   Gearbox sprocket
2   Alternator rotor
3   Starter clutch and motor assemblies
4   Oil sump cover and sump filter
5   Oil pump drive gear and pump assembly
6   Primary driven and drive gears
7   Clutch plates and hub
8   Clutch pushrod
9   Oil filter
10   Cylinder head cover and breather cover
11   Cylinder head and barrel
12   Camshafts
13   Piston assemblies
14   Cam chain tensioner assembly
15   Gearchange pedal and shaft
16   CDI pulse generator assembly and ATU

It should be noted that where a number of the above items require attention, it can often be easier to remove the engine/gearbox unit and carry out the dismantling work with the unit positioned on a work surface.

3 Operations with engine/gearbox unit removed from frame

As previously described, the crankshaft, gearbox and counter balance shaft are housed within a common casing. Any work that needs to be carried out on these assemblies and their associated bearings will necessitate removal of the engine/gearbox unit from the frame so that the crankcase halves can be separated.

4 Removing the engine/gearbox unit from the frame

1   Place the machine on its centre stand making sure that it is standing firmly. Although by no means essential it is useful to raise the machine a number of feet above floor level by placing it on a long bench or horizontal ramp. This will enable most of the work to be carried out in an upright position, which is eminently more comfortable than crouching or kneeling in a

puddle of sump oil.

2   Place a suitable receptacle below the crankcase and drain off the engine oil. The sump plug is located in the centre of the finned sump cover which itself is located just forward of the exhaust system balancer pipe. The oil will drain at a higher rate if the engine has been warmed up previously, thereby heating and thinning the oil. Approximately 2.0 litres (250 models) or 2.6 litres (400 models) should drain out. Remove the oil filler cap to assist oil drainage. Loosely refit the filler cap to prevent loss and refit and tighten the sump plug, remembering to fit a new sealing washer before doing so, on completion of the oil draining.

3   Release the seat from its rear mounting points and pull the seat up off the mounting points and rearwards to release it from the forward frame mounting. On machines equipped with a seat fairing, the seat may be released simply by turning the key in the centrally positioned lock located just to the rear of the pillion passenger grab rail. On machines equipped with a conventional type of seat, release the seat by unscrewing and removing the two dome-headed bolts, one each side of the seat.

4   Before attempting to release the fuel tank from its frame mountings, position the fuel tap lever in the 'On' position and release the two spring clips that hold the fuel and vacuum hoses to the fuel tap stubs. Carefully pull each hose off its stub; if necessary, using a small screwdriver to help release the hose. Once the fuel hose is detached, allow any fuel contained in the hose to drain into a small clean container.

5   On machines fitted with a 'peardrop' style tank, release the plastic trim from around the front of the tank by unscrewing the crosshead screw retaining each half of the trim. Move to the rear of the tank and remove the single tank retaining bolt. Lift the rear of the tank up and pull it rearwards to release the tank from the forward rubber mountings. Lift the tank clear of the machine and store it carefully in a safe place to avoid any risk of damage to the paint finish. Unclip each of the sidepanels and store them and the seat next to the fuel tank.

6   Position the receptacle used to catch the engine oil beneath the oil filter housing and remove the filter cap by unscrewing and removing the three retaining nuts and washers. Allow the small amount of oil contained within the housing to drain into the receptacle and withdraw the spring and filter from their housing.

7   Move to the right-hand side of the machine and pull the battery breather pipe off its retaining stub at the top of the battery. Release the battery retaining strap and withdraw the battery far enough out of its tray to allow the negative (-) and positive (+) leads to be disconnected from their terminals. Remove the battery and store it carefully. It is advisable to service the battery at a convenient time whilst the engine is removed; full details of servicing procedures are contained within Chapter 6 of this Manual.

8   On completion of battery removal disconnect the following electrical leads:

a)   The leads from the CDI pulse generator and oil pressure switch at the block connector located just forward of the top of the battery tray.

b)   The leads from the alternator at the push connectors located within the rubber cover situated between the top frame tubes just forward of the rear mudguard.

c)   The neutral indicator switch lead at the single push connector located adjacent to the HT coil on the frame top tube.

d)   On those models fitted with a gear position indicator disconnect the switch leads at the block connector.

Note the colour coding of the wiring when disturbing the single bullet connections and if considered necessary, mark the leads for reference when refitting. Release the leads from any frame clips, etc and ensure that they will not be pulled or damaged during engine removal.

9   Remove the horn by first disconnecting the electrical leads whilst noting their positions and then releasing the mounting bracket from the frame by unscrewing and removing the single

retaining bolt. Detach the tachometer drive cable from the cylinder head cover by unscrewing the knurled retaining ring and allow the cable to hang clear of the engine.

10 Pull the HT lead suppressor caps from the sparking plugs and tuck both leads out of the way on the frame top tube. Loosen the sparking plugs now as some difficulty may be experienced later when the cylinder head is removed if the plugs have partially seized in their threads. Detach the breather pipe from the breather cover by releasing the spring clip and pulling the pipe from its retaining stub.

11 Commence removal of the carburettor assembly by detaching the throttle cable from the carburettor throttle operating pulley. To do this, loosen both the locknuts that retain the cable end adjuster to the carburettor mounted retaining bracket. Rotate the throttle twistgrip so that the throttle is fully open and with the flat of a small screwdriver, hold the pulley in the fully open position whilst releasing the twistgrip to provide enough slack in the cable to allow the nipple to be released from its location in the pulley. Slide the cable end adjuster sideways to free it from the retaining bracket and tuck the cable out of the way on the frame top tube.

12 Loosen the clamps between the carburettors and air filter housing and release the housing from the three mounting points located on each inner side panel. Note when removing the retaining screws that one screw on each side has electrical leads located beneath its head; the position of these leads must be noted for reference when refitting. Note also that there are bushes located beneath all the screw heads and these must be removed before the housing can be slid rearwards to clear the carburettor mouths. It will also be necessary to ease each side panel outwards so that it clears the housing. Some difficulty was experienced in freeing the housing from the carburettor mouths and moving it back far enough to clear completely the carburettor assembly, but with the aid of an assistant and a certain amount of patience the job can soon be completed.

13 Finally, to remove the carburettor assembly from the machine, loosen the clamps between the carburettors and cylinder head and pull the carburettor assembly rearwards to clear the inlet manifolds and then to the right to clear the machine.

14 Moving to the left-hand side of the machine, unscrew and remove the gearchange lever securing bolt and pull the lever clear of its splined shaft. Remove the left-hand footrest by unscrewing its two securing bolts and remove the gearchange sprocket cover by unscrewing and removing its five securing screws. At this point the cover may be either left attached to the clutch cable and secured to a point on the frame clear of the engine or detached from the clutch cable by bending back the cable nipple retaining tab within the cable end retainer with the flat of a small screwdriver and moving the nipple down and sideways to release it from the retainer. With the nipple released, the adjuster locknut can be loosened and the adjuster unscrewed from the cover to allow the cable to be pulled clear.

15 Using the flat of a large screwdriver, carefully knock back the lock washer from where it is bent over one of the flats of the gearbox sprocket retaining nut. Place the machine in gear, apply the rear brake and undo the retaining nut. In practice it was found that the nut was secured in position by a thread locking compound and had been tightened with some force; it was therefore necessary to recruit the help of an assistant to apply the rear brake whilst the nut was released. To avoid damage to the nut, the machine and to one's knuckles it is strongly recommended that a socket of the correct size is used in conjunction with an extension bar.

16 Before removing the nut and lock washer, mark the face of the sprocket so that it is not inadvertently reversed whilst refitting. Place a sheet of clean paper or a piece of clean rag beneath the chain run and remove the nut, lock washer and sprocket together with the chain. Detach the sprocket from the chain and allow the chain to fall onto the paper or rag.

17 Remove the two crosshead screws that retain the chromed starter motor cover to the upper crankcase and disconnect the electrical lead from the starter motor. It was found that both these screws are very tight and required the use of an impact driver to free them. Move the lead clear of the engine and secure it in position.

18 Remove the right-hand footrest assembly by unscrewing and removing its two retaining bolts. Detach the exhaust system by first removing the bolts securing the exhaust pipe clamps to the cylinder head; this must be done first as it prevents the weight of the complete system from being allowed to impose an unacceptable strain on the cylinder head stud threads once the rest of the mounting points are undone. On systems where the silencers are detachable from the exhaust pipes the silencers may be left attached to the machine if so desired. In this case the retaining clamps securing the silencers to the pipes should be fully loosened before removing the two balancer pipe to crankcase securing bolts and pulling the pipe assembly forward and down to clear the machine. Each silencer is retained to its frame attachment mounting by a silencer bolt, this bolt may be loosened slightly to allow the front of the silencer to be pivoted downwards to clear the engine during the following engine removal procedure.

19 On exhaust systems where the silencers are permanently attached to the pipes, the system must be removed from the machine in two halves. With the cylinder head connections already detached as previously described, fully loosen the balancer pipe retaining clamp(s) and with an assistant supporting one side of the system, support the opposite side. Release each silencer from its frame mounting by removing the securing bolt and move the complete system forward to clear the cylinder head before lowering it to clear the machine. At this point the two halves of the system may be separated by pulling them apart whilst twisting slightly so that they separate at the balancer pipe connection.

20 Before freeing the engine from its frame mountings, carry out a final check around the engine and adjacent frame components to ensure that there is no chance of cables, electrical leads, etc becoming entangled with the engine during removal. It is advisable to protect the front and lower frame tubes with strips of rag or foam rubber. The engine/gearbox unit is a close fit in the frame and the chances of lifting it out of position without it coming into contact with the paint finish are remote to say the least.

21 The help of an assistant is invaluable, although not completely necessary, in supporting the engine whilst it is freed from its frame mountings and in lifting the engine out of the frame. If this job is to be attempted without assistance, it is necessary that a strong wooden box or similar platform is positioned next to the engine on the right-hand side of the machine so that when the engine is freed from its mountings it can be lifted straight onto the box. The suggested method of lifting the engine out of the frame without assistance is to stand on the left-hand side of the machine, facing forward, and to place the right hand over the frame top tubes in order to grasp the underside of the rear of the right-hand crankcase cover whilst placing the left hand on the forward left-hand corner of the cylinder head. The right hand may then be used to take the weight of the engine unit whilst the left hand prevents it from toppling forward.

22 Commence freeing the engine from the frame by removing the three bolts holding the front mounting plates in position and removing the plates. Move to the right-hand side of the machine and remove the lower mounting plate by removing the single nut. Withdraw the long bolt from the left-hand side of the machine, and remove the two bolts and lock washers to allow the mounting plate to be dropped clear of the frame.

23 With an assistant supporting the engine, or with a wooden block placed firmly underneath the crankcase, remove the bolt passing through the lower of the two rear mountings. Remove the upper rear mounting plate by unscrewing the single nut and two bolts and lock washer and drawing the long bolt out from the left-hand side of the machine whilst taking care to retain the cylindrical spacer and plain washer. Take the weight of the engine and carefully manoeuvre it out of the frame towards the right-hand side of the machine.

4.2 Remove the sump plug to drain the engine oil...

4.6 ...followed by the oil filter housing cap

4.10 Release the spring clip to detach the breather pipe

4.14 The clutch cable may be released from its retainer

4.17 Disconnect the electrical lead from the starter motor

4.22a Free the engine from the frame by first removing the front mounting plates...

4.22b ...followed by the right-hand lower mounting plate...

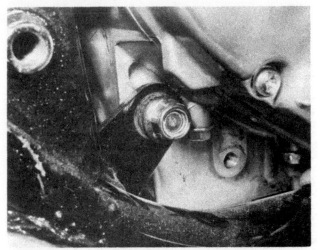
4.23a ...the bolt passing through the lower of the two rear mountings...

4.23b ...and finally, the upper rear mounting plate...

4.23c ...taking note of the cylindrical spacer and washers

5 Dismantling the engine/gearbox unit: general

1 Before commencing work on the engine unit, the external surfaces should be cleaned thoroughly. A motorcycle engine has very little protection from road grit and other foreign matter, which will find its way into the dismantled engine if this simple precaution is not taken. One of the proprietary cleaning compounds, such as 'Gunk' or 'Jizer' can be used to good effect, particularly if the compound is worked into the film of oil and grease before it is washed away. In the USA, 'Gumout' degreaser is an alternative product. Special care is necessary when washing down to prevent water from entering the now exposed parts of the engine unit.

2 Never use undue force to remove any stubborn part unless specific mention is made of this requirement. There is invariably good reason why a part is difficult to remove, often because the dismantling operation has been tackled in the wrong sequence.

3 Mention has already been made of the benefits of owning an impact driver. Most of these tools are equipped with a standard $\frac{1}{2}$ inch drive and an adaptor which can take a variety of screwdriver bits. It will be found that most engine casing screws will need jarring free due both to the effects of assembly by power tools and an inherent tendency for screws to become pinched in alloy castings. If an impact screwdriver is not available, it is often possible to use a crosshead screwdriver fitted with a T-handle as a substitute.

4 A cursory glance over many machines of only a few years' use, will almost invariably reveal an array of well-chewed screw heads. Not only is this unsightly, it can also make emergency repairs impossible. It should be borne in mind that there are a number of types of crosshead screwdrivers which differ in the angle and design of the driving tangs. To this end, it is always advisable to ensure that the correct tool is available to suit a particular screw.

5 Before commencing dismantling, make arrangements for storing separately the various sub-assemblies and ancillary components, to prevent confusion on reassembly. Where possible, replace nuts and washers on the studs or bolts from which they were removed and refit nuts, bolts and washers to their components. This too will facilitate straightforward reassembly.

6 Identical sub-assemblies, such as valve springs and collets or rocker arms and pins etc should be stored separately, to prevent accidental transposition and to enable them to be fitted in their original locations.

6 Dismantling the engine/gearbox unit: removing the cylinder head cover and camshafts

1 Support the engine firmly on a clean work surface and

detach the chromed end caps from each side of the cylinder head cover by unscrewing the two crosshead screws retaining each of the caps in position. On 250cc models remove the breather cover from the top of the cylinder head cover by unscrewing its four retaining bolts and lifting it clear. In practice it was found that the gasket beneath the cover caused it to stick to its location on the cylinder head cover and some force was necessary to free it. The method used to free the cover without damaging it was to place a hardwood wedge against the side of the cover and move it around the cover periphery whilst tapping the end of the wedge sharply with a mallet. On no account should the flat of a screwdriver or a similar item be forced between the mating surfaces in an attempt to lever the cover from position. Removal of the breather cover prior to cylinder head cover removal is not required on 400cc models.

2   Free the tachometer drive gear assembly from the front of the cylinder head cover by unscrewing the single crosshead screw and removing the retaining plate to allow the assembly to be pulled from position. Remove the cylinder head cover by unscrewing its retaining bolts and easing it clear of its two locating dowels. Unscrew the retaining bolts a little at a time at first and in a diagonal sequence to avoid any risk of distortion to the cover. It should be noted for reference when refitting that the bolts are of differing lengths, the two longer bolts passing through the locating dowels. Where necessary, tap around the gasket face with a soft-faced mallet to aid separation.

3   Before attention can be given to the camshafts and chain, the automatic chain tensioner must be detached from the rear of the cylinder block. The recommended sequence for detaching the tensioner is to commence by loosening the locknut retaining the stop screw in the left-hand side of the tensioner body. Turn the stop screw in so that it tightens against the spring loaded plunger contained within the tensioner body and retighten the locknut. This will prevent the plunger from shooting out of the tensioner body directly it is detached from the cylinder block. Free the tensioner assembly from the cylinder block by removing the two mounting bolts.

4   Having removed the tensioner assembly, the camshafts may be removed individually, without separating the cam chain. It will be noticed that no matter in what position the engine is placed, at least one cam lobe will be depressing one pair of valves and springs to some extent. To prevent uneven stress to the camshafts when removing the camshaft bearing caps, Suzuki recommend that a large self-grip wrench be used to hold the camshaft down. A G-clamp of suitable size will make a substitute if a wrench is not available. Fit the wrench as shown in the accompanying figure, ensuring that it is so placed that slipping is not possible. Loosen the bearing cap securing bolts in a diagonal sequence and in small even increments at first until they can be withdrawn and the caps removed. Note that each cap is marked with a letter set inside a triangle. A matching symbol is cast in the cylinder head to enable the caps to be refitted in their original locations. The point of the triangle must face forward when the caps are refitted. Note also the fitting of two small locating dowels to each of the cam bearing caps. Finally, displace the clamping tool to free the camshaft and repeat the procedure to free the second camshaft.

5   If a suitable clamping tool is not available, it is acceptable to slacken the bearing cap bolts without restraining the camshafts, providing extreme caution is exercised. Again, the bolts must be loosened evenly, in a diagonal sequence and a little at a time, so that neither the camshaft nor the bearing caps are allowed to tilt.

6   Lift the cam chain off one sprocket and remove the camshaft complete with sprocket. Repeat the procedure for the other camshaft. It is not necessary to remove either of the sprockets from the camshafts unless either of the components require renewal.

7   If a top-end overhaul only is anticipated, the cam chain should not be allowed to fall down into the chain tunnel, as retrieval can be very difficult. Insert a long bar through the chain so that it rests on the cylinder head or use a length of stout wire secured to an adjacent stud or bolt hole.

6.2 Remove the tachometer drive gear retaining plate

6.3 Free the tensioner assembly from the cylinder block

6.4 Each camshaft bearing cap is marked as shown

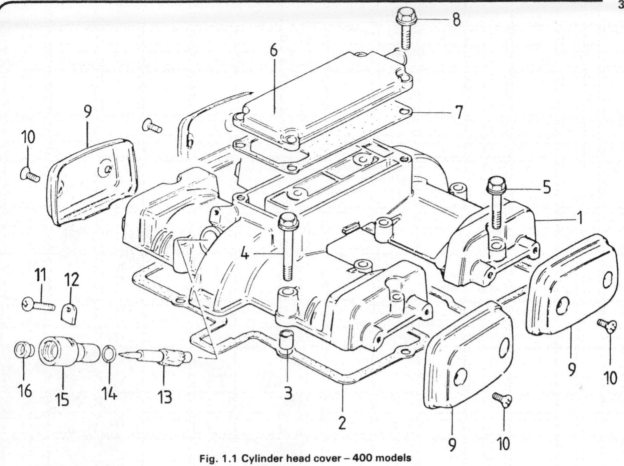

Fig. 1.1 Cylinder head cover – 400 models

1 Cylinder head cover	7 Breather cover gasket	12 Retaining plate
2 Cover gasket	8 Bolt – off	13 Tachometer drive gear
3 Locating dowel – 2 off	9 End cap – 4 off	14 O-ring
4 Bolt – off	10 Screw – 8 off	15 Drive gear housing
5 Bolt – off	11 Screw	16 Oil seal
6 Breather cover		

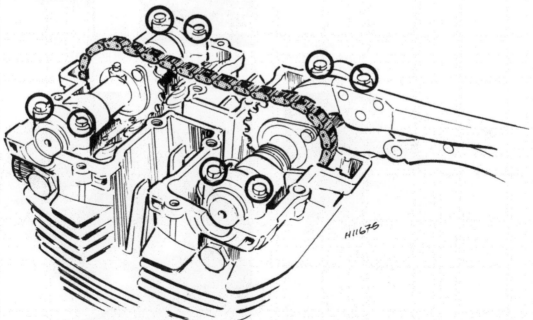

Fig. 1.2 Method of camshaft retension when removing bearing caps

7 Dismantling the engine/gearbox unit: removing the cylinder head, cylinder block, pistons and rings

1   Withdraw the cam chain guide from its location between the two pairs of exhaust valves and commence removal of the cylinder head by unscrewing the single 6 mm bolt located between the exhaust ports. Slacken the eight cylinder head retaining nuts, each by half a turn to start with and then repeating the sequence until all the nuts can be run off their studs. The nuts must be slackened in a diagonal sequence to avoid any risk of distortion occurring to the cylinder head. Note that the numbers cast into the casing adjacent to the nuts refer to the tightening sequence; this sequence should be reversed when slackening the nuts. It was found in practice that a $\frac{3}{8}$ inch drive adaptor fitted with a thin-walled socket of the correct size was the only tool (with the possible exception of a box spanner) with which it was possible to gain proper access to, and therefore slacken, the nuts. Once freed of its stud, each nut can be removed from its recess within the cylinder head by using a pair of long-nose pliers; a similar method should be used to remove the copper washer located beneath each nut.

2   Lift the cylinder head, if necessary using a soft-faced mallet to break the seal between the cylinder head and gasket. Strike only those parts of the casting which are well supported by lugs or webs. **Under no circumstances** should levers be used to raise the head; this will only result in broken cooling fins. Where required, prevent the cam chain from falling down its tunnel into the crankcase by rearranging the temporary securing wire or rod. After lifting the cylinder head from position, remove the two locating dowels and the gasket.

3   Rotate the crankshaft until the two pistons are at approximately equal positions in the bore. Again using a rawhide mallet, separate the cylinder block from the base gasket, taking care not to damage the cooling fins.

4   Slide the cylinder block up and off the pistons taking care to support each piston as the cylinder block becomes free. If a top end overhaul is being carried out, place a clean rag in each crankcase mouth before the lower edge of each cylinder frees the rings. This will preclude any small particles of broken ring falling into the crankcase. It will no doubt be appreciated that the job of removing the cylinder block will be made much easier if an assistant is present to both support the pistons as they leave their bores and to guide the cam chain through its tunnel. Endeavour to lift the cylinder block squarely so that the pistons do not bind in their bores.

5   Note the fitted positions of the oil supply jets located on the outer edge of each crankcase mouth and remove them from their locations. Renew the O-ring on each jet and store them safely until required for reassembly. Remove the cylinder base gasket and locating dowels.

6   Before removing the pistons, note that each one has an arrow on its crown. The arrow should point towards the front of the engine if the piston has been correctly fitted. Mark the fitted position of each piston with an R or L inside its skirt directly after it is disconnected from the connecting rod. This will aid subsequent fitting of the pistons in their correct locations.

7   Proceed to remove each piston by using the flat of a small screwdriver to prise one of the gudgeon pin circlips out of position, then press the gudgeon pin out of the small-end eye through the piston boss. If the pin is a tight fit, it may be necessary to warm the piston so that the grip on the gudgeon pin is released. A rag soaked in warm water and wrapped around the piston should suffice. The piston may be detached from the connecting rod once the gudgeon pin is clear of the small-end eye.

8   If the gudgeon pin is still a tight fit after warming the piston, it can be lightly tapped out of position with a hammer and soft metal drift. **Do not** use excess force and make sure the connecting rod is supported during this operation, or there is a risk of it bending.

9   With the piston free of the connecting rod, remove the second circlip and fully withdraw the gudgeon pin from the piston. Place the piston and gudgeon pin aside for further attention. On no account reuse the circlips, they should be discarded and new ones fitted during rebuilding.

7.1 Note the single 6 mm bolt located between the exhaust ports...

7.5 ...and the oil supply jets located on the edge of each crankcase mouth

7.7 Prise each gudgeon pin circlip from position

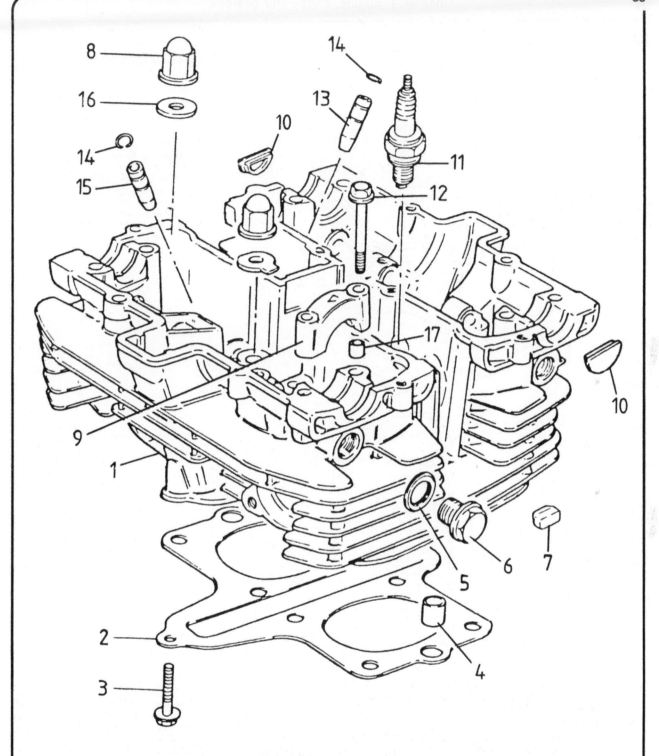

Fig. 1.3 Cylinder head – all models

1	Cylinder head	7	Damping rubber – 2 off	13	Inlet valve guide – 4 off
2	Cylinder head gasket	8	Nut – 8 off	14	Locating ring – 8 off
3	Bolt	9	Bearing cap – 4 off	15	Exhaust valve guide – 4 off
4	Locating dowel – 2 off	10	Camshaft end plug – 4 off	16	Copper washer – 8 off
5	Sealing washer – 4 off	11	Sparking plug	17	Locating dowel – 8 off
6	Rocker spindle plug – 4 off	12	Cap securing bolt – 8 off		

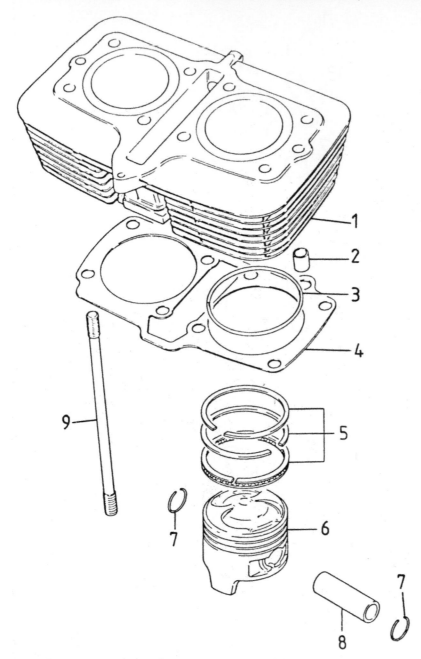

Fig. 1.4 Cylinder barrel – all models

1   Cylinder barrel
2   Locating dowel
3   O-ring – 2 off
4   Cylinder base gasket
5   Piston ring set – 2 off
6   Piston – 2 off
7   Circlip – 4 off
8   Gudgeon pin – 2 off
9   Stud – 8 off

8   Dismantling the engine/gearbox unit: removing the alternator assembly and starter motor

1   Free the electrical leads to the alternator stator from the cable clip on the neutral indicator switch (or gear position switch) and remove the switch from its housing by unscrewing the two crosshead retaining screws. Store the switch in a clean, dry place until it is required for reassembly. Note the pin and spring located beneath the switch; these too should be stored safely.

2   The left-hand crankcase cover complete with the alternator stator assembly can be removed by unscrewing its retaining screws. These screws should be loosened evenly and in a diagonal sequence to prevent any risk of distortion to the cover. Using a soft-faced mallet, tap carefully around the mating face of the cover to free it from the crankcase. Lift the cover clear of the alternator rotor and remove the gasket and the single

locating dowel from the crankcase mating surface. There is no need to remove the stator assembly from its location within the crankcase cover unless the unit has been proved to be defective; in which case it can be easily removed by releasing its three crosshead retaining screws and lock washers and removing the three electrical lead retaining plates before lifting the complete assembly out of its location. It was found in practice that the retaining screws were extremely tight and required the use of an impact driver to free them. Once removed, the cover and stator assembly should be stored in a safe, dry place to avoid accidental damage occurring to the coils.

3   Withdraw the starter motor idler gear and shaft. The starter motor itself is secured by two bolts passing through a flange in the end cap, into the crankcase. Remove the bolts and ease the starter motor across towards the right-hand side until it hits the chamber wall. The motor can then be lifted up at the right-hand end and away from the engine. If difficulty is encountered in moving the starter motor initially, a wooden lever may be used

between the gear casing wall and starter motor front cover. **NEVER** strike the starter motor shaft as this may damage the reduction gear within the motor.

4   Remove the bolt and lock washer retaining the alternator rotor to the crankshaft end. The rotor centre boss is provided with two milled flats to which a spanner may be fitted to prevent rotor rotation during bolt removal. If a spanner cannot be fitted over these flats without its coming into contact with and therefore damaging the edge of the rotor, lock the crankshaft in position by placing a close-fitting tommy bar through the small-end eye of one of the connecting rods so that when the crankshaft is turned by means of the rotor retaining bolt the ends of the bar abut on the mating surface of the crankcase mouth. Care must be taken to place an effective form of padding between the bar and mating surface, soft wood blocks being ideal for the purpose.

5   The alternator rotor is a tapered fit on the crankshaft end, and as such must be drawn from position. The centre of the rotor has an internal thread povided to take a slide hammer. This tool consists of a long, headed rod upon which a free sliding weight is fitted. After screwing the rod firmly into the rotor centre the weight is slid along the rod with force until it strikes the head. This action will separate the two tapers.

6   If a suitable slide hammer is not available, the rear wheel spindle may be used to draw the rotor off its taper by using it in conjunction with a spacer cut from a length of metal dowel. The dowel should be of as large a diameter as possible and of such a length that once inserted into the centre hole of the rotor its upper end is a good $\frac{1}{4}$ inch below the top of the rotor centre boss. Ensure that the threaded end of the wheel spindle is clean of all road dirt before fitting it into the rotor boss until its end abuts onto that of the dowel. Take care to check the thread compatibility of the boss and spindle when inserting the spindle and ensure that at least four turns can be made on the spindle before it comes into contact with the dowel, otherwise there is a risk of the thread being stripped. Insert a tommy bar through the hole in the end of the spindle and tighten the spindle whilst locking the crankshaft in position by the method described in paragraph 4. Sharply tap the end of the spindle; this should free the rotor from the crankshaft. If the rotor does not free at the first attempt, check tighten the spindle and repeat the procedure. Once freed, the rotor assembly may be removed from the crankshaft end and stored with the stator.

7   Before proceeding further, remove the oil seal retaining plate (where fitted) from around the gearbox sprocket shaft and clutch pushrod by knocking the tab away from the head of each retaining bolt and unscrewing the bolts.

8.1 Remove the neutral indicator switch from its housing

8.2 The alternator stator assembly is contained within the left-hand crankcase cover

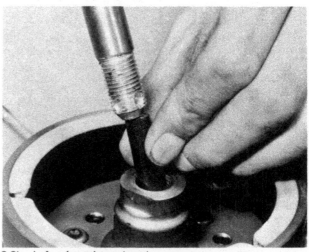

8.6a Remove the alternator rotor retaining bolt and washer...

8.6b ...before inserting a dowel spacer...

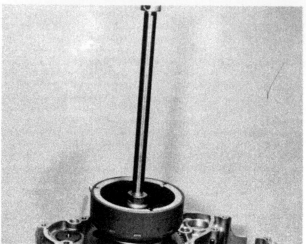

8.6c ...and fitting and tightening the wheel spindle to free the rotor

8.7 Remove the oil seal retaining plate – where fitted

9 Dismantling the engine/gearbox unit: removing the ignition pick-up assembly

1   The ignition pick-up assembly is located beneath the small circular cover on the forwardmost half of the right-hand crankcase cover. Once the cover is removed, the pulse generator unit may be detached from the crankcase cover by removing its two mounting screws, detaching the electrical lead from the oil pressure switch and lifting the unit from position whilst at the same time detaching the electrical cable from its retaining clips along the crankcase wall. Note that it is a good idea to mark accurately the generator unit baseplate in relation to the crankcase cover before disturbing it. This will facilitate setting-up of the assembly during the refitting procedure.
2   Remove the automatic timing unit (ATU) by placing an open-ended spanner across the flats of the crankshaft turning nut and holding the nut in position whilst unscrewing the centre bolt located within the nut itself. Withdraw the centre bolt, together with the nut and lift the ATU out of its housing. Note the locating pin set in the crankshaft end.

10 Dismantling the engine/gearbox unit: removing the clutch assembly

1   To gain access to the clutch assembly, remove the right-hand crankcase cover by unscrewing the two retaining screws situated within the ATU housing followed by the retaining screws around the periphery of the cover. These screws should be loosened a little at a time at first and in a diagonal sequence in order to avoid any risk of the cover becoming distorted.
2   Remove the cover gasket and its two locating dowels and proceed to remove the clutch assembly by unscrewing the spring retaining bolts. Take care to avoid placing any undue strain upon the pressure plate by unscrewing the bolts in small increments and in a diagonal sequence until they can be removed along with their plate washers. The pressure plate can now be withdrawn from the hub along with the clutch springs to expose the bearing and operating boss located behind it. Remove the bearing and withdraw the operating boss. Note that 400 models may have a shim located between the bearing and pressure plate.
3   Remove the plain and friction plates from their location within the clutch drum and place them in the order of removal on a clean work surface so that they may be inspected and if serviceable, refitted in the same order. Knock the centre nut tab

washer away from the flat of the nut and place a close-fitting socket or box spanner over the nut. Prevent the clutch shaft from rotating by loosely refitting the gearbox sprocket and its retaining nut and washer and locking the sprocket in position by placing a short length of final drive chain between the sprocket and crankcase as shown in the accompanying photograph. With the engine unit properly supported, loosen the centre nut and remove it, the tab washer, the clutch hub, the large thrust washer and the clutch drum assembly from the splined shaft. Remove the gearbox sprocket and withdraw the clutch operating pushrod from the clutch end of the shaft. On those models fitted with an oil seal retaining plate (photo 8.7) it is possible to renew the clutch pushrod oil seal at this stage. After removal of the plate the seal can be carefully levered from location. On all other models (without the retaining plate) the seal can only be renewed after the crankcases have been separated.

11 Dismantling the engine/gearbox unit: removing the gear selector external components and the oil pump assembly

1   Slacken the pivot screw, which passes through the gear change drum stopper arm, sufficiently to allow the arm to be lifted up away from the change drum and released so that the roller end clears the end of the drum. The pivot bolt may now be unscrewed fully without danger of the return spring tearing the bolt out of the last few threads in the casing. Detach the return spring from the anchor plate in the casing in order to free the arm fully.
2   Depress the main change arm so that the pawls clear the change drum end and then withdraw the gearchange shaft complete with the arm. Remove the three or four crosshead screws that hold the bearing retainer plate in position over the mainshaft and remove the plate. In practice these screws were found to be extremely tight and required the use of an impact driver to free them.
3   To remove the oil pump, the oil pump driven gear must first be removed. Displace the circlip which retains the gear and slide the gear off the pump drive pin. Note the fitting of a washer, which should also be removed, behind the driven gear.
4   Remove the three screws which retain the oil pump in position and pull the oil pump from place. These crosshead screws will almost certainly again be very tight, and require removal with an impact driver. With these screws removed, pull the oil pump from place. Note the two O-rings fitted to the rear of the pump; these should be renewed as a matter of course upon reassembly. Finally, take great care when removing the pump to retain the drive pin.

9.1 Detach the lead from the oil pressure switch

9.2 Withdraw the centre bolt and crankshaft turning nut to free the ATU

10.2 Take care to avoid placing undue strain upon the clutch pressure plate when removing the spring retaining bolts

10.3 Use a short length of final drive chain to lock the gearbox sprocket in position

11.3 Displace the circlip which retains the oil pump driven gear

12 Dismantling the engine/gearbox unit: removing the sump cover and separating the crankcase halves

1   Before attempting to remove the sump cover, the engine unit should be inverted and held over a suitable receptacle to allow any oil remaining within the unit to drain out through the crankcase mouths. On completion of the oil draining, keep the engine unit inverted and position it on wooden blocks on the work surface so that no strain is placed on the cylinder block retaining studs.

2   Commence removal of the sump cover by unscrewing its retaining bolts. These bolts should be loosened a little at a time at first and in a diagonal sequence to prevent any risk of the cover becoming distorted. Before removing the bolts, note for reference when refitting, the position of the electrical cable retaining clips located beneath two of the bolt heads. It will be found that even with all the retaining bolts removed, the cover will be stuck fast in position and will need to be tapped quite sharply around its mating face with a soft-faced mallet in order to free it.

3   Inspect the gauze filter plate for contamination and if necessary remove it for cleaning in accordance with the instructions given in Chapter 2.

4   Remove the two Allen bolts from within the oil filter housing. Suzuki recommend that a special T-type tool (No 09914-25811) be used for this purpose but it was found that a standard Allen key was sufficient, although a small diameter tube (or box spanner) had to be placed over the key 'handle' in order to form an extension so that enough leverage could be placed on the key in order to free the bolts.

5   Because the bolts holding the two halves of the crankcase together are of varying lengths, it is advisable to make up a template from a sheet of card so that when each bolt is removed it may be inserted into its respective hole in the template and thus its position noted for reference when refitting. To avoid any risk of distortion to either of the crankcase halves, slacken the bolts a little at a time at first and in a diagonal sequence. Note the position of the earth lead fitted underneath one of the bolt heads. Carry out a check to ensure that all bolts have been removed and reposition the engine so that a single retaining bolt located on the right-hand side of the starter motor housing can be removed.

6   With the engine once again inverted and blocked in position, sharply tap around the mating faces of the crankcase halves with a soft-faced mallet so as to break the seal between them. A prising point is provided at both the front and rear mating faces of the crankcase halves. These points provide a means by which the crankcase halves may be levered apart. Suzuki recommend that a special tool (No 09912-34510) be inserted into the recess provided and used to ease the crankcase halves apart, although it was found in practice that the flat of a large screwdriver was perfectly adequate for the job. Take great care when using the screwdriver as a lever not to place too great a strain on the crankcase castings; if all the bolts have been removed and the seal between the mating faces broken as described previously, it will only be necessary to use a moderate amount of force to cause the crankcase halves to separate.

7   Once the crankcase halves have been fully separated, remove the four locating dowels from the mating faces and the two small rubber retaining blocks from their location in the crankcase casting at the base of the cam chain tensioner blade.

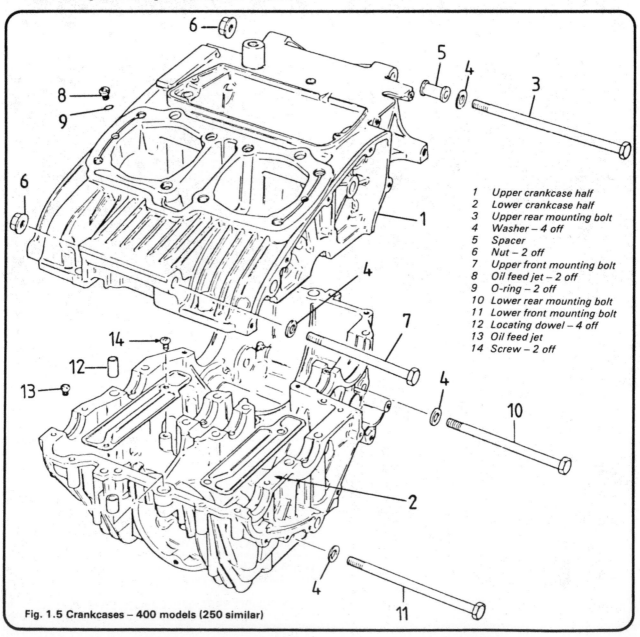

1   Upper crankcase half
2   Lower crankcase half
3   Upper rear mounting bolt
4   Washer – 4 off
5   Spacer
6   Nut – 2 off
7   Upper front mounting bolt
8   Oil feed jet – 2 off
9   O-ring – 2 off
10   Lower rear mounting bolt
11   Lower front mounting bolt
12   Locating dowel – 4 off
13   Oil feed jet
14   Screw – 2 off

Fig. 1.5 Crankcases – 400 models (250 similar)

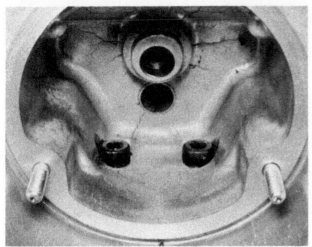

12.4 Remove the two Allen bolts from within the oil filter housing

12.6 One of the prising points provided as an aid to separation of the crankcase halves

13 Dismantling the engine/gearbox unit: removing the counterbalance shaft, crankshaft assembly and gearbox components

1 Before removing the counterbalance shaft from the crankcase, check that the shaft gear pinion outer face has a punch mark which corresponds to a similar mark on the outer face of the crankshaft drive pinion. If these marks are not visible, mark each gear pinion at the point where they meet so that they may be aligned in the same position when refitted; this is to ensure the counterbalance shaft is positioned exactly 180° out of phase with the crankshaft.

2 Lift the counterbalance shaft out of the crankcase, rotating it at the same time so that its gear pinion clears that of the crankshaft. Grasp the crankshaft assembly with both hands and lift it out of the crankcase, together with the cam chain. If the crankshaft is firmly seated it may be tapped lightly with a soft-faced mallet in order to free it. Take great care whilst using this procedure.

3 Lift the two gearbox shafts individually out of the crankcase, together with their pinions, seals and bearings. Place each shaft on a clean work surface ready for inspection and cover them with a piece of clean rag or paper to prevent any ingress of dirt or dust into the assemblies. Note the positions of the bearing

location half clips in the crankcase and very carefully prise them from position. Note also the mainshaft bearing locating pin and if possible, draw it out of its recess within the crankcase bearing location and store it with the half clips to prevent loss. If any bearing shells become dislodged from either casing half during dismantling they should be refitted in their original positions until the examination stage. This applies equally on 400cc models to the two crescent shaped thrust bearings fitted to the outer faces of the centre bearing webs in the upper casing half.

4 Note carefully the position of each of the three selector forks, marking them with tape if necessary. Remove the crosshead retaining screw from the end of the selector fork shaft and withdraw the shaft from the crankcase. Lift out the freed selector forks and refit them on the shaft in their original positions so as to facilitate subsequent assembly.

5 Move to the drive pin end of the gearchange drum and remove the single bolt retaining the guide plate in position. Lift away the guide plate and reposition the crankcase half so that the detent plunger retaining bolt can be removed from its location behind the starter motor housing. Once the retaining bolts, sealing gasket, spring and plunger are removed, the gearchange drum may be withdrawn from the casing. Finally, pull the cam chain tensioner blade from its location within the upper crankcase half.

6 On 400cc machines two baffle plates are fitted to the floor of the crankcase lower half. These need not be removed.

13.1 Check the alignment marks of both the counterbalance shaft and crankshaft pinions

13.3a Note the positions of the bearing location half clips...

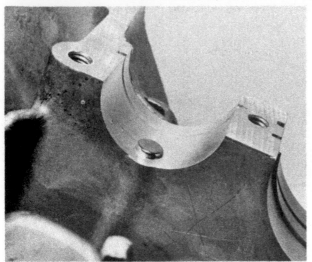

13.3b ...and the mainshaft bearing locating pin

13.4 The selector fork shaft is retained in position by a single retaining screw

13.5a Remove the second of the guide plate retaining bolts...

13.5b ...followed by the detent plunger retaining bolt...

13.5c ...the plunger spring...

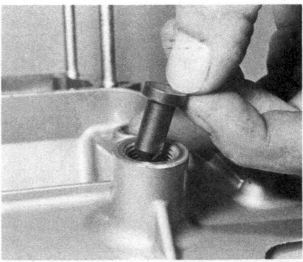

13.5d ...and the plunger itself

14 Examination and renovation: general

1 Before examining the individual components of the dismantled engine/gearbox unit for wear, it is essential that they should be cleaned thoroughly. Use a paraffin/petrol mix to remove all traces of old oil and sludge that may have accumulated within the engine.

2 Examine the crankcase castings for cracks or other signs of damage. If a crack is discovered, it will require professional repair.

3 Examine carefully each part to determine the extent of wear, if necessary checking with the tolerance figures listed in the Specifications Section of this Chapter, or accompanying the text.

4 Use a clean, lint-free rag for cleaning and drying the various components, otherwise there is risk of small particles obstructing the internal oilways.

5 Should any studs or internal threads require repair, now is the appropriate time to attend to them. Where internal threads are stripped or badly worn, it is preferable to use a thread insert, rather than tap oversize. Most dealers can provide a thread reclaiming service by the use of Helicoil thread inserts. They enable the original component to be re-used.

15 Examination and renovation: main and big-end bearings

1 The models covered in this Manual are fitted with shell type bearings on the crankshaft and big-end assemblies. Bearing shells are relatively inexpensive and it is prudent to renew the entire set of main bearing shells when the engine is dismantled completely, especially in view of the amount of work which will be necessary at a later date if any of the bearings fail. Always renew all four sets of main bearings together.

2 Wear is usually evident in the form of scuffing or score marks in the bearing surface. It is not possible to polish these marks out in view of the very soft nature of the bearing surface and the increased clearances that will result. If wear of this nature is detected, the crankshaft must be checked for ovality as described in the following Section.

3 Failure of the big-end bearings is invariably accompanied by a pronounced knock within the crankcase. The knock will become progressively worse and vibration will also be experienced. It is essential that bearing failure is attended to without delay because if the engine is used in this condition there is a risk of breaking a connecting rod or even the crankshaft, causing more extensive damage.

4 Before the big-end bearings can be examined the bearing caps must be removed from each connecting rod. Each cap is retained by two high tensile nuts. Before removal, mark each cap in relation to its connecting rod so that it may be replaced correctly. As with the main bearings, wear will be evident as scuffing or scoring and the bearing shells must be replaced as complete sets. If the bearings appear to be in good order measurement of the clearances should be made as described in Section 17. Note that once the bearing caps are detached from the connecting rods, the studs will remain firmly attached to the rods. Under no circumstances attempt to displace the studs as this is likely to result in their becoming misaligned, therefore making accurately refitting of the bearing caps difficult if not impossible.

5 When selecting bearing shells to fit the big-end bearings, note the number etched on the machined side of the connecting rod where the cap meets the rod itself. This number is the code for the connecting rod big-end inner diameter. Note also the corresponding code number marked on the innermost face of each flywheel. This number is the code for the crankpin overall diameter. Compare these numbers with the table accompanying this text and cross refer to obtain the correct colour coding for the bearing shell required. This colour coding indicates the bearing thickness and part number and should be quoted when ordering a replacement item.

6 A similar method to the above is used to select the correct main bearings. Again the bearing shells are colour coded to denote the bearing thickness and part number; this colour coding being obtained by cross-referring the crankcase bearing housing inner diameter code letter with the code letter for the crankshaft journal outside diameter. The code letter for the crankcase bearing housing may be found stamped on the rearmost face of the upper crankcase half, whereas the code letter for the crankshaft journal is marked on the flywheel adjacent to the journal. Both main bearing and big-end bearing shells are colour marked on one edge.

Main bearing and counterbalance bearing selection table

		Crankshaft and Balancer shaft		
	Code	A	B	C
Crankcase	A	Green	Black	Brown
	B	Black	Brown	Yellow

Big-end bearing selection table

		Crank pin		
	Code	1	2	3
Conrod	1	Green	Black	Brown
	2	Black	Brown	Yellow

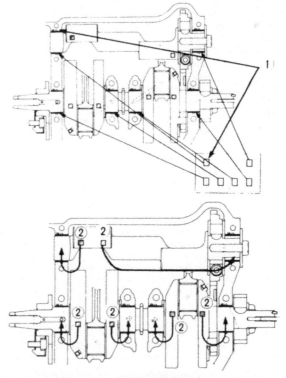

Fig. 1.6 Main and balance shaft bearing code number positions

1 *Crankcase housing inside diameters*
2 *Shaft journal outside diameters*

15.5a Note the number etched on the connecting rod...

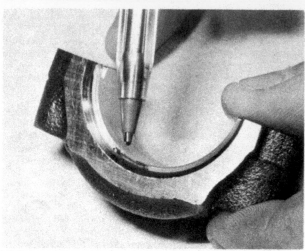

15.5b ...together with the colour coding on the bearing shell

16 Examination and renovation: connecting rods

1   It is unlikely that any of the connecting rods will become damaged during normal usage unless an unusual occurrence such as a dropped valve causes the engine to lock. This may well bend the connecting rod in that cylinder. Carelessness when removing a tight gudgeon pin can aso give rise to a similar problem. It is not advisable to straighten a bent connecting rod; renewal is the only satisfactory solution.

2   The small-end eye of the connecting rod is unbushed. If the gudgeon pin is found to be a slack fit in the eye, check the outside diameter of the pin and the internal diameter of the small-end eye and compare the measurements obtained with those given in the Specifications at the beginning of this Chapter. If either component is worn beyond the service limits given, it must be replaced with a new item. Always check that the oil hole in the small-end eye is not blocked since if the oil supply is cut off, the bearing surfaces will wear very rapidly.

3   Measure the clearance between the thrust face of the crankshaft flywheel and the side face of the connecting rod big-end by inserting a feeler gauge between the two. If the measurement obtained exceeds that given in the Specifications Section of this Chapter, the connecting rod will have to be removed from the crankshaft and measured across its big-end width. Any excessive wear will necessitate renewal of the connecting rod. If the width of the big-end is found to be within limits, then the width of the crank pin must be suspect. Measure between each flywheel thrust face and compare the measurement obtained with that given in the Specifications Section. If the width of the crankpin is found to be in excess of the measurement given in the Specifications, it is possible that the crankshaft assembly will need to be renewed. Before making a final decision on this matter, return the crankshaft assembly, together with the connecting rods, to an official Suzuki service agent who will confirm whether renewal is necessary.

17 Examination and renovation: crankshaft assembly

1   If wear has necessitated the renewal of the big-end and/or main bearing shells, the crankshaft should be checked with a micrometer to verify whether ovality has occurred. Suzuki give no figures for the maximum amount of ovality allowed but if any amount of ovality is found it is advisable to obtain the advice of an official Suzuki service agent.

2   Mount the crankshaft by supporting both ends on V-blocks or between centres of a lathe and check the run-out at the centre main bearing surfaces by means of a dial gauge. The run-out will be half that of the gauge reading indicated. A measured run-out of more than 0.05 mm (0.002 in) indicates the need for crankshaft renewal. It is wise, however, before taking such drastic (and expensive) action, to consult a Suzuki specialist.

3   The clearance between any set of bearings and their respective journal may be checked by the use of plastigage (press gauge). Plastigage is a strip of plastic material that can be compressed between two mating surfaces. The resulting width of the material when measured with the gauge provided will give the amount of clearance. For example, if the clearance in the big-end bearing was to be measured, plastigage should be used in the following manner.

4   Cut a strip of plastigage to the width across the bearing to be measured. Place the plastigage strip across the bearing journal so that it is parallel with the crankshaft. Place the connecting rod complete with its half shell on the journal and then carefully refit the bearing cap complete with half shell onto the connecting rod bolts. Fit and tighten the retaining nuts to the correct torque and then loosen and remove the nuts and the bearing cap. When checking the main bearing clearances the crankcase halves must be assembled and the bolts tightened in a similar manner. To save time, measure the clearance of each bearing simultaneously.

5   Measure the width of the compressed plastigage at its widest point by comparing it with the scale on the side of the envelope in which the plastigage is supplied. The maximum allowable clearances are listed in the Specifications at the beginning of this Chapter. Clearances may also be checked by direct measurement of each journal and bearing using external and internal micrometers.

6   The crankshaft has drilled oil passages which allow oil to be fed under pressure to the working surfaces. Blow the passages out with a high pressure air line to ensure they are absolutely free. Blanking plugs in the form of small steel balls are fitted in each flywheel, to close off the outer ends of the passages. Check that these balls, which are peened into place, are not loose. A plug coming free in service will cause oil pressure loss and resultant bearing and journal damage.

7   When refitting the connecting rods and shell bearings, note that under no circumstances should the shells be adjusted with a shim, 'scraped in' or the fit 'corrected' by filing the connecting rod and bearing cap or by applying emery cloth to the bearing surface. Treatment such as this will end in disaster; if the

bearing fit is not good, the parts concerned have not been assembled correctly. This advice also applies to main bearing shells.

8   Oil the bearing surfaces before reassembly takes place and make sure the tags of the bearing shells are located correctly. Check also that the oil hole in each bearing shell is aligned with the oil hole in the base of the connecting rod.

9   When fitting each connecting rod to the crankshaft, ensure that the oil hole drilled in the shoulder of the rod faces to the rear. After the initial tightening of the connecting rod nuts, check that each connecting rod revolves freely, then tighten to a torque setting of 21.5 – 24.5 lbf ft (3.0 – 3.4 kgf m). Check once again for ease of rotation.

10   Crankshaft endfloat is determined by the width across the flanks of the centre main bearing supports in the upper crankcase half and the distance between the machined faces of the two centre crankshaft webs. On 250 cc models, place the crankshaft in position in the upper crankcase half and, using a feeler gauge, check the thrust clearance between the machined face of one flywheel web and the adjacent bearing support wall. If the clearance exceeds 0.35 mm (0.0138 in) take direct measurements of the crankshaft and crankcase dimensions to determine which component exceeds the service limit.

Service limits
Centre bearing width	53.90 – 54.00 mm
	(2.122 – 2.166 in)
Flywheel to flywheel width	54.05 – 54.15 mm
	(2.128 – 2.132 in)

To restore the correct clearance it is recommended by Suzuki that the worn component is renewed. This, however, is an expensive cure. It is possible that a competent light engineering specialist might modify the cases to take thrust bearings of the type fitted as standard to the 400cc models.

11   On 400cc models the crankshaft endfloat is controlled by removable crescent-shaped thrust bearings located in recesses in the flanks of the centre main bearing supports. Ease the bearings from position and measure their thickness with a micrometer. If the thickness is below the service limit of 2.85 mm (0.112 in) they should be renewed to restore the endfloat clearance.

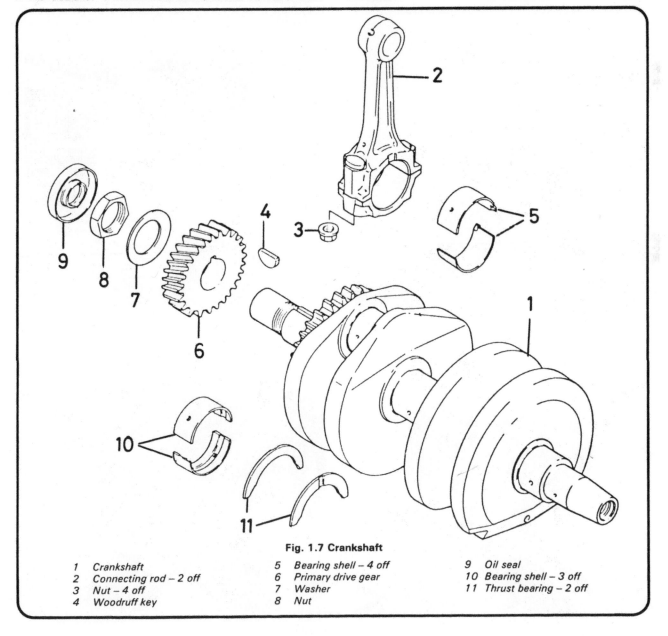

Fig. 1.7 Crankshaft

1	Crankshaft	5	Bearing shell – 4 off	9	Oil seal
2	Connecting rod – 2 off	6	Primary drive gear	10	Bearing shell – 3 off
3	Nut – 4 off	7	Washer	11	Thrust bearing – 2 off
4	Woodruff key	8	Nut		

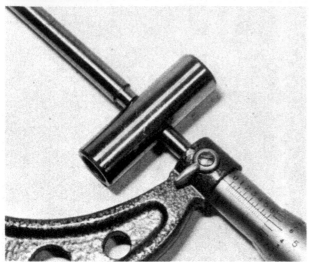

16.2 Measure the outside diameter of the gudgeon pin

16.3a If the width of the connecting rod big-end is found to be within limits...

16.3b ...check the measurement between each flywheel thrust face

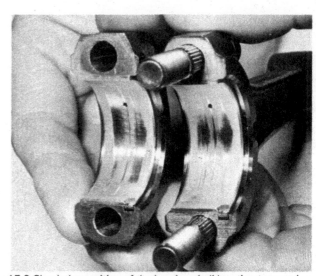

17.8 Check the position of the bearing shell locating tags and oil holes...

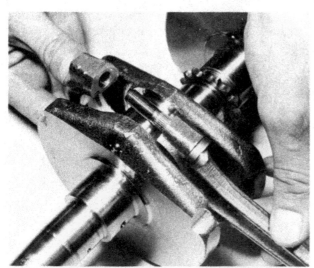

17.9a ...fit the connecting rod to the crankshaft...

17.9b ...and tighten the securing nuts to the correct torque loading

17.10a Measuring distance across centre main bearings

17.10b Measuring distance between machined flywheel faces

18 Examination and renovation: counterbalance shaft

1   The counterbalance shaft runs on two shell bearings which may be checked in a similar manner to that described in Section 17 for the crankshaft main bearings. Reference to the figures accompanying the text of Section 15 should also be made when identifying the bearings.

2   Inspect the condition of the shaft pinion and also the drive pinion on the crankshaft with which it meshes. Examine for excessive wear and broken or chipped teeth. If renewal of the drive pinion on the crankshaft is considered to be necessary, the crankshaft assembly should be taken to an official Suzuki service agent who will give further advice on the matter. The construction of the balance shaft assemblies fitted to the 250 and 400 models differs in that the assembly fitted to the 400 models includes a shock absorber assembly between the shaft pinion and the shaft whereas on the 250 models the pinion is a direct fit on the shaft. Service each assembly as follows.

400 models

3   Grasp the balance shaft pinion firmly and attempt to turn it whilst holding the shaft steady. If looseness or rattling of the three shock absorber springs can be detected, the balance shaft should be dismantled and the springs removed.

4   The balance shaft pinion may be pulled from position using a two or three-legged sprocket puller. Grip the shaft between the protected jaws of a vice and loosen the centre bolt approximately two turns. Leave the bolt in place as a thrust point for the puller centre. Draw the pinion off slowly, undoing the centre bolt at intervals to give room for movement. To prevent the three shock absorber springs from flying out in an uncontrolled manner, wrap a rag around the assembly until the pinion has been removed. Any springs remaining in place may be carefully prised from position, using the flat of a small screwdriver to do so. Place each spring, together with its two pads, to one side ready for inspection and reassembly.

5   The shock absorber centre boss is a tight interference fit on the shaft and is located in position by a Woodruff key. Removal is not normally required because if the boss becomes worn it must be renewed complete with the shaft. These components are not supplied separately.

6   Commence reassembly of the shock absorber unit by inspecting each spring for damage or loss of tension and its pads for damage or deterioration. The minimum service length for the springs is 14.2 mm (0.56 in). If any spring has 'set' to less than this length the springs should be renewed as a set.

7   Place each spring and set of pads in position in the recessed centre boss. Position the pinion and push it fully home onto the springs. Note that a punch mark is provided on the outer face of the pinion. This punch mark must be aligned with a similar mark on the centre boss face. Inspect the bearing surface of the spacer for any sign of damage or excessive wear before lightly lubricating it with engine oil and refitting it to the shaft. Refit the large end washer. Before fitting the end bolt, clean its threads and coat them with a thread locking compound. Fit the bolt and tighten it to the specified torque loading.

250 models

8   If the balance shaft pinion is seen to be damaged, it should be removed from the shaft by using the procedure given in paragraph 4 of this Section. Do not omit to refit the Woodruff key when fitting the replacement pinion and follow the procedure given in the preceding paragraph when refitting the spacer, end washer and bolt.

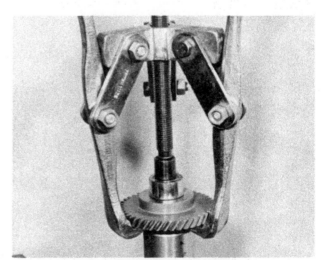

18.8 Pull the balance shaft pinion from position with a tool similar to that shown above (250 models)

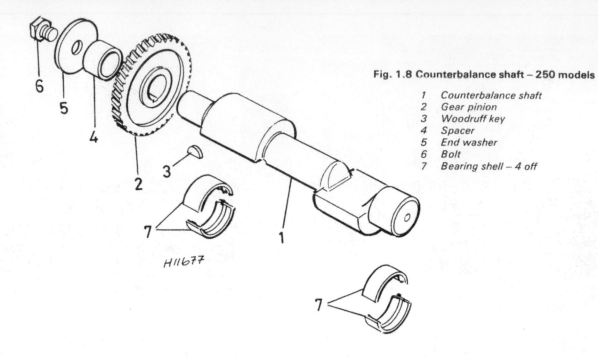

Fig. 1.8 Counterbalance shaft – 250 models

1   Counterbalance shaft
2   Gear pinion
3   Woodruff key
4   Spacer
5   End washer
6   Bolt
7   Bearing shell – 4 off

H11677

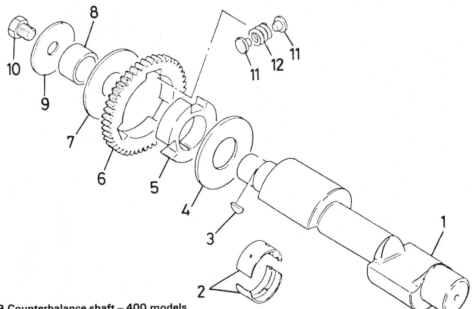

Fig. 1.9 Counterbalance shaft – 400 models

1   Counterbalance shaft
2   Bearing shell – 4 off
3   Woodruff key
4   Washer
5   Shock absorber centre boss
6   Gear pinion
7   Washer
8   Spacer
9   End washer
10 Bolt
11 Shock absorber pad – 6 off
12 Spring – 3 off

19 Examination and renewal: oil seals

1   Oil seals are fitted to both the crankshaft and gearbox mainshaft assemblies. In time it is possible that any one of these seals will wear or harden with age thus allowing oil to find its way past. Although the seals themselves are relatively inexpensive, a considerable amount of dismantling work is necessary should one fail in service. For this reason it is advisable to renew all oil seals in the event of a full engine overhaul. In any case, a seal which looks remotely suspect should be renewed.

20 Examination and renovation: cylinder block

1   The usual indication of badly worn cylinder bores and pistons is excessive smoking from the exhausts, high crankcase compression which causes oil leaks, and piston slap, a metallic rattle that occurs when there is little or no load on the engine. If the top of the cylinder bore is examined carefully, it will be found that there is a ridge at the front and back, the depth of which will indicate the amount of wear which has taken place. This ridge marks the limit of travel of the top piston ring.
2   Since there is a difference in cylinder wear in different directions, side to side and back to front measurements should be made. Take measurements at three different points down the length of the cylinder bore, starting at a point just below the wear ridge and following this with measurements half way down the bore and at a point just above the lower edge of the bore. If any of these measurements exceed the service limit given in the Specifications Section of this Chapter, the cylinder must be rebored and fitted with an oversize piston. Never rebore one cylinder without reboring the other to the same size as this will only result in the balance of the engine being disturbed with the resultant increase in vibration and wear rate.
3   Oversize pistons are available in two oversizes of 0.5 mm (0.020 in) and 1.0 mm (0.040 in).
4   Check that the surface of the cylinder bore is free from score marks or other damage that may have resulted from an earlier engine seizure or a displaced gudgeon pin. A rebore will be necessary to remove any deep scores, irrespective of the amount of bore wear that has taken place, otherwise a compression leak will occur.
5   Clean the cylinder block to cylinder head mating surface and check it for distortion with a straight-edge and a set of feeler gauges. The straight-edge should be laid diagonally across the mating surface from corner to corner and any clearance between its lower edge and the mating surface checked with the gauges. Repeat the procedure with the straight-edge laid between the opposite corners and then at several positions in between the four corners. If the largest clearance found exceeds 0.10 mm (0.004 in), Suzuki recommend that the cylinder block be replaced with a new item. It is worth noting however, that if the amount of distortion found is only slightly greater than the limit given, it could well be worth seeking the advice of a competent motorcycle engineer who can advise on whether skimming the mating surface flat is possible without the subsequent risk of the pistons coming into contact with the valve heads once the engine is reassembled and started.
6   Finally, make sure the external cooling fins of the cylinder block are not clogged with oil or road dirt which will prevent the free flow of air and cause the engine to overheat.

21 Examination and renovation: pistons and piston rings

1   Attention to the pistons and piston rings can be overlooked if a rebore is necessary, since new components will be fitted.
2   If a rebore is not necessary, examine each piston carefully.

Reject pistons that are scored or badly discoloured as the result of exhaust gases by-passing the rings.
3   Remove all carbon from the piston crowns, using a blunt scraper, which will not damage the surface of the piston. A scraper made from soft aluminium alloy or hardwood is ideal; never use a hard metal scraper with sharp edges as this will almost certainly gouge the alloy surface of the piston. Clean away carbon deposits from the valve cutaways and finish off with metal polish so that a smooth, shining surface is achieved. Carbon will not adhere so readily to a polished surface.
4   Small high spots on the back and front areas of the piston can be carefully eased back with a fine swiss file. Dipping the file in methylated spirits or rubbing its teeth with chalk will prevent the file clogging and eventually scoring the piston. Only very small quantities of material should be removed, and never enough to interfere with the correct tolerances. Never use emery paper or cloth to clean the piston skirt; the fine particles of emery are inclined to embed themselves in the soft aluminium and consequently accelerate the rate of wear between bore and piston.
5   Measure the outside diameter of the piston about 15 mm (0.6 in) up from the skirt at right angles to the line of the gudgeon pin. To determine the piston/cylinder barrel clearance, subtract the maximum piston measurement from the minimum bore measurement. If the clearance exceeds the service limit given in the Specifications, the piston should ideally be renewed (given that the cylinder bore is within limits). This however, is seeking perfection, and an additional clearance of perhaps 0.025 mm (0.001 in) will not reduce engine performance dramatically.
6   Refer to the figures given in the Specifications Section of this Chapter and check that the gudgeon pin bosses are not worn beyond the service limit or the circlip grooves damaged in any way. Check also that the ring grooves are not enlarged and that the ring to groove clearances for the top and second rings do not exceed the limits given.

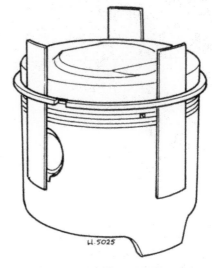

Fig. 1.10 Method of removing and replacing piston rings

7   Piston ring wear can be measured by inserting the rings in the bore from the top and pushing them down with the base of the piston so that they are square with the bore and close to the bottom of the bore where the cylinder wear is least. Place a feeler gauge between the ring ends. If the clearance exceeds the service limit the ring should be renewed. The expander band of the oil control ring cannot be measured. In practice, if wear of the two side rails exceeds the limit, the three components should be renewed. It is advised that, provided new rings have not been fitted recently, a complete set of rings be fitted as a matter of course whenever the engine is dismantled. This action will ensure maintenance of compression and performance. If

new rings are to be fitted to cylinder bores which are in good condition and do not require a rebore, it is essential to have the surface of the bores honed lightly. This operation is known as glazebusting and as the name suggests, it removes the mirror smooth surface which has been produced by the previous innumerable up and down strokes of the piston and rings. If the glaze is not removed, the new rings will glide over the surface, making the running-in process unnecessarily protracted. Note that new rings should not be fitted to a part-worn cylinder bore, as the resulting wear ridge may break the top ring, particularly at high engine speeds. The resulting debris could have very expensive results.

8   Check that there is no build up of carbon either in the ring grooves or the inner surfaces of the rings. Any carbon deposits should be carefully scraped away. A short length of old piston ring fitted with a handle and sharpened at one end to a chisel point is ideal for scraping out encrusted piston ring grooves.

9   Oversize top and second piston rings are stamped with a number on their upper edge adjacent to the end gap. The number 50 denotes an oversize ring of 0.5 mm whereas the number 100 denotes an oversize ring of 1.0 mm. Oversize oil ring spacers are identified by the following colour codes.

Blue denotes 0.5 mm oversize
Yellow denotes 1.0 mm oversize
Red denotes a standard sized spacer

The oil ring side rail oversizes are determined by measurement across the outside diameter.

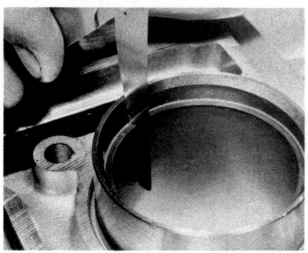

21.7 Measure piston ring wear as shown

22 Examination and renovation: cylinder head and valves

1   It is considered good practice to remove all carbon deposits from the combustion chambers before removing the valves for inspection and grinding-in. Use a blunt ended chisel or scraper so that the surfaces are not damaged. Finish off with a metal polish to achieve a smooth, shining surface. If a mirror finish is required, a high speed felt mop and polishing soap may be used. A chuck attached to a flexible drive will facilitate the polishing operation.

2   Obtain eight marked containers so that the appropriate valve components can be kept separate once removed. A valve spring compression tool must be used to compress each set of valve springs in turn, thereby allowing the split collets to be freed from the valve stem and the spring retaining plate, the springs, the spring seat and the oil seal to be removed from the cylinder head along with the valve itself. As each valve is removed, check that it will pass through the guide bore without

resistance. After high mileages have been covered it is possible that the collet groove will have spread and if this increased diameter is pulled through the guide bore, it will enlarge it. If any resistance is encountered, relieve the high spots with fine abrasive paper until the valve can be removed easily. Mark each valve so that it can be refitted in the correct combustion chamber. There is no danger of inadvertently replacing an inlet valve in an exhaust position, or vice-versa, as the valve heads are of different sizes. Do not mark the valves by centre punching them on the valve head. This method is not recommended on valves, or any other highly stressed components, as it will produce high stress points and may lead to early failure. Tie-on labels, suitably inscribed, or a spirit-based marker, are ideal for the purpose.

3   Using a micrometer, measure the diameter of the valve stem at various points along its length. If any one of the measurements obtained is less than the service limit given in the Specifications Section of this Chapter, the valve should be renewed. Check the amount of valve stem runout by supporting the valve stem below two V-blocks and slowly rotating it whilst measuring the amount of runout with a dial gauge. Use a similar method to measure the amount of radial runout of the valve head and compare both sets of measurements obtained with the service limits given in the Specifications. In either case, if these measurements are beyond limits, then the valve must be renewed.

4   The valve stem to guide clearance can be measured with the use of a dial gauge and a new valve. Place the valve into the guide and measure the mount of shake with the tip of the gauge resting against the edge of the valve stem head. Take two measurements, one at 90° to the other, and renew the valve guide if the amount of wear indicated is greater than the service limit given in the Specifications.

5   To remove an old valve guide, place the cylinder head in an oven and heat to it about 100°C (212°F). The old guide can now be tapped out from the cylinder side. The correct drift should be shouldered with the smaller diameter the same size as the valve stem and the larger diameter slightly smaller than the OD of the valve guide. If a suitable drift is not available a plain brass drift may be utilized with great care. Even heating is essential, if warpage of the cylinder head is to be avoided. Before removing old guides scrape away any carbon deposits which have accumulated on the guide where it projects into the port. Removal of carbon will ease guide movement and help prevent broaching of the guide bore in the cylinder head. If in doubt, seek the advice of a Suzuki specialist. Note that each valve guide is fitted with a metal locating ring which must be discarded along with the guide. It is essential that the cylinder head is properly supported during valve guide removal and that the rocker arms are not allowed to interfere with the guide as it is tapped out of the casing. If necessary, remove each rocker arm and spindle by following the procedure given in Section 23 of this Chapter.

6   Ream out the valve guide holes in the cylinder head with an 11.20 mm (0.44 in) reamer, clean away any swarf and lubricate the surface of each hole. Failure to keep the hole lubricated during insertion of the guide may result in damage to the guide or head. Ensure the cylinder head is well supported on a clean flat worksurface during insertion of the guides and is heated to the same heat as for removal. Fit a new locating ring to each new guide and drive each guide into its hole until the locating ring abuts against its recess in the cylinder head. It should be noted that whereas the valve guides removed from the head and fitted as original equipment are of differing shapes, the replacement items are identical in shape. Following renewal of the guides, each valve seat may have to be recut to centre the seat with the guide axis. Carry out the following check to detemine if this is necessary.

7   Before attempting to fit the valves into their guides, ream out the guide bores with a 5.50 mm (0.216 in) reamer and clean away any metal particles from the bore surfaces and the head casting. Lightly lubricate each valve stem with oil and evenly coat the valve seat with Engineers Blue before fitting

each valve into its guide and pressing it down onto its seat. Lightly tap each valve onto its seat in order to allow the Engineers Blue to form a clear impression of the seating contact. Remove each valve and examine the impression left by the Engineers Blue on the valve face. This impression must be continuous and its width within the limits 0.9 – 1.1 mm (0.035 – 0.043 in).If this is not the case, recut the valve seat by using the following procedure. Note that the valve seats must also be recut if their surfaces have become so pitted that the damage cannot be removed by using the valve grinding procedure listed in the following paragraphs.

8   In view of the high cost of the cutting tools required to recut the valve seats and of the expertise required, it is strongly suggested that this work be entrusted to a Suzuki service agent or a competent motorcycle engineer. Never resort to excessive grinding because this will only pocket the valves in the head and lead to reduced engine efficiency. If, however, it is decided to attempt this work at home, proceed as follows.

9   The valve seat cutting set supplied by Suzuki (tool no 09916-21110) contains cutter attachments with 45° and 70° angles. Fit the 45° attachment to the tool. Insert the solid pilot into the valve guide and rotate it slowly whilst pushing it into the guide until it seats correctly. Place the tool into the pilot and rotate it once to cut the valve seat. Remove the tool and recheck the valve seat contact area with Engineers Blue. Repeat this procedure until the contact area is seen to be continuous. If the contact area is correct but the valve seat is still pitted, repeat the cutting procedure until all signs of pitting are removed, It should be noted however, that if too much of the valve seat is cut away, correct valve adjustment will be made impossible due to the end of the valve stem becoming too close to the rocker arm.

10 On completion of recutting the valve seat, inspect its width to see if it is within the limits 0.9 – 1.1 mm (0.035 – 0.043 in). Check also that the seat contact area with the valve is in the position indicated by the figure accompanying this text; ie in the centre of the valve face. If the contact area is found to be too low, it may be raised and its width narrowed by carefully using the 75° cutter on the tool. This may well result in the valve seat becoming too narrow, in which case the 45° cutter will have to be used to return the seat to its correct width. Alternatively, if the contact area is found to be too high, use the 45° cutter to lower it.

Fig. 1.11 Valve seat renovation

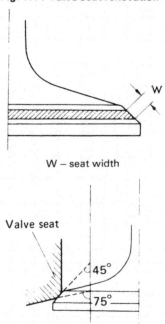

W – seat width

Seat recutting angles

11 Once the correct valve seat width and position is achieved, lightly skim the seat surface with the 45° cutter in order to remove any burrs caused by the cutting procedure. The finished seat surface should be matt in appearance and have a smooth finish, thus providing the ideal surface for correct bedding in of the valve once the engine is started. It is not necessary to grind the valve in on completion of the recutting procedure.

12 If it is found that the valve seat does not need to be recut and is free of all but the lightest pitting, then the valve may be reground into the seat by using the following procedure. Valve grinding is a relatively simple task; commence by smearing a trace of fine valve grinding compound (carborundum paste) on the valve seat and apply a suction tool to the head of the valve. Oil the valve stem and insert the valve in the guide so that the two surfaces to be ground in make contact with one another. With a semi-rotary motion, grind in the valve head to the seat, using a backward and forward action. Lift the valve occasionally so that the grinding compound is distributed evenly. Repeat the application until an unbroken ring of light grey matt finish is obtained on both valve and seat. This denotes the grinding operation is now complete. Before passing to the next valve, make sure that all traces of the valve grinding compound have been removed from both the valve and its seat and that none has entered the valve guide. If this precaution is not observed, rapid wear will take place due to the highly abrasive nature of the carborundum base. Note that no matter which method of seating the valve is employed, the valve periphery must be checked to determine whether its thickness has worn beyond the service limit of 0.5 mm (0.02 in). If, on completion of the valve seating process this proves to be the case, then the valve must be renewed.

13 Suzuki state that refacing of the valve stem end is permissible, provided the length between the stem end and the upper lip of the valve collet retaining groove is not reduced to less than 3.60 mm (0.14 in), in which case the valve must be renewed. Upon refitting a refaced valve, refer to the figure accompanying this text whilst checking that the end of the valve stem protrudes above the upper faces of the valve collets.

14 Examine the condition of the valve collets and the groove on the valve stem in which they seat. If there is any sign of damage, new parts should be fitted. Check that the spring retaining plate is not cracked. If the collets work loose or the plate splits whilst the engine is running, a valve could drop into the cylinder and cause extensive damage.

15 Check the free length of each of the valve springs. The springs have reached their serviceable limit when they have compressed to the limit readings given in the Specifications Section of this Chapter.

16 Clean the cylinder head (to cylinder block) mating surface and check it for distortion with a straight-edge and a set of feeler gauges. The straight-edge should be laid diagonally across the mating surface from corner to corner and any clearance between its lower edge and the mating surface checked with the gauges. Repeat the procedure with the straight-edge laid between the opposite corners and then at several positions in between the four corners. If the largest clearance found exceeds 0.10 mm (0.004 in), Suzuki recommend that the cylinder head be replaced with a new item. It is worth noting however, that if the amount of distortion found is only slightly greater than the limit given, it could well be worth seeking the advice of a competent motorcycle engineer who can advise on whether skimming the mating surface flat is possible without the subsequent risk of the pistons coming into contact with the valve heads once the engine is reassembled and started.

17 If the cylinder head is seen to be only slightly warped, the mating surface may be lapped on a surface plate or a sheet of plate glass. Place a sheet of 400 or 600 grit abrasive paper on the surface plate. Lay the cylinder head, gasket-face down, on the paper and gently rub it with an oscillating motion to remove any high spots. Lift the head at frequent intervals to inspect the progress of the operation, and take care to remove the

minimum amount of material necessary to restore a flat sealing surface. It should be remembered that most cases of cylinder head warpage can be traced to unequal tensioning of the cylinder head nuts and bolts by tightening them in incorrect sequence or by using incorrect or unmeasured torque settings.
18 Finally, make sure that the external cooling fins of the cylinder head are not clogged with oil or road dirt which will prevent the free flow of air and cause the engine to overheat.
19 Reassemble the valve and valve springs by reversing the dismantling procedure. Fit new oil seals to each valve guide and oil both the valve stem and the valve guide with a molybdenum disulphide based lubricant prior to reassembly. Take special care to ensure the valve guide oil seal is not damaged when the valve is inserted. As a final check after assembly, give the end of each valve stem a light tap with a hammer, to make sure the split collets have located correctly. Note that each spring must be fitted with its close coil end nearest the cylinder head.

22.2 Free the split collets by compressing the valve springs

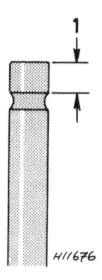

Fig. 1.12 Valve stem end refacing measurement

1   Not less than 3.60 mm (0.14 in)

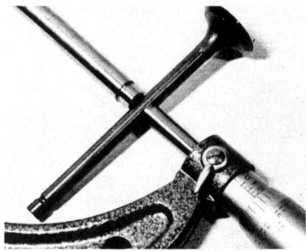

22.3 Using a micrometer, measure the diameter of the valve stem

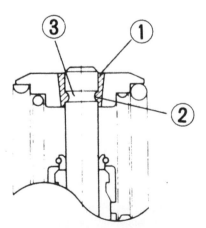

Fig. 1.13 Correct filing of a refaced valve stem

1   Valve retaining collets
2   Groove in valve stem
3   Valve stem

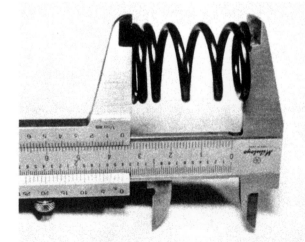

22.15 Measure the free length of each valve spring

22.16 Check the cylinder head for distortion with a straightedge and feeler gauge

22.19a Fit a new oil seal to each valve guide...

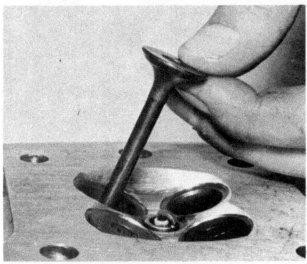
22.19b ...before lubricating and inserting the valve

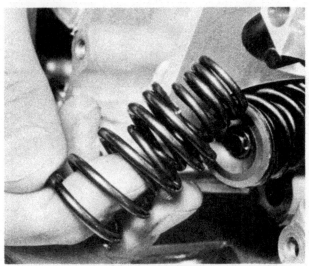
22.19c ...followed by the valve springs (close-coils in the shown position)...

22.19d ...an undamaged spring retaining plate...

22.19e ...and the split collets, which must be located correctly

23 Examination and renovation: rocker arms and spindles

1   It is unlikely that excessive wear will occur in either the rocker arms or the rocker spindles unless the flow of oil has been impeded or the machine has covered a very high mileage. A clicking noise from the rocker area is the usual symptom of wear in the rocker gear, which should not be confused with a somewhat similar noise caused by excessive valve clearance.

2   If any shake is present and the rocker arm iş loose on its shaft, excessive wear between the two components can be suspected. Remove each rocker spindle from its location within the head casting by removing the plug and sealing washer from the cylinder head, unscrewing the spindle stop bolt and inserting it into the threaded end of the spindle. The bolt may then be used to withdraw the spindle. Pull the rocker arm assembly from its location and inspect its bearing surface with that of the spindle for signs of lack of lubrication. If such signs are apparent, the fault in the lubrication system must be found and cured before progressing further.

3   Measure the overall diameter of each rocker spindle and the bore diameter of each rocker arm and compare the measurements obtained with those given in the Specifications Section of this Chapter. Renew each component as necessary.

4   Closely inspect each rocker arm for signs of cracking around its bore ends and between each adjuster screw threaded eye. Ensure that the arm has been thoroughly cleaned in a degreasing agent before it is inspected as this will help to clarify any hairline cracks that may be present in the casting surface.

5   Inspect the surface of each rocker arm where it makes contact with the cam lobe. If signs of cracking, scuffing or break through in the case hardened surface are in evidence, then the arm must be renewed. Check also the thread of each adjusting screw, the threads of the rocker arm into which they fit and the thread of each locknut. If any of these threads show signs of damage, the component must be renewed. The hardened tip of each adjusting screw must also be in good condition. Finally, inspect the rocker arm springs for signs of fatigue or failure and renew each one as necessary.

6   Reassembly of the rocker components is carried out by reversing the dismantling procedure. Ensure that the hole drilled in the spindle to accommodate the stop bolt is correctly aligned with the hole in the head casting before refitting the bolt and tightening it to the specified torque loading. Do not omit to fit the washer beneath the head of the stop bolt and lubricate the surface of the spindle with clean engine oil prior to inserting it into the cylinder head. Refit a new sealing washer beneath the plug head before fitting it in position.

23.2a Remove the rocker spindle end plug from the cylinder head...

23.2b ...and use the spindle stop bolt...

23.2c ...as a means of withdrawing the spindle from its location

23.5 Inspect the condition of the rocker arm contact surface and spring

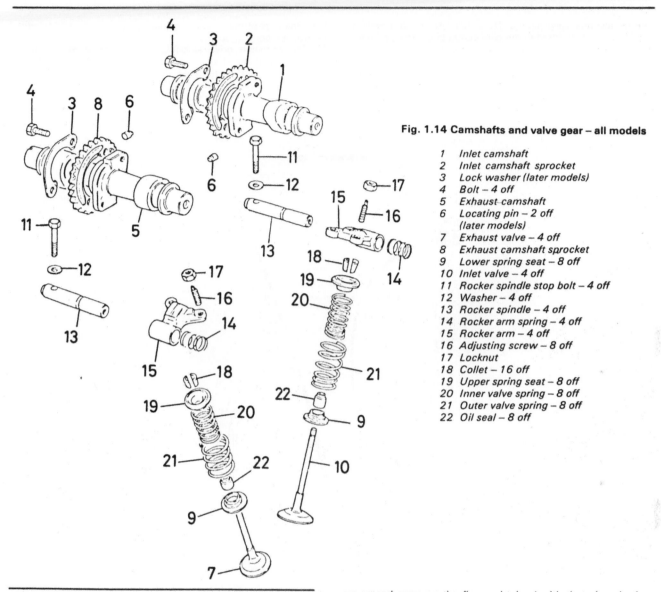

Fig. 1.14 Camshafts and valve gear – all models

1   Inlet camshaft
2   Inlet camshaft sprocket
3   Lock washer (later models)
4   Bolt – 4 off
5   Exhaust camshaft
6   Locating pin – 2 off
    (later models)
7   Exhaust valve – 4 off
8   Exhaust camshaft sprocket
9   Lower spring seat – 8 off
10 Inlet valve – 4 off
11 Rocker spindle stop bolt – 4 off
12 Washer – 4 off
13 Rocker spindle – 4 off
14 Rocker arm spring – 4 off
15 Rocker arm – 4 off
16 Adjusting screw – 8 off
17 Locknut
18 Collet – 16 off
19 Upper spring seat – 8 off
20 Inner valve spring – 8 off
21 Outer valve spring – 8 off
22 Oil seal – 8 off

24 Examination and renovation: camshafts, camshaft bearings and camshaft drive sprockets

1   The camshafts should be examined visually for wear, which will probably be most evident on the ramps of each cam and where the cam contour changes sharply. Also check the bearing surfaces for obvious wear and scoring. Cam lift can be checked by measuring the height of the cam from the bottom of the base circle to the top of the lobe. If the measurement is less than the service limit given in the Specifications at the beginning of this Chapter, the opening of the particular pair of valves will be reduced thus resulting in poor engine performance. If any one lobe surface on either camshaft is found to be at fault, that camshaft should be renewed in order to restore performance.
2   Each camshaft bears directly on the cylinder head material and that of the bearing caps, there being no separate bearings. Check the bearing surfaces for wear and scoring. The clearance between the camshaft bearing journals and the aluminium bearing surfaces may be checked using plastigage (press gauge) material in the same manner as described for crankshaft bearing clearance in Section 17 of this Chapter. If the clearance is greater than given for the service limit the recommended course is to determine the degree of wear on each camshaft journal by measuring across its diameter with a micrometer or vernier

gauge and compare the figure obtained with that given in the Specifications. If journal wear is found to be minimal, the internal diameter of the cylinder head bearing surfaces may be assumed to have worn beyond limits, thus resulting in the need to replace the cylinder head with a serviceable item. Before committing oneself to purchasing this relatively expensive component, double check the bearing clearances and the degree of wear on the camshaft journals before confirming one's worst suspicions by measuring the internal diameter of the cylinder head bearing surfaces with the bearing caps fitted and their securing bolts tightened to the correct torque loading.
3   If bad scuffing is evident on the camshaft bearing surfaces, due to a lubrication failure, the only remedy is to renew the cylinder head and bearing caps and the camshaft if it transpires that it has been damaged also. Make absolutely sure that the cause of lubrication failure has been found and rectified before fitting any new components.
4   Examine the camshaft chain sprockets for hooked worn, or broken teeth. If any damage is found, the camshaft sprocket in question should be renewed. Each sprocket is retained on the camshaft flange by two socket screws (hexagon-headed bolts on later models). When refitting either sprocket note that each is marked IN or EX as are the camshafts. It is important that the sprockets are fitted on the correct camshaft and in the position shown in the accompanying illustration. Incorrect assembly will

prevent accurate valve timing. Note that later models have a locating pin incorporated in the cam centre boss. This pin must be properly located in the cam boss before the sprocket is fitted over it.

5   Fit a new lock washer (later models with hexagon-headed bolts); apply a small amount of locking fluid to the cleaned thread of each securing bolt before fitting and tightening to the correct torque loading. Finally, where a lock washer is fitted,

bend the tabs of the washer against the flats of the bolt heads to lock the bolts in position.

6   The camshaft drive sprocket is an integral part of the crankshaft and therefore if damage is evident, the crankshaft must be renewed. Fortunately, this drastic course of action is rarely necessary since the parts concerned are fully enclosed and well lubricated, working under ideal conditions.

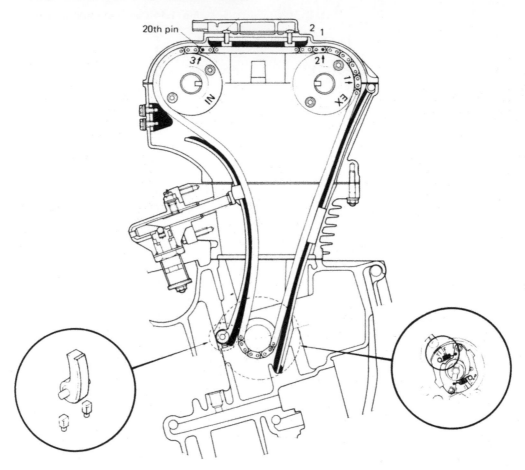

Fig. 1.15 Correct fitting of camshaft and cam chain

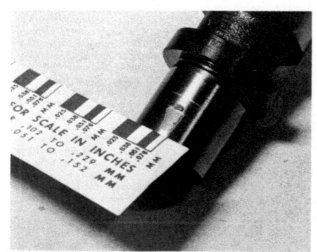

24.2 Use a Plastigage strip to determine camshaft bearing clearance

24.4 Note the camshaft sprocket markings, the locating pin and the lock washer

25 Examination and renovation: camshaft chain and chain tensioner mechanism

1 Inspect the cam chain for obvious signs of damage, such as broken or missing rollers or fractured links. Some indication of the amount of chain wear may be gained by checking the extent of adjustment remaining on the automatic tensioner assembly. If the plunger has moved towards the end of the stroke, it may be assumed that the chain is near the end of its useful life. Wear of the chain may be measured by washing it in petrol, then compressing it endwise so that the free play in both runs is taken up fully. Anchor one end, then pull on the chain so that it stretches as far as possible. If the extension measured over a 20 pitch length exceeds 157.80 mm (6.213 in), the chain must be renewed. Although the cam chain works in almost ideal conditions, being fully lubricated and enclosed, wear will develop after an extended mileage. If there is any doubt as to the chain's condition, it should be renewed, as breakage will cause extensive engine damage.

2 Inspect the cam chain tensioner mechanism by first loosening the locking screw on the tensioner unit body to free the push rod and then rotating the adjuster knob fully anti-clockwise so that the push rod may be moved in and out. Check that the push rod slides freely. If push rod movement is not perfectly smooth, remove the rod and inspect it for signs of seizure, damage to the surface finish and any indication that it may be bent. The rod must be renewed if it is either bent or its surface finish

damaged.

3 Carry out a second check by once again turning the adjuster knob fully anti-clockwise against the resistance of the spring beneath it and then allowing it to return to its original position whilst checking for signs of erratic movement. If the return movement of the adjuster knob is seen to be faulty, return the whole tensioner assembly to an official Suzuki service agent who will be able to confirm whether or not the complete tensioner assembly should be renewed.

4 Renew both the O-ring and gasket washer located beneath the lock nut of the locking screw and inspect both the push rod and adjuster knob return springs for signs of corrosion, fatigue and failure, renewing each as considered necessary.

5 Inspect the chain contact surface of both the cam chain tensioner blade and the cam chain guide blades. If the rubber has been badly scored by the chain or is coming away from the steel backing, the blade in question should be renewed.

6 Note that the smaller of the two guide blades may be removed from its location within the cylinder head cover by removing its two securing screws, the heads of which are located beneath the breather pipe cover. The 400 models covered in this Manual also incorporate a guide block which should be inspected in a similar manner to the blades and which can be removed by unscrewing its retaining screws. Always clean the thread of each retaining screw and coat it in a thread locking compound before refitting either the guide blade or guide block.

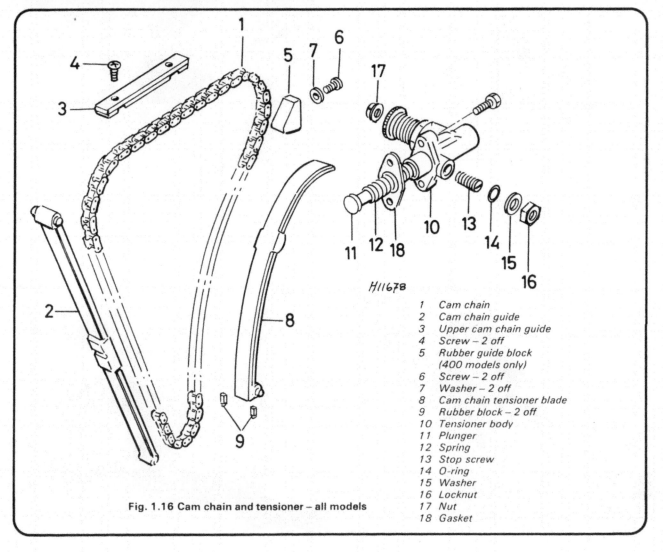

H11678

1 Cam chain
2 Cam chain guide
3 Upper cam chain guide
4 Screw – 2 off
5 Rubber guide block
   (400 models only)
6 Screw – 2 off
7 Washer – 2 off
8 Cam chain tensioner blade
9 Rubber block – 2 off
10 Tensioner body
11 Plunger
12 Spring
13 Stop screw
14 O-ring
15 Washer
16 Locknut
17 Nut
18 Gasket

Fig. 1.16 Cam chain and tensioner – all models

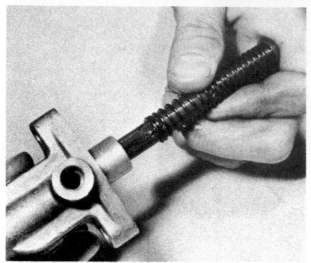

25.2 Remove and inspect the cam chain tensioner pushrod

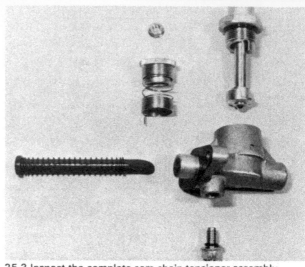

25.3 Inspect the complete cam chain tensioner assembly...

25.4a ...paying particular attention to the adjuster retainer O-ring...

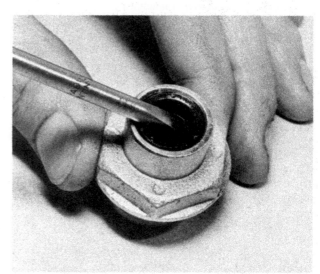

25.4b ...and seal, which may be removed as shown

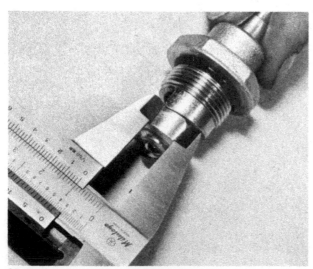

25.4c Align the adjuster retainer and rod end...

25.4d ...before locating the return spring...

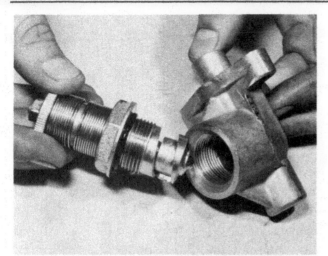

25.4e ...and inserting the complete assembly into the tensioner body

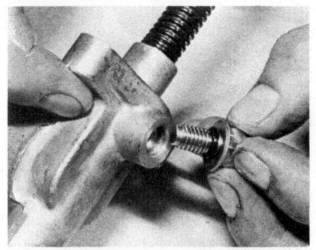

25.4f Renew the O-ring and gasket washer beneath the pushrod locknut

25.6 The smaller cam chain guide blade is retained within the cylinder head cover by two screws

remove the component parts from within the unit, press down on the spring retaining cap and remove the circlip from its retaining groove within the top of the unit. Withdraw the retaining cap by hooking it out of the unit with the flat of a small screwdriver. Use a similar method to remove the spring and the plunger.

2   It is very unlikely that the components within the unit will have become in any way damaged unless the unit has become severely contaminated. If contamination is apparent within the unit, remove the casing from within the crankcase and wash it and the internal components in clean petrol or paraffin, taking care to observe the necessary safety precautions whilst doing so. It is best to discard the washer fitted to the base of the unit and replace it with a new item.

3   Inspect the spring for damage and signs of fatigue or failure. If either the spring, the end cap or plunger or the unit casing is in any way damaged, the complete unit will have to be renewed. If in the slightest doubt as to the condition of the unit, seek the advice of an official Suzuki service agent; do not under any circumstances take a chance by refitting a suspect unit.

4   If the component parts are found to be in a satisfactory condition, lightly lubricate each part in clean engine oil before refitting it into the unit casing. Check that once the circlip is refitted, it is located correctly in its retaining groove. If the unit was removed from its location within the crankcase, it should now be wrapped in a piece of clean rag and placed to one side ready for refitting during the engine reassembly sequence.

26 Examination and renovation: tachometer drive assembly

1   The worm drive to the tachometer is an integral part of the exhaust camshaft which meshes with a pinion attached to the cylinder head cover. If the worm is damaged or badly worn, it will be necessary to renew the camshaft complete.

2   The driveshaft and pinion are a single part retained in the cylinder head in a bush housing which is secured by a plate and screw. Renewal is therefore straightforward. It is unlikely that wear will develop on either the drive or driven pinion as both are well lubricated and lightly loaded.

3   It must be noted that the driveshaft and pinion assembly can only be refitted once the cylinder head cover has been refitted to the cylinder head. Fitting the cylinder head cover with the driveshaft already installed may well cause breakage of the teeth on both the driveshaft and cam pinions. Renew both oil seals on the assembly prior to installation.

27 Examination and renovation: oil pressure relief valve

1   The oil pressure relief valve is located within the lower crankcase half and may be examined in or out of its location. To

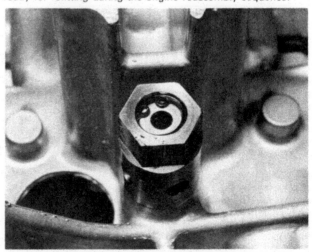

27.1a Dismantle the oil pressure relief valve by removing the circlip...

27.1b ...hooking out the spring retaining cap...

27.1c ...withdrawing the spring...

27.1d ...and displacing the plunger

27.2 A contaminated relief valve casing must be removed for cleaning

28 Examination and renovation: gearbox components

1   The gearbox fitted to both the 250 and 400 models is a substantial assembly and should not normally suffer a great deal of wear. Light general wear may be expected after a high mileage has been covered but the usual causes of accelerated wear or damage can invariably be traced to misuse or poor lubrication. Initial examination can be carried out with the shafts intact, and should be directed at the pinion teeth. Damaged teeth will be self-evident and will of course demand renewal of the component concerned.

2   Look for signs of general wear at the points of contact between the gear teeth. These should normally present a smooth, highly polished profile if in good condition. Pitting of the hardened faces will necessitate renewal of the affected pinions. This type of damage will often be most pronounced on the intermediate (2nd, 3rd and 4th gear) pinions and can be caused by water contamination due to condensation. This is only likely where the machine has been used for frequent short trips which may have prevented the engine unit reaching full operating temperature. Alternatively, neglected oil changes and the resulting thinned and dirty oil can have the same consequences. Pinions with chipped or pitted teeth must always be renewed, there being a real danger of breakage if re-used. In

view of the risk of extensive engine damaged that this presents, do not be tempted to 'economise' at this point.

3   The gearbox bearings should be cleaned in a high flash-point solvent, or petrol (gasoline) if care is taken to avoid the obvious fire risk, and then checked for play or roughness when the bearing is rotated. Renew all bearings that show signs of damage or old age, noting that a worn bearing can cause accelerated wear of the gearbox components and may lead to poor gear selection.

4   If the general examination described above has indicated a likelihood of general wear, the gearbox shafts should be dismantled for further examination. Always deal with one shaft at a time to avoid confusion, and ensure that each component is kept scrupulously clean. The photographic sequence which accompanies this text shows the step-by-step reassembly order required for the components fitted to the GS250 T (UK) model. The illustrations accompanying this text show exploded views for both the 250 and 400 models and should be used as a reference at all times during dismantling and reassembly. In order to eliminate the risk of misplacement, make rough sketches as the clusters are dismantled. Also strip and rebuild as soon as possible, to reduce any confusion which might occur at a later date.

5   When dismantling the layshaft (output shaft) on 250 models note that the gearbox sprocket spacer may be secured

in position by a thread locking compound. If removal of the layshaft left-hand bearing is found to be necessary the bearing and spacer must be displaced from the shaft simultaneously using a legged puller or a fly-press. To facilitate removal of gear pinions from the mainshaft of 250 models the correct sequence must be used. Start by displacing from its groove the circlip which lies between the 6th gear pinion and the 3rd/4th gear pinion. Move the circlip and the adjacent washer along the shaft so that they are hard up against the 3rd/4 th gear pinion face. The 6th gear pinion and 2nd gear pinion can now be slid inwards along the shaft to allow removal of the circlip previously obscured by the outer edge of the 2nd gear pinion. Removal of the pinions, washers and circlips can now be made in a logical sequence. When reasembling the mainshaft the sequence must be reversed to allow correct positioning of the outer circlip.

6   When removing the gear pinions from the layshaft of 400 models note that 3rd gear pinion is retained by a special locking washer system. To allow removal of the pinion, first withdraw the splined locking washer from the shaft and then rotate slightly the second splined washer so that its splines align with the spline channels in the shaft. The second washer can now be slid from position. On reassembly the locking washers must be fitted by reversing the dismantling sequence. On 400 models the mainshaft 2nd gear pinion, which is the outermost pinion on the shaft, is a tight interference fit on the shaft end and is also secured by a locking compound. To allow safe removal, and also reassembly, the 2nd pinion must be displaced using a fly-press. This is more easily accomplished if the 6th gear pinion is drawn off at the same time.

7   After dismantling both gearshafts check that each pinion moves freely on its shaft, but without undue free play. Check for blueing of the shaft or the gearbox pinions as this can indicate overheating due to inadequate lubrication. The dogs on each pinion should be checked for damage or rounded edges. Such damage can lead to poor gear engagement and will require renewal if extensively damaged. Look for signs of hairline cracks around the pinion bosses and dogs.

8   When rebuilding the gearbox shaft assemblies ensure that all washers and circlips are fitted in their correct positions, where necessary employing the special assembly sequences discussed in paragraphs 5 and 6. It is recommended that all circlips, and any washers found to be damaged, are renewed. On 400 models the 2nd gear pinion must be refitted onto the mainshaft end using a press. Prior to this coat the inside bore of the pinion with a locking compound. To ensure correct registering of the gears the fitted overall length of the mainshaft assembly measured between the outer faces of the 2nd gear pinion and the integral 1st gear pinion must be within the range

114.7 – 114.8 mm (4.516 – 4.520 in). See the accompanying illustration. On completion of assembly check that the 6th gear pinion is free to rotate and has not been contaminated by locking fluid.

9   Before refitting the pinions to their respective shafts, lightly smear the splines of the shaft with a molybdenum disulphide grease. Before refitting the gearbox sprocket spacer to the layshaft coat its inside surface with a thread locking compound, ensuring that both the inside surface of the spacer and the contact surface of the shaft are degreased before doing so. Finally, ensure tht the lip of each oil seal is well lubricated with clean engine oil before inserting the shaft through it; on 250 models the O-ring should also be lubricated. It is recommended that seals and O-rings be renewed as a matter of course.

10 The importance of cleanliness during reassembly of a gearbox cannot be overstressed, any piece of grit or dirt which finds its way into the assembly will begin to cause damage directly the rebuilt machine is started, thereby undoing all the work carried out in the renovation and rebuilding sequences and causing extra expense in the meantime. Lubrication of each component part prior to assembly is also essential as dry bearing surfaces will not receive lubrication directly the engine is started.

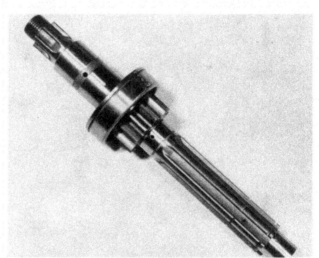

28.8a The gearbox mainshaft and drive side bearing (250 models)

28.8b Commence mainshaft assembly by fitting the 5th gear pinion...

28.8c ...and retaining it in position with the washer and circlip

28.8d Fit the 3rd/4th gear pinion, its retaining circlip and the washer...

28.8e ...followed by the 6th gear pinion...

28.8f ...the 2nd gear pinion...

28.8g ...and the retaining ring

28.8h Correctly locate the circlip and washer behind the 6th gear pinion...

28.8i ...before lubricating and fitting the end bearing

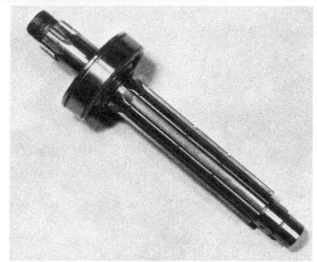

28.8j The gearbox layshaft with bearing (250 models)

28.8k Commence layshaft assembly by fitting the 2nd gear pinion...

28.8l ...and retaining it with the washer and circlip

28.8m Fit the 6th gear pinion, its retaining circlip and washer...

28.8n ...followed by the 3rd gear pinion...

28.8o ...the 4th gear pinion, with its washer and retaining circlip...

28.8p ...the 5th gear pinion...

28.8q ...the 1st gear pinion...

28.8r ...and finally, the thrust washer followed by the end bearing

28.8s Lubricate and fit the O-ring to the layshaft end...

28.8t ...before fitting the gearbox sprocket spacer and oil seal

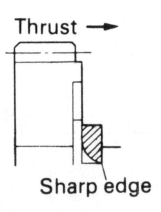

Thrust →

Sharp edge

Fig. 1.17 Correct fitting of gear cluster circlips

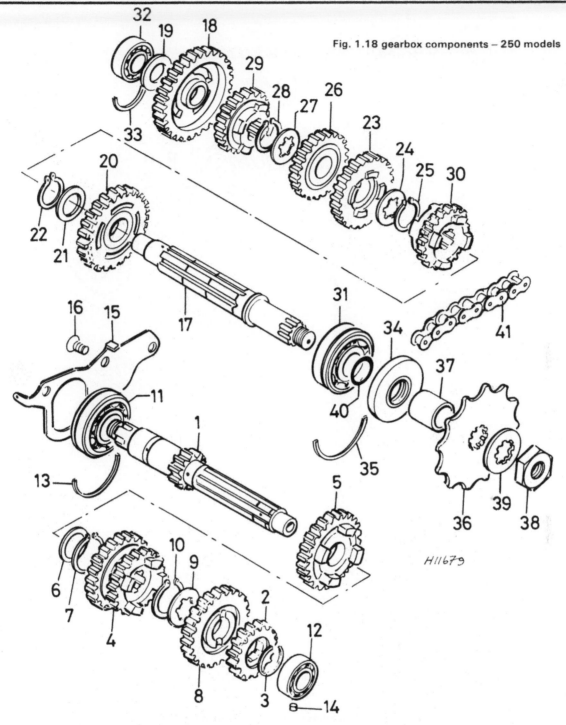

Fig. 1.18 gearbox components – 250 models

H11679

1 Mainshaft	12 Mainshaft left-hand bearing	22 Circlip	31 Layshaft left-hand bearing
2 Mainshaft 2nd gear pinion	13 Bearing half ring	23 Layshaft 3rd gear pinion	32 Layshaft right-hand bearing
3 Circlip	14 Pin	24 Splined washer	33 Bearing half ring
4 Mainshaft 3rd and 4th gear pinion	15 Bearing retainer	25 Circlip	34 Oil seal
5 Mainshaft 5th gear pinion	16 Screw – 4 off	26 Layshaft 4th gear pinion	35 Bearing half ring
6 Washer	17 Layshaft	27 Splined washer	36 Final drive sprocket
7 Circlip	18 Layshaft 1st gear pinion	28 Circlip	37 Spacer
8 Mainshaft 6th gear pinion	19 Washer	29 Layshaft 5th gear pinion	38 Nut
9 Splined washer	20 Layshaft 2nd gear pinion	30 Layshaft 6th gear pinion	39 Tab washer
10 Circlip	21 Washer		40 O-ring
11 Mainshaft right-hand bearing			41 Final drive chain

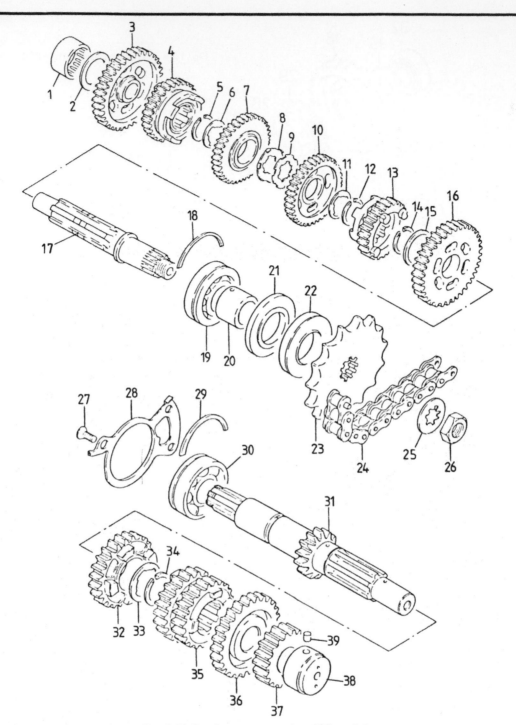

Fig. 1.19 Gearbox components – 400 models

1 Layshaft right-hand bearing	15 Washer	28 Bearing retainer
2 Washer	16 Layshaft 2nd gear pinion	29 Bearing half ring
3 Layshaft 1st gear pinion	17 Layshaft	30 Mainshaft right-hand bearing
4 Layshaft 5th gear pinion	18 Bearing half ring	31 Mainshaft
5 Circlip	19 Layshaft left-hand bearing	32 Mainshaft 5th gear pinion
6 Splined washer	20 Spacer	33 Splined washer
7 Layshaft 4th gear pinion	21 Oil seal	34 Circlip
8 Tabbed splined washer	22 Oil seal	35 Mainshaft 3rd and 4th gear pinion
9 Splined washer	23 Final drive sprocket	36 Mainshaft 6th gear pinion
10 Layshaft 3rd gear pinion	24 Final drive chain	37 Mainshaft 2nd gear pinion
11 Splined washer	25 Tab washer	38 Mainshaft left-hand bearing
12 Circlip	26 Nut	39 Pin
13 Layshaft 6th gear pinion	27 Screw – 3 off	
14 Circlip		

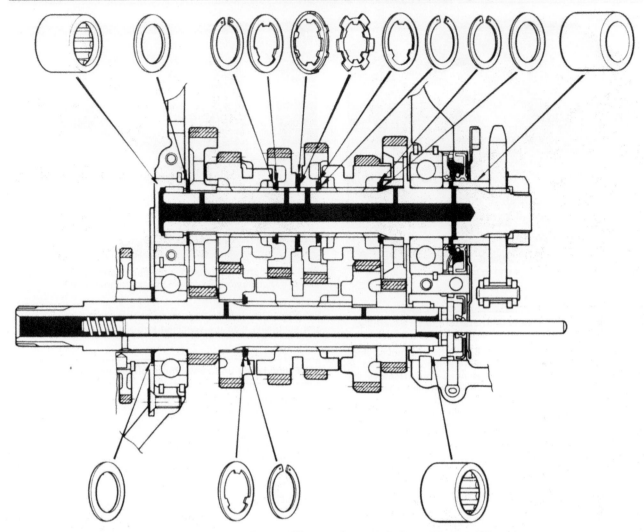

Fig. 1.20 Gear shaft assemblies washer and circlip positions – 400 models

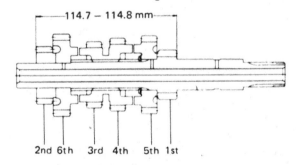

|←――― 114.7 – 114.8 mm ―――→|

2nd 6th   3rd   4th   5th 1st

Fig. 1.21 Mainshaft assembled length – 400 models

29 Examination and renovation: gear selector mechanism

1   Inspect the gear selector fork support shaft for wear or scoring, renewing it if obviously damaged. Check for straightness by rolling it across a surface plate or a sheet of plate glass. If the shaft is even slightly bent, it must be renewed as it will cause difficulty in selecting gears and will make the gear change particularly heavy.

2   Examine the selector forks carefully, noting any signs of wear on the fork ends where they engage with the groove in the gear pinion. Refer to the Specifications Section of this Chapter for details of fork end thickness, gear pinion groove width and fork end to pinion groove clearance. Measure the fork end to groove clearance by inserting the fork into the pinion groove and sliding a feeler gauge into the groove alongside the fork. Compare the measurement obtained with the service limit given in the Specifications. If the service limit is exceeded, measure the fork end thickness and pinion groove width and reject the fork, the pinion, or both components as necessary. It is important that the forked end is not bent in any way. Check the fit of the forks on the support shaft. These should be a light sliding fit with no appreciable free play. Any movement at the bore will be greatly magnified at the fork end, leading to imprecise gear selection.

3   Examine the tracks in the selector drum in conjunction with the selector fork pins which run in them. In practice, the tracks will not wear appreciably, even over high mileages, but the small pins on the selector forks may show signs of flattening, in which case they should be renewed to reduce play. Note that the pins are integral with the fork, and thus the entire fork will have to be renewed if the pins are worn.

4   Check the tension of the gearchange pawl, gearchange arm and drum stopper arm springs. Weakness in the springs will lead to imprecise gear selection. Check the condition of the gear stopper arm roller and on the pins in the change drum end with which it engages. It is unlikely that wear will take place here except after considerable mileage.

29.2 Check the selector fork end to pinion groove clearance

29.3 Do not attempt to remove the gearchange drum bearing unless it is unserviceable, in which case remove as shown

30 Examination and renovation: starter clutch assembly

1   The starter clutch assembly will have been removed from the crankshaft end along with the alternator rotor and is attached to the rotor by three large Allen bolts. Remove the starter driven sprocket from its location within the starter clutch to expose the heads of these bolts.

2   There is no need to detach the starter clutch from the alternator rotor unless excessive wear on the roller bearing faces of either the shim plate or the casing is suspected. Inspect the condition of the sprocket boss where it comes into contact with the rollers. If the surface of the boss is scored or pitted or shows signs of excessive wear, then the sprocket should be replaced with a serviceable item. The sprocket should also be replaced if its teeth are chipped or badly worn. Suzuki do not list the bush contained within the sprocket as a separate item, nor do they give any indication as to the amount of wear allowed on the bush. If the bush seems excessively worn and the crankshaft surface with which it comes into contact is marked, return both items to an official Suzuki service agent who will be able to give advice on the matter.

3   Check the condition of the three engagement rollers, springs and pushrods. Each separate assembly may be removed from its location within the casing by carefully easing the roller from position with the flat of a small screwdriver. Carefully inspect each roller and pushrod for signs of deterioration in the finish of their bearing surfaces and renew as necessary. Inspect each spring for signs of fatigue or failure and if possible, compare their length with that of a new item to determine whether or not they have taken a 'set' to a shorter length. If any one spring is found to be defective, it is advisable to renew all the springs as a set.

4   If it is found necessary to renew the rollers due to excessive wear on their bearing surfaces, then it is advisable to closely inspect the bearing faces of both the shim plate and starter clutch casing. This may be achieved initially by simply cleaning the roller location of any residue of oil and directing a strong light into the location; any excessive wear will show quite clearly, necessitating the need for the shim plate to be removed from the casing for further investigation and subsequent renewal.

5   If it has been found necessary during the course of examining and renovating the starter clutch assembly to detach the assembly from the alternator rotor, then ensure that each Allen bolt is thoroughly degreased, along with its corresponding thread in the rotor casting, before reassembly takes place. Coat the thread of each bolt with thread locking compound before

inserting it through the starter clutch assembly, into the rotor and tightening it to the specified torque loading.

31 Examination and renovation: clutch assembly

1   After an extended period of service the clutch linings will wear and promote clutch slip. The limit of wear measured across each friction plate and the standard measurement is given in the Specifications. When the overall width reaches the limit, the friction plates must be renewed, preferably as a complete set. Check the warpage of each friction plate by laying it on a sheet of plate glass and attempting to insert a feeler gauge beneath it. The maximum allowable warpage is 0.2 mm (0.008 in).

2   The plain plates should not show any excess heating (blueing). Check the warpage of each plate using plate glass or surface plate and a feeler gauge. The maximum allowable warpage is 0.1 mm (0.004 in).

3   Check the free length of each clutch spring with a vernier gauge. After considerable use the springs will take a permanent set thereby reducing the pressure applied to the clutch plates. The service limit for each spring free length is given in the Specifications Section.

4   Check the condition of the slots in the outer surface of the clutch hub and the inner surfaces of the outer drum. In an extreme case, clutch chatter may have caused the tongues of the friction plates to make indentations in the slots of the outer drum, or the tongues of the plain plates to indent the slots of the clutch hub. These indentations will trap the clutch plates as they are freed and impair clutch action. If the damage is only slight, the indentations can be removed by careful work with a file and the burrs removed from the tongues of the clutch plates in similar fashion. More extensive damage will necessitate renewal of the parts concerned. The recommended service limit for friction plate tongue width is given in the Specifications Section.

5   Check the condition of the clutch hub spacer, the thrust bearing, the operating boss and the pressure plate. Wear in the bearing will be indicated by roughness felt in the rollers as they are rotated and by cracking of the bearing cage. Inspect the pressure plate for signs of cracking due to uneven loosening and tightening of the spring retaining bolts.

6   The clutch release mechanism in the clutch cover does not normally require attention, provided that it is greased from time to time. If the unit fails, it should be renewed as a complete assembly.

30.2 Inspect the starter clutch sprocket boss and bush for wear or damage

30.3a Carefully ease each starter clutch roller from its location...

30.3b ...to gain access to its plunger and spring

31.2 Check the warpage of each clutch plate as shown

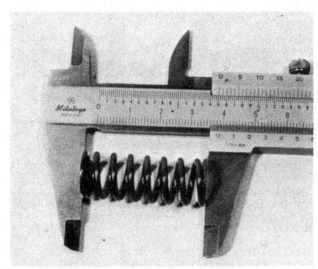

31.3 Check the free length of each clutch spring

31.4 The friction plate tongue width should not be worn beyond the service limit

32 Examination and renovation: primary drive and oil pump drive pinions

1   Inspect the primary drive pinion and the pinion affixed to the clutch outer drum, for worn or chipped teeth. It is probable that if damage or wear is found on one pinion, similar problems will have occurred on the other.
2   Check the driven pinion shock absorber springs for breakage or excessive slack (400 models only). If movement can be felt between the pinion and the clutch outer drum renewal is required. The primary driven pinion, shock absorber and the clutch outer drum are supplied as a complete unit. If one component is damaged the whole unit must be renewed.
3   Under no circumstances should wear or breakage of the pinion teeth be ignored as failure of the components will most certainly result within a relatively short space of time thereby causing problems in transmission and possible damage to other moving components. If the primary drive pinion is to be removed from the crankshaft end and the same item refitted, mark the outer face of the pinion so that it may be refitted in the same position. Closely inspect the condition of the key and its keyways once the pinion is removed. The key itself may easily be renewed but expert advice will have to be sought if it is considered necessary to recut the keyways.
4   The oil pump drive pinion is located against the rear face of the clutch drum; it too should be inspected for worn or chipped teeth and its condition compared with that of the driven pinion on the oil pump. If damage or wear is apparent, both pinions should be renewed as at set. The drive pinion is easily removed from the clutch drum boss by first removing its retaining circlip and then pulling it from position.

33 Examination and renovation: crankcase covers

1   The right and left-hand crankcase covers and the inspection covers are unlikely to become damaged unless the machine is dropped or involved in an accident. Cracks in a casing can be repaired by specialist aluminium welding, providing the damage is not too extensive and care is taken to prevent distortion.
2   The covers are lightly polished before leaving the factory. Badly scratched covers can be refurbished using a single cut file treated with chalk to prevent clogging and finished off with fine emery paper and metal polish or aluminium cleaner.

34 Engine reassembly: general

1   Before reassembly of the engine/gearbox unit is commenced, the various component parts should be cleaned thoroughly and placed on a sheet of clean paper, close to the working area.
2   Make sure all traces of old gaskets have been removed and that the mating surfaces are clean and undamaged. One of the best ways to remove old gasket cement is to apply a rag soaked in methylated spirit. This acts as a solvent and will ensure that the cement is removed without resort to scraping and the consequent risk of damage.
3   If a gasket becomes bonded to the surface through the effects of heat and age, a new sharp scalpel blade should be used with great care to effect removal. Keep the cutting edge of the scalpel blade flat on the mating surface so that there is no chance of it damaging the soft aluminium alloy of the casting.
4   Gather together all the necessary tools and have available an oil can filled with clean engine oil. Make sure that all new gaskets and oil seals are to hand, also all replacement parts required. Nothing is more frustrating than having to stop in the middle of a reassembly sequence because a vital gasket or replacment has been overlooked.
5   Make sure that the reassembly area is clean and that there is adequate working space. Refer to the torque and clearance

settings wherever they are given. Many of the smaller bolts are easily sheared if overtightened. Always use the correct size screwdriver or bit for the crosshead screws, never an ordinary screwdriver or punch. If the existing screws show evidence of maltreatment in the past it is advisable to renew them as a complete set.

35 Reassembling the engine/gearbox unit: fitting the upper crankcase components

1   With the upper crankcase half inverted and positioned on wooden blocks so that no strain is imposed on the cylinder block retaining studs, commence reassembly of the engine/gearbox unit by sliding the gearchange drum into the casing whilst ensuring that its bearing surfaces are thoroughly lubricated with engine oil. Retain the drum in position by refitting the detent plunger, its spring and its retaining bolt. Note that the sealing washer fitted beneath the bolt head must be renewed before the bolt is fitted. Note also that the gearchange drum must be rotated so that it is in the neutral position when the detent plunger is installed.
2   If, on examination, the change pins were found to be worn and the pins and retainer plates have been removed, do not refit them at this stage. The pins should be fitted in a special sequence, as described in Section 37 of this Chapter.
3   Lubricate and insert the selector fork shaft, locating it with each of the three selector forks as it is pushed home. Ensure that the forks are refitted in the positions noted during removal. The photograph accompanying this text shows the exact fitted position. The guide pin on each fork must be pointing rearwards and engaged with the appropriate channel in the change drum.
4   Having refitted the selector fork shaft, fit and tighten the shaft securing screw after having degreased its thread and coated it with thread locking compound.
5   If the mainshaft bearing locating pin was removed from its recess within the crankcase bearing location, it should now be refitted. Refit the bearing location half clips in the casing grooves and lower each gear shaft assembly into position, ensuring that the half clips and pin fit correctly in the bearing recesses. Lubricate the gear shaft assemblies with clean engine oil.
6   Relocate the cam chain tensioner blade in its location within the crankcase and push the two small rubber blocks into position over its retaining spigots. Check that the main bearing and counterbalance bearing shells in both upper and lower crankcases are properly positioned with the tags located in the housing recesses. On 400cc models ensure that the thrust bearings are fitted each side of the centre main bearing support wall. Lubricate the crankshaft bearings with clean engine oil and carefully lower the crankshaft assembly into position in the crankcase whilst ensuring that the cam chain passes cleanly through the crankcase tunnel to rest on a piece of clean rag or card placed on the work surface beneath.
7   Lubricate the counterbalance shaft bearings with clean engine oil and lower the shaft into position. Ensure that the shaft gear pinion meshes with the crankshaft pinion in such a way that the two previously noted alignment marks meet each other.

36 Reassembling the engine/gearbox unit: fitting the lower crankcase components and joining the crankcase halves

1   If the oil pressure relief valve has been removed from its location within the lower crankcase half, it should now be refitted. Ensure that the seating washer is in good condition before screwing home the unit and tightening it to the specified torque loading.
2   Carefully inspect the small oil feed nozzle located in the layshaft bearing location for signs of clogging due to contamination. If in doubt, remove the nozzle and blow it through with compressed air to ensure that it is clear before refitting it in

position. Once this is done, carry out a final inspection of the crankcase halves and associated components for cleanliness and correct assembly before joining the crankcase halves together as follows.

3 Carefully clean the crankcase halves mating surfaces. Refit the four hollow locating dowels, tapping them into position carefully so as not to distort them. If the dowels have become slightly burred they should be cleaned up with a small file.

4 With the mating surfaces properly degreased, carefully smear a thin, even layer of jointing compound onto the surface of one crankcase half. Take great care not to apply the compound on or near any of the bearing surfaces, keeping it about 2 mm away from the bearing edge. Suzuki recommend that Suzuki Bond No. 4 be used to make a joint. A good quality non-hardening compound will make a suitable substitute.

5 Suzuki recommend that their jointing compound be given 10 minutes in which to set before pressing the crankcase halves together. It is suggested that this period of waiting is best spent checking that all the bearing surfaces have been well lubricated with clean engine oil. Bear in mind that when the rebuilt engine is first started there will be a delay of several seconds before the oil begins to circulate, so careful oiling of vulnerable areas is essential.

6 Lower the lower crankcase half into position, using a soft-faced mallet to carefully tap it fully home. Check that all the shafts are able to rotate freely before refitting the securing bolts into the lower crankcase half in their previously noted positions. Tighten these bolts a little at a time and in even increments,

following the numerical sequence stamped on the casing adjacent to each bolt, commencing with No 1. All of these bolts (Nos 1 to 12) should be finally tightened to the torque figures given in the Specifications Section of this Chapter. Do not forget to refit the earth lead in its previously noted position.

7 Check once again that all the shafts rotate freely before refitting and tightening the two Allen bolts located within the oil filter housing. Reposition the engine unit and fit and tighten the single 6 mm bolt located within the starter motor housing to the specified torque loading.

8 Re-invert the engine unit and block it in position. If the gauze filter plate has been removed for cleaning, it should now be refitted and secured in position with its three crosshead retaining screws. Ensure that the mating surfaces of the sump cover and the lower crankcase half have been cleaned of all old gasket material and jointing compound and are properly degreased. Smear an even coating of Suzuki Bond No 4 (or equivalent) over the area of sump cover mating surface indicated in the accompanying photograph and over the corresponding area of crankcase mating surface. Fit the new gasket into position on the crankcase and place the sump cover in position on top of it. Refit the sump cover retaining bolts into their respective locations, remembering to position the electrical cable retaining clips in their previously noted positions beneath the bolt heads. Tighten the retaining bolts evenly and in a diagonal sequence, noting the final torque loading given in the Specifications Section of this Chapter.

35.1 Slide the gearchange drum into the casing

35.3 Engage each selector fork guide pin in its change drum channel

35.5 Lower each gear shaft assembly into position

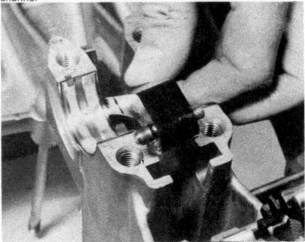

35.6a Relocate the cam chain tensioner blade end in its location...

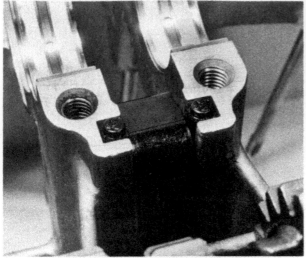

35.6b ...and push the two rubber blocks into position

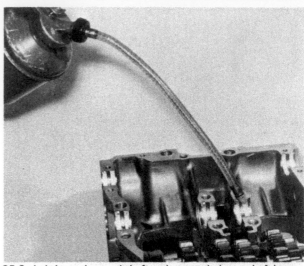

35.6c Lubricate the crankshaft and counterbalance shaft bearings with clean engine oil...

35.6d ...and fit the components into the crankcase

36.6a Place the lower crankcase half in position...

36.6b ...and tighten the securing bolts in the numerical sequence indicated

36.8a Smear an even coating of Suzuki Bond No 4 over the indicated crankcase and sump cover mating surfaces before fitting a new gasket

36.8b Correctly position the two electrical cable retaining clips beneath the bolt heads

37 Reassembling the engine/gearbox unit: refitting the oil pump assembly and gear selector external components

1   Refit the bearing retainer plate into position over the mainshaft end. Degrease the threads of the plate securing screws and coat them with a thread locking compound before fitting and tightening them with an impact driver.
2   The oil pump must be refitted in the primary drive casing as a complete unit, either before or after the driven gear is fitted. The driven gear is retained by a circlip on the shaft end, and is located by a drive pin which passes through the shaft, engaging with a recess in the rear face of the pinion.
3   Place a new O-ring in each of the casing recesses against the wall to which the pump is secured. Omission of the O-rings will lead to lubrication failure. Position the oil pump and fit and tighten the three mounting screws. It is recommended that locking fluid be used on the mounting screws.
4   Position the change drum guide plate in the casing. Degrease the threads of the plate retaining bolt which locates nearest to the change drum end and coat the threads with a thread locking compound. Insert the bolt and tighten it finger-tight. Grease the splined end of the gearchange arm so that when it passes through the oil seal in the left-hand wall of the gearbox, the sealing lip will not be damaged. Insert the gearchange shaft, complete with the pawl arm and return spring, into the primary drive chamber. Note that the arms of the change arm centraliser spring must locate one either side of the peg in the casing. Push the assembly fully home, simultaneously pulling back the pawl arm so that it clears the end of the change drum and engages with the change pins. If the change pins and retaining plate were removed for pin renewal, the replacement pins can now be fitted, using the correct positioning procedure detailed in paragraph 6.
5   Refit the change drum stopper arm pivot bolt after having degreased its threads and coated them with a thread locking compound. The pivot bolt is of the shouldered type and serves also as the change drum guide plate front securing bolt. Fully tighten both the pivot bolt and the rearmost change drum guide plate securing bolt and check that the stopper arm is free to move and has not been trapped by the bolt shoulder. Reconnect the arm return spring with the anchor lug on the bearing retainer plate.
6   Note that of the six change pins, one is of differing dimensions, having a relieved portion at one end. It is against

this recess that the stopper arm roller abuts when the gearbox is in neutral. Check that the gearbox is still in the neutral position and then insert the neutral pin in the change drum pin hole nearest to the stopper arm roller. Push the roller hard up against the recess so that the two components self-align. Insert the five remaining pins in the holes provided. Position the pin retainer plate on the end of the change drum so that the punched depression engages with the neutral pin and secure the plate in position by fitting and tightening the central retaining screw. The threads of the screw must be degreased and coated with a thread locking compound before the screw is fitted.

38 Reassembling the engine/gearbox unit: fitting the primary drive pinion and clutch assembly

1   Fit the Woodruff key into the keyway in the crankshaft end. Slide the pinion into place so that the key on the shaft engages with the keyway in the pinion. Lock the crankshaft in position by placing a close-fitting tommy bar through the small-end eye of one of the connecting rods and abutting the ends of the bar against pads placed over the crankcase mouth mating surfaces. Fit the special washer over the crankshaft end, concave side against the pinion, and fit and tighten the retaining nut to the specified torque loading.
2   Check that the pushrod oil seal is in place. On models fitted with a retaining plate over the seal (photo 40.7) it is possible to install the seal at this stage; drift it into place, ensuring that the drift bears only on the seal's hard outer edge. Models without the seal retaining plate should have the seal fitted prior to joining the crankcase halves. Lightly grease the clutch pushrod and insert it into the mainshaft from the clutch end. Note that the scrolled end of the rod must be nearest the clutch. On 400 models only, refit the heavy washer over the mainshaft end. If the oil pump drive pinion was removed, it must now be fitted to the rear face of the clutch drum before continuing further. On 250 models only ensure that the two notches in the face of the pinion align correctly with the corresponding tongues in the rear face of the clutch drum before pushing it into position and securing it with the circlip. The pinion fitted to 400 models is prevented from turning about the clutch drum boss by a drive pin; ensure that this pin is properly located before pushing the pinion into position and fit the pinion so that the raised boss faces towards the clutch drum.
3   Position the clutch drum over the shaft end so that it meshes with the primary drive pinion. Likewise, the oil pump drive pinion must be meshed with the driven gear.
4   Place the thrust washer over the shaft end and refit the clutch hub onto the shaft splines. Fit a new tab washer and then fit and tighten the centre nut to the specified torque loading. Use the same procedure for tightening the nut as was used for loosening. After torque loading the nut, bend up the tab washer against one of its flats to lock it in position.
5   Lightly grease the thrust bearing and the operating boss before refitting them to the shaft end. If a shim was removed from the outer face of the bearing, it too should be lightly greased and refitted. Refit the plain and friction plates, one at a time, into the clutch drum in the order noted during removal. If in doubt as to the sequence of plate refitting, refer to the figures accompanying this text.
6   Place the pressure plate in position over the outermost friction plate and insert the clutch springs in position in its recesses. Before fitting each spring retaining bolt and plate washer, degrease the thread of the bolt and coat it with a thread locking compound. Fit the bolts finger-tight at first, tightening them in small increments and in a diagonal sequence until they are fully tightened to the specified torque loading. Failure to observe this simple tightening sequence will almost certainly result in a damaged pressure plate.

37.1 Refit the bearing retainer plate over the mainshaft end

37.3a Place a new O-ring in each of the casing recesses...

37.3b ...before fitting the oil pump in position

37.4 Refit the change drum guide plate and gearchange shaft into position, noting the centraliser spring location

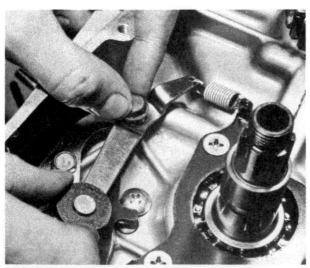

37.5a Refit the stopper arm pivot bolt...

37.5b ... whilst ensuring that its end is correctly located in the change drum end

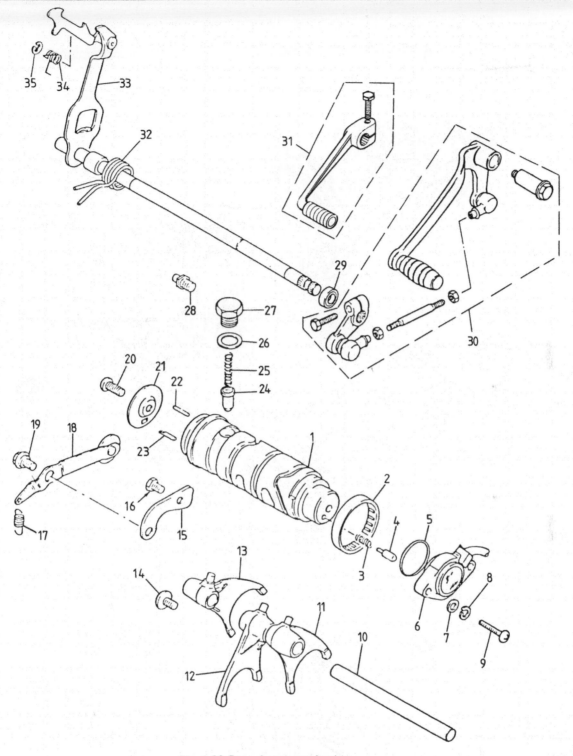

Fig. 1.22 Gear change mechanism

1	Gear change drum	10	Selector fork shaft	20	Screw	29	Oil seal
2	Change drum bearing	11	Selector fork	21	Pin retaining plate	30	Gearchange lever
3	Spring	12	Centre selector fork	22	Change pin – 5 off		assembly – 400 models
4	Contact pin	13	Selector fork	23	Change pin	31	Gearchange lever –
5	O-ring	14	Screw	24	Detent plunger		250 models
6	Neutral indicator/gear	15	Anchor plate	25	Spring	32	Shaft return spring
	position switch	16	Bolt	26	Sealing gasket	33	Gear change shaft
7	Washer – 2 off	17	Return spring	27	Bolt	34	Centraliser spring
8	Spring washer – 2 off	18	Stopper arm	28	Spring anchor	35	E-clip
9	Screw – 2 off	19	Bolt				

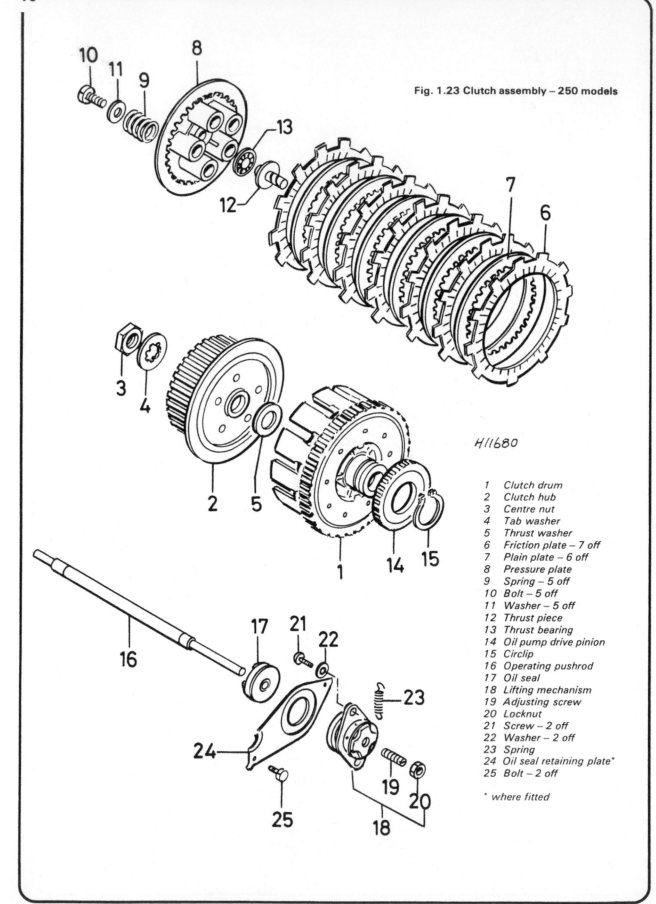

Fig. 1.23 Clutch assembly – 250 models

H11680

1   Clutch drum
2   Clutch hub
3   Centre nut
4   Tab washer
5   Thrust washer
6   Friction plate – 7 off
7   Plain plate – 6 off
8   Pressure plate
9   Spring – 5 off
10 Bolt – 5 off
11 Washer – 5 off
12 Thrust piece
13 Thrust bearing
14 Oil pump drive pinion
15 Circlip
16 Operating pushrod
17 Oil seal
18 Lifting mechanism
19 Adjusting screw
20 Locknut
21 Screw – 2 off
22 Washer – 2 off
23 Spring
24 Oil seal retaining plate*
25 Bolt – 2 off

* where fitted

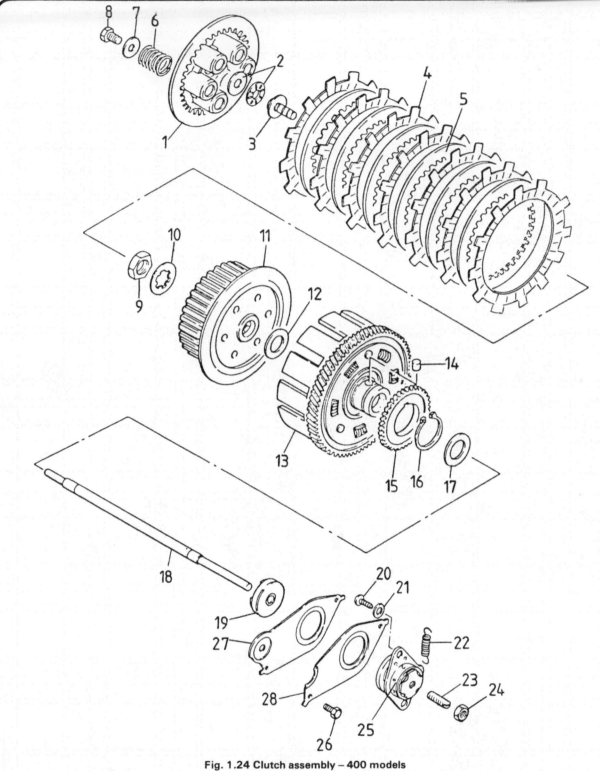

Fig. 1.24 Clutch assembly – 400 models

1 Pressure plate	11 Clutch hub	21 Washer – 2 off
2 Thrust bearing	12 Thrust washer	22 Spring
3 Thrust piece	13 Clutch drum	23 Adjusting screw
4 Friction plate – 7 off	14 Roller	24 Locknut
5 Plain plate – 6 off	15 Oil pump drive pinion	25 Lifting mechanism
6 Spring – 6 off	16 Circlip	26 Bolt – 2 off
7 Washer – 6 off	17 Thrust washer	27 Oil seal retaining plate*
8 Bolt – 6 off	18 Operating pushrod	28 Retaining plate*
9 Centre nut	19 Oil seal	
10 Tab washer	20 Screw – 2 off	*where fitted

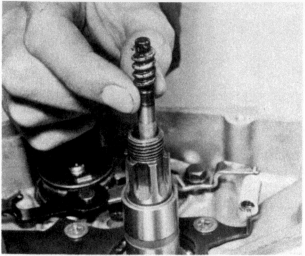

38.2a The scrolled end of the clutch pushrod must be nearest the clutch

38.2b Align the notches on the oil pump drive pinion with the tongues on the clutch drum (250 models only)...

38.2c ...before securing the pinion with the circlip

38.3 Position the clutch drum over the shaft end...

38.4a ...followed by the thrust washer...

38.4b ...and the clutch hub

38.4c Bend up the tab washer to lock the centre nut in postion

38.5a Lightly grease the operating boss...

38.5b ...and the clutch thrust bearing

38.5c Fit the plain and friction plates into the clutch drum...

38.6a ...followed by the pressure plate...

38.6b ...and the clutch springs

38.6c Tighten the spring retaining bolts in small increments in a diagonal sequence

39 Reassembling the engine/gearbox unit: refitting the ignition pick-up assembly and oil pressure switch

1   Refit the right-hand crankcase cover locating dowels and ensure that both the crankcase and cover mating surfaces are properly cleaned and degreased. Place a new gasket over the crankcase mating surface, checking that the holes in the portion of gasket below the primary drive pinion and balance shaft end are correctly aligned with the oil passage in the crankcase.
2   Before refitting the crankcase cover, lightly smear the lip of the new seal with clean engine oil to obviate the risk of its being damaged as it is passed over the crankshaft end. Push the cover into position over the locating dowels and gently tap around its periphery with a soft-faced mallet to seat it on the gasket. Refit the retaining screws into the periphery of the cover and tighten them evenly and in a diagonal sequence. Refit and tighten the two retaining screws located within the ATU housing.
3   If, for any reason, the oil pressure switch has been removed from the ATU housing, it should now be refitted. Degrease the threads of the switch and coat them with a thread locking compound before screwing the switch into position and tightening it to the specified torque loading.

4   Before refitting the ATU, note the locating pin in the crankshaft end and the slot in the rear face of the ATU. Align the pin and groove before pushing the ATU into place and ensure that the pin is properly located in the groove once the ATU is in position. Refit the centre bolt and crankshaft turning nut. Hold the nut against its location on the outer face of the ATU and tighten the bolt finger-tight. Place an open-ended spanner across the flats of the crankshaft turning nut and hold the nut steady whilst tightening the centre bolt to the specified torque loading.
5   Place the pulse generator unit over the ATU and screw in its two mounting screws finger-tight. Reconnect the electrical lead to the oil pressure switch and rotate the electrical cable through its retaining clips along the crankcase wall, making sure the rubber locating grommet is correctly inserted into the crankcase cover recess before doing so. If the pulse generator unit baseplate was marked in relation to the crankcase cover during the removal sequence, realign the reference marks before fully tightening both its mounting screws. Alternatively, align the mark cast in the baseplate with that cast in the portion of cover adjacent to the rearmost mounting screw.
6   If necessary, refit and tighten the crankcase cover oil level plug. Do not omit to place a new sealing washer beneath the head of the bolt.

39.1a Place a new gasket over the right-hand crankcase cover locating dowels...

39.1b ...ensuring that the holes in the gasket (arrowed) are aligned with the oil passage in the crankcase...

39.2 ...before pushing the cover into position

39.4 Note the locating pin in the crankshaft end...

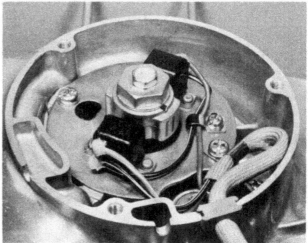

39.5a ...before fitting the ATU and pulse generator unit into position

39.5b The electrical leads from the pulse generator unit must be correctly routed

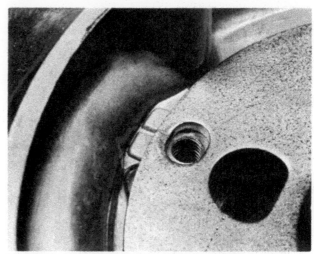

39.5c The pulse generator unit baseplate alignment marks

40 Reassembling the engine/gearbox unit: refitting the starter motor and alternator assembly

1   Lubricate and fit a new O-ring to the starter motor boss before inserting the motor into the casing. Push the motor fully home before fitting and tightening the two retaining bolts to the specified torque loading. Note that the threads of these bolts should be degreased and coated with a thread locking compound before the bolts are fitted.

2   Carefully degrease both the tapered bore of the alternator rotor and the tapered end of the crankshaft before pushing the alternator rotor/starter clutch assembly into position on the crankshaft end. Degrease the threads of the rotor retaining bolt and coat them with a thread locking compound before fitting the bolt together with a new spring washer into the crankshaft end. Lock the crankshaft in position by using the method described for rotor removal and tighten the retaining bolt to the specified torque loading.

3   Lubricate the starter motor idler gear stub shaft with clean engine oil and refit both the shaft and gear pinion into the casing. Ensure that the gear teeth mesh correctly with those of the starter clutch and the starter motor shaft.

4   Refit the single locating dowel into the crankcase mating surface and ensure that the mating surfaces of the crankcase

cover and crankcase are both clean and degreased. Fit the new gasket to the crankcase, taking great care to ensure that the holes in the gasket correspond exactly with the oil passage in the area of mating surface directly below the alternator rotor.

5   Check that the rubber guide grommet through which the electrical leads to the alternator stator pass is correctly located in the cover recess before placing the cover on the crankcase. Push the cover into position over the locating dowel and gently tap it around its periphery with a soft-faced mallet to seat it on the gasket. Refit and tighten the cover retaining screws, taking care to tighten them evenly and in a diagonal sequence to prevent any risk of distortion to the cover.

6   Check that the neutral indicator switch (or gear position switch) contact pin and spring are correctly located in the gearchange drum end and that a new O-ring is located in the crankcase groove. Refit the switch and secure it in position with the two crosshead retaining screws. Do not omit to refit the lead retaining clip beneath the screw head nearest the alternator. Route the stator leads beneath the retaining clip and check that both the switch and stator leads are routed through the recess in the top edge of the crankcase.

7   Finally, fit a new oil seal retaining plate (where fitted) over the clutch pushrod end and layshaft end and secure it in position by fitting and tightening the retaining bolts. Lock the bolts in position by bending the plate tabs against their heads.

40.1a Lubricate and fit a new O-ring to the starter motor boss...

40.1b ...before inserting the starter motor into the casing

40.2a Push the alternator rotor/starter clutch assembly onto the crankshaft end...

40.2b ...and secure it in position with the bolt and new spring washer...

40.5 ...before fitting the starter idler gear and shaft, a new gasket, the locating dowel and cover

40.6 Do not omit to fit the neutral indicator (or gear position) switch contact pin and spring and a new O-ring

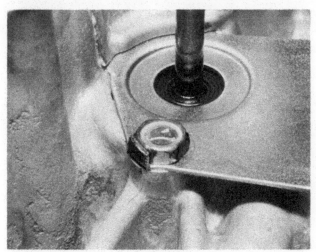

40.7 Lock the oil seal retaining plate (where fitted) bolts by bending up the plate tabs as shown

41 Reassembling the engine/gearbox unit: refitting the pistons, piston rings, cylinder block and cylinder head

1   Fit the piston rings to each piston, commencing with the three-piece oil control ring. The first of the three parts to be fitted should be the corrugated spacer band. Ensure that the band ends are not allowed to overlap when the spacer is in place in the ring groove. Fit the oil control ring side rails one at a time; these may be fitted in either position as no top or bottom ring designations are given.
2   The two compression rings are of differing type and cross-section. The upper ring has a chrome plated face which is slightly curved. The 2nd ring is not chrome plated and has a tapered face. As a visual guide, the colour of the top ring will be seen to be appreciably lighter than that of the 2nd ring. When fitting the two compression rings, ensure that both are fitted with the N mark facing upwards.
3   Take great care when fitting the piston rings in their correct relative positions, otherwise breakage of the rings will most certainly occur. With a little practice the ring ends can be eased apart with the thumbs and the ring lowered into position. Alternatively, a piston ring expander can be used. This is a plier-like tool that does much the same thing as described above. One of the safest (and cheapest) methods is to use three thin steel strips such as old feeler gauges to ease the ring into place.
4   Before refitting the pistons to the connecting rod ends, pad the crankcase mouths with clean rag to prevent the ingress of foreign matter during piston and cylinder barrel replacement. It is only too easy to drop a circlip while it is being inserted into the piston boss, which will necessitate a further strip down for its retrieval.
5   If old pistons are being re-used, refer to the left and right-hand markings on the pistons, which were made during dismantling, and fit the pistons onto their original connecting rods, with the arrow embossed on each piston crown facing forwards. If the gudgeon pins are a tight fit in the piston bosses, warm each piston first to expand the metal. Do not forget to lubricate the gudgeon pin, small-end eye and the piston bosses before reassembly.
6   Use new circlips, **never** re-use old circlips. Check that each circlip has located correctly in its groove. A displaced circlip will cause severe engine damage. Note that the circlips should be fitted with the gap well away from the recess in the end of the piston boss.
7   Ensure that the mating faces of the crankcase and cylinder block are both clean and degreased. Refit the oil supply jets and

O-rings into their locations in the crankcase mating surface and refit the locating dowels. On 400 models only, fit new O-rings to the cylinder bore spigots and push them fully home so that they seat correctly in the grooves provided. Carefully ease a new gasket over the cylinder block retaining studs, cam chain tensioner blade and cam chain, and push it down onto the crankcase mating surface.
8   As for removal, the following procedure will be completed with far greater ease if an assistant is at hand. Arrange the piston ring gaps as shown in the accompanying diagram, in order to maintain the best sealing characteristics. Using clean engine oil, lubricate thoroughly the cylinder bores. Lift the cylinder block up onto the studs and support it there whilst the camshaft chain is threaded through the tunnel between the bores. This task is best achieved by using a piece of stiff wire to hook the chain through, and pull up through the tunnel. The chain must remain in engagement with the crankshaft drive sprocket.
9   The cylinder bores have a generous lead in for the pistons at the bottom, and although it is an advantage on an engine such as this to use the special Suzuki ring compressor, in the absence of this it is possible to gently lead the pistons into the bores working across from one side. Position the pistons so that one will enter its bore before the other and take great care not to put too much pressure on the fitted piston rings. When the pistons have finally engaged, remove the rag padding from the crankcase mouths and lower the cylinder block still further until it seats firmly on the base gasket.
10   Care should be taken at all times during the above operation to anchor the camshaft chain so that it does not drop back into the crankcase. This precaution should also be taken when refitting the cylinder head.
11   Check that the mating surfaces of the cylinder block and cylinder head are both clean and degreased. Lubricate the cylinder bores with liberal amounts of clean engine oil. Refit the two locating dowels into their locations in the cylinder block and carefully fit the new gasket into position over the retaining studs, cam chain and tensioner blade before pressing it down onto the mating surface.
12   Ease the cylinder head down its retaining studs whilst guiding the cam chain and tensioner blade through the tunnel. Re-anchor the chain and loosely fit the cylinder head retaining nuts and copper washers. Check that the cylinder head is properly seated on the barrel by gently tapping it down with a soft-faced mallet and proceed to tighten fully the nuts in the indicated sequence, a little at a time until they reach the specified torque loading figure.
13   Move to the front of the cylinder head and refit the single 6 mm bolt into its location between the exhaust ports. Tighten the bolt to the specified torque loading.
14   Insert the cam chain guide blade into the cylinder head central tunnel so that the lower end engages in the recess in the guide seat and the upper end locates with the recesses each side of the tunnel. Make a special check that the lower end of the blade is restrained; if it is allowed to float in operation the cam chain may become damaged.

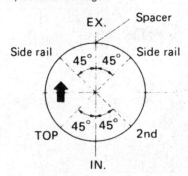

Fig. 1.25 Piston ring gap arrangement

41.5a The arrow embossed on each piston crown must face forwards when the piston is fitted

41.5b Lubricate the gudgeon pin before fitting

41.6 Each circlip must be fitted with its gap away from the recess in the piston boss

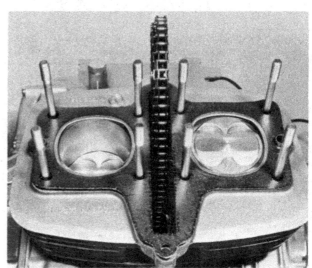

41.11 Carefully fit the new cylinder head gasket...

41.12a ...before easing the cylinder head into position

41.12b Do not omit to fit copper washers over the cylinder head retaining studs...

41.12c ...before fitting and tightening the retaining nuts to the correct torque loading

41.14 Ensure that the cam chain guide blade is correctly located

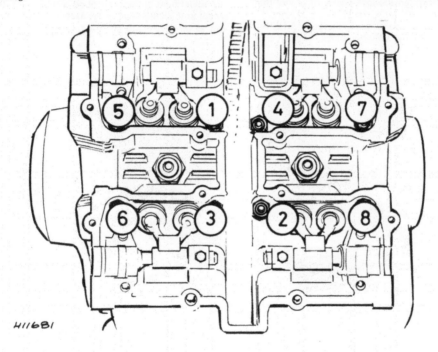

H11681

Fig. 1.26 Cylinder head nut tightening sequence

42 Reassembling the engine/gearbox unit: refitting the camshafts and timing the valves

1 Because the cam chain is endless, the camshaft replacement and valve timing operations must be made simultaneously. Fit a spanner to the engine turning hexagon, located beneath the ATU central retaining bolt head, and rotate the crankshaft forwards until the 'T' mark on the 'R' side of the ATU is in **exact** alignment with the index mark on the static plate. In this position No 1 cylinder (RH cylinder) is at TDC.

2 Note that whilst the crankshaft is being turned, the cam chain must be hand fed into the cylinder head tunnel so that it is prevented from becoming snagged or bunched up and from becoming detached from the crankshaft sprocket. Do not, under any circumstances, attempt to rotate the crankshaft by placing a spanner on the ATU retaining bolt.

3 Before fitting either of the camshafts, lubricate the com-

plete area of their journals with a high quality molybdenum disulphide lubricant, taking care not to leave any dry spots. Lubricate the journal bearings in the cylinder head and bearing caps with clean engine oil.

4 Select the exhaust camshaft (marked EX) and feed it through the cam chain. Pull the forward run of the chain taut whilst taking care not to rotate the crankshaft. Mesh the cam sprocket to the chain so that when the camshaft is lying across the bearing housings, the 1 marked arrow is pointing flush with the upper mating surface of the cylinder head. To check, refer to Fig. 1.15.

5 Insert the inlet camshaft through the cam chain. To mesh the sprocket correctly, count the chain roller pins from the exhaust camshaft to the inlet camshafts, start with the pin directly above the 2 marked arrow on the exhaust camshaft sprocket and count to the 20th pin along the chain. Mesh the 3 marked arrow on the inlet sprocket with the 20th pin. Provided that the crankshaft has not moved during this pro-

cedure, the valve timing is now correct.

6   The two bearing caps retaining each camshaft can now be refitted in their original locations, noting the symbol cast in both the cap and cylinder head. Ensure that the point of the triangle on each cap faces forward and that the cap locating dowels are fitted. Great care should be taken when tightening down the bearing cap bolts. If, due to the position of a camshaft, a camlobe is in contact with a rocker arm and thereby preventing correct seating of the shaft, the bolts should be tightened evenly and diagonally, a little at a time, allowing neither the caps nor the camshaft to tilt. Fully tighten the bolts to the specified torque loading and recheck the valve timing.

43 Reassembling the engine/gearbox unit: refitting and adjusting the cam chain tensioner

1   Hold the cam chain tensioner in one hand and whilst restraining the plunger rod, slacken the locking screw a few turns. Push the plunger inwards fully, simultaneously rotating the knurled adjuster wheel anti-clockwise. Continue turning until the plunger is fully retracted and the knurled adjuster has moved as far as possible. Tighten the lock screw to secure the pushrod.

2   Fit a new gasket to the tensioner body flange and fit the completed unit to the rear of the cylinder block. Tighten the securing bolts evenly to the specified torque loading. If any difficulty is experienced in pushing the tensioner fully into the cylinder block, turn the crankshaft slowly clockwise so that the full amount of slack is present in the rearmost run of the camshaft chain.

3   With the tensioner fitted to the engine, unscrew the locking screw $\frac{1}{4} - \frac{1}{2}$ a turn so that the plunger is free to move forwards under tension from the spring. Without allowing further rotation of the locking screw, tighten the locknut.

4   The cam chain tensioner is now set for automatic adjustment in service. To check whether the unit is functioning correctly, rotate the engine backwards to take up all the slack in the rear run of the chain. Whilst turning the engine backwards, rotate the knurled adjuster wheel slowly in an anti-clockwise direction as far as possible. Now turn the engine in a forward direction, which will have the effect of slackening the chain on the rear run. The knurled adjuster wheel should be seen to rotate in a clockwise direction as the plunger moves out and automatically tensions the chain.

5   **WARNING.** After initial adjustment of the cam chain tensioner, the tensioner will continue to function automatically. **DO NOT** under any circumstances rotate the knurled adjuster wheel either clockwise or anti-clockwise except when making this adjustment in the prescribed manner. Rotation of the wheel except at this stage will cause excessive chain tightness and will lead to tensioner and chain damage.

42.4 Select and fit the exhaust camshaft through the cam chain...

42.5 ...followed by the inlet camshaft which must be correctly aligned

42.6a Refit each camshaft bearing cap onto its two locating dowels...

42.6b ...and tighten the retaining bolts evenly, a little at a time

44 Reassembling the engine/gearbox unit: checking and adjusting the valve clearances

1   Place a spanner on the engine turning hexagon, located beneath the ATU central retaining bolt head, and rotate the crankshaft until the 'T' mark on the 'R' side of the ATU is in alignment with the index mark on the static plate and the notches in the camshaft ends are pointing outwards as shown in the table accompanying this text. With the engine set in this position, the left-hand pair of exhaust valves may be checked for clearance as follows.

2   Using a feeler gauge, check the clearance between each rocker arm adjuster and the end of the valve stem with which it makes contact. The clearance between each adjuster and stem should be identical to that listed in the Specifications at the beginning of this Chapter.

3   If the clearances are found to be incorrect, adjustment can be made by slackening each locknut concerned and turning the adjuster until the correct setting is obtained. Hold the adjuster still and tighten the locknut. Then recheck the setting before passing to the next valve. The feeler gauge should be a light sliding fit between the adjuster and the valve stem.

4   Note that valve clearances should always be adjusted with the engine COLD, otherwise a false reading will be obtained. Badly adjusted tappets normally give a pronounced clicking noise from the vicinity of the cylinder head.

5   Once the clearances of the left-hand pair of exhaust valves are found to be satisfactory, refer again to the table accompanying this text and rotate the crankshaft through 180° until the camshaft notches are in the positions indicated. The right-hand pair of exhaust valves may now be checked for clearance by following the previously noted procedure.

6   Continue to check the clearances of the two pairs of inlet valves by turning the crankshaft in further increments of 180° whilst referring to the table for the respective positions of the camshaft notches and the pair of valves to check.

7   On completion of checking the valve clearances, place a new gasket over the ignition pick-up assembly cover to right-hand crankcase cover mating surface, place the cover in position and secure it with the three retaining screws.

44.3a Turn the rocker arm adjuster whilst checking the valve clearance with a feeler gauge

44.3b Hold the adjuster whilst tightening its locknut

45 Reassembling the engine/gearbox unit: refitting the cylinder head cover and oil filter

1   Check that the smaller of the two cam chain guide plates is securely fitted in the roof of the cylinder head cover and that on the 400 models, the guide block is also fitted. Note that the tachometer drive assembly must be removed from the cover prior to the cover being placed in position over the camshafts.

2   Clean and degrease the mating surfaces of the cylinder head and cover and push the two locating dowels into position in the cylinder head. Place the new gasket in position on the cylinder head mating surface after having first relocated the semi-circular seals at each end of the camshaft chambers.

3   Check that the oil drain plug was refitted and tightened as described during the engine removal sequence. Fit a new seal to the oil filter cap, lightly greasing it before fitting it in the cap

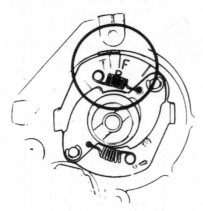

ATU alignment mark

	Notch position	
Cam Position	Intake Camshaft	Exhaust Camshaft
Ⓐ L. EX.	◖▪	◗▪
Ⓑ R. EX.	◖▪	◖▪
Ⓒ L. IN.	◖▪	◖▪
Ⓓ R. IN.	◖▪	◖▪

Camshaft notch positions

Fig. 1.27 Valve clearance alignment marks

groove to lessen the risk of it dropping out of position during positioning of the cap. Refit the spring with a new oil filter into the filter housing and secure the cap in position by refitting and tightening the three retaining nuts and washers. The nuts should be tightened evenly and in small increments so that the spring remains in its location in the cap and filter end. This simple precaution will also ensure that the edges of the cap around the retaining stud holes are not subjected to excessive stress.

4   Pour 50cc (1.76/1.69 Imp/US fl oz) of clean engine oil into each of the cylinder head pockets surrounding the four camshaft lobes. Place the cylinder head cover into position over the two locating dowels and tap it gently with a soft-faced mallet to seat it on the gasket. Insert the retaining bolts into the cover, remembering that the two longer bolts must pass through the dowel locations. Tighten the bolts evenly and in a diagonal sequence until the specified torque loading is reached.

5   Clean and degrease the mating surfaces of the cylinder head cover and breather cover and place a new gasket on the cover mating surface. Refit the breather cover and fit and tighten its four retaining bolts to secure it in position.

6   Before refitting the chromed end caps to each side of the cylinder head cover, clean and degrease the threads of each of the cap retaining screws and coat the threads in a thread locking compound.

7   Finally, insert the tachometer drive gear assembly into its location in the cylinder head cover, having first lightly lubricated its oil seals with clean engine oil. Secure the assembly in the cover by refitting the retaining plate and fitting and tightening the single crosshead retaining screw.

46 Refitting the engine/gearbox unit into the frame

1   It is worth checking at this stage that nothing has been omitted during the rebuilding sequences. It is better to discover any left-over components at this stage rather than just before the rebuilt engine is to be started.

2   Installation is, generally speaking, a reversal of the removal sequence. Pad the front and lower frame tubes as for removal and lift the engine into position in the frame. Refit the engine plates and mounting bolts, taking care to refit the cylindrical spacer and plain washer to the long bolt passing through the upper rear mounting plate. Do not fully tighten any of the mounting bolts or nuts until all the plates and bolts have been fitted. Refer to the Specifications Section at the beginning of this Chapter and tighten the mounting bolts and nuts to the torque figures specified.

3   With the engine firmly secured in the frame, mesh the gearbox sprocket with the final drive chain and slide the sprocket onto the splined end of the layshaft. Take note of the mark made on the sprocket during removal to ensure that it is not refitted in the reverse position. If difficulty is experienced in fitting the sprocket and chain to the shaft, loosen the rear wheel spindle securing nut and move the wheel forward to give some play in the chain. Place a new lock washer over the shaft end and fit and tighten the retaining nut to the specified torque loading. Lock the nut in position by bending the lock washer up against one of its flats. A similar method to that used for removal will have to be employed in order to prevent the layshaft from turning as the sprocket retaining nut is tightened. Note that the nut has a recess machined in one of its sides and that this recess must face the sprocket.

4   If necessary, refit the clutch operating cable into the gearbox sprocket cover, using a reversal of the procedure given for removal. Clean the mating surface of the cover and place it in position over the sprocket. Insert the cover securing screws and tighten them evenly and in a diagonal sequence. Refit the gearchange lever to its required position on the splined shaft

and refit both footrest assemblies.

5   Refit the exhaust system to the machine, using a reverse sequence to that given for removal whilst noting the torque figures given in the Specifications Section of this Chapter. Always renew the sealing rings before inserting the pipe ends into the cylinder head.

6   Reconnect the electrical lead to the starter motor and refit the chromed cover to its location on the upper crankcase. Fit and tighten the two cover retaining screws whilst ensuring that the lead is correctly routed between it and the crankcase.

7   As for removal, the help of an assistant is desirable when refitting the carburettor assembly. Follow the procedure given for removal in the reverse sequence, taking care to refit electrical leads to their original locations beneath the air filter housing retaining screw heads. Check also that the throttle cable is correctly routed and that all connections are correctly located and tightened so as to prevent any likelihood of fuel or air leakage.

8   Prior to refitting the sparking plugs, ensure that their electrodes are clean and correctly gapped. Reconnect all electrical connections, taking care to note any marks made for reference during engine removal. Refit and reconnect the horn and reconnect the tachometer drive cable. Once all the control cables, electrical leads, etc have been reconnected, carry out a final check to ensure that they are not likely to chafe on cycle components or come into contact with hot engine castings.

9   Reconnect the battery leads, observing the polarity markings. Ensure that the connections are clean and free from corrosion; to protect the connections from corrosion, smear them with a liberal coating of petroleum jelly. Do not use ordinary grease. Reconnect the breather pipe to its retaining stub at the top of the battery and ensure that it is routed correctly and is not trapped or kinked.

10   After the fuel tank has been refitted, reconnect the fuel and vacuum hoses to the fuel tap. Utilise the primary function of the tap and carefully check both ends of the fuel hose for any signs of fuel leakage. On no account should fuel be allowed to come into contact with hot engine castings; if this is allowed to happen fire may result causing serious personal injury.

11   After refitting the seat unit, carry out a final check around the engine/gearbox unit and the associated cycle components to ensure that all component parts have been correctly refitted. Carry out an inspection of the work area and any containers used to determine whether or not any components have been omitted from the reassembly sequence and if satisfied, proceed to prepare the engine for start-up.

45.2a Relocate the camshaft chamber seals...

45.2b ...followed by the cylinder head cover locating dowels and a new gasket

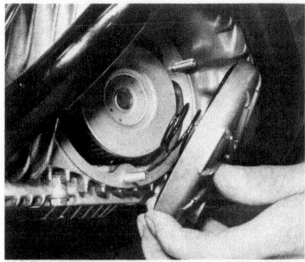

45.3 Fit the oil filter element, spring and cap with new O-ring to the crankcase

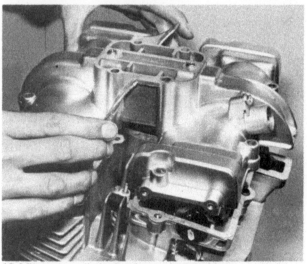

45.4 Place the cylinder head cover into position

46.3a Note the recess machined in the gearbox sprocket retaining nut...

46.3b ...before fitting the nut and locking it in position with the tab washer

46.4 The clutch cable is retained by bending the tag in its holder

46.5 Always renew the exhaust pipe to cylinder head sealing rings

46.8 Ensure that the tachometer drive cable retaining ring is fully tightened

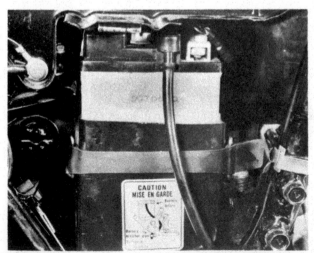

46.9 The battery breather pipe must be correctly routed

47 Final adjustments and preparation

1   Commence preparation of the engine/gearbox unit by referring to the Specifications at the beginning of Chapter 2 and pouring the correct amount of SAE 10W/40 engine oil into the engine through the filler point on the right-hand crankcase cover. After the initial start-up wait a few minutes for the oil to settle within the engine and check its level with the dipstick attached to the filler cap. If necessary, add more oil to bring it to the correct level.

2   Adjust the clutch cable and lifting mechanism as follows. Loosen the locknut on the adjuster screw and rotate the screw inwards until it can be felt to abut against the end of the pushrod. To gain the necessary running clearance unscrew the screw by $\frac{1}{4} - \frac{1}{2}$ a turn and then tighten the locknut. The cable should be adjusted so that there is 4 mm (0.16 in) play measured between the handlebar stock and lever before the clutch commences lifting. Cable adjustment can be carried out by loosening the locknut of the adjuster located at the gearbox sprocket cover end of the cable and rotating the adjuster until the correct amount of play is achieved. Lock the adjuster in position by tightening the locknut whilst holding the adjuster to prevent it from turning and carry out any fine adjustment on the cable by rotating the knurled adjuster ring at the handlebar stock.

3   To check that the throttle cable is correctly adjusted, locate the adjuster at the carburettor assembly retaining clip and pull upwards on the cable to check that there is 0.5 mm (0.02 in) of free play between the cable end and the adjuster. If necessary, carry out adjustment by loosening the adjuster locknut and turning the adjuster. Once the desired amount of free play is achieved, hold the adjuster steady and retighten the locknut.

4   If it was found necessary to loosen the rear wheel spindle securing nut in order to move the wheel forward to aid refitting of the gearbox sprocket, refer to the instructions given in Chapter 5 for the relocation of the wheel and the retensioning of the final drive chain. Once the wheel has been correctly positioned carry out any necessary adjustment on the rear brake and stop lamp switch.

48 Starting and running the rebuilt engine

1   Utilise the primary function of the fuel tap and allow a few seconds for the carburettor float chambers to fill. Check that the machine is in neutral, start the engine and return the tap to the 'On' position as soon as the engine is running. Push the choke knob fully in as soon as the engine will run evenly and keep it running at a low speed for a few minutes to allow oil pressure to build up and the oil to circulate. If the oil pressure indicator lamp is not extinguished, stop the engine immediately and investigate the lack of oil pressure.

2   If the engine refuses to start, carry out a check for a spark at the sparking plug electrodes and check that fuel is finding its way to the combustion chambers; compression should be apparent and the engine should rotate freely once the starter button is pressed. Under no circumstances engage the starter motor for more than five seconds at a time as this will cause overheating problems in the motor and its wiring harness.

3   Once running, the engine may tend to smoke through the exhaust initially, due to the amount of oil used when assembling the various components. The excess of oil should gradually burn away as the engine settles down.

4   Check the exterior of the machine for oil leaks or blowing gaskets. Make sure that each gear engages correctly and that all the controls function effectively, particularly the brakes. This is an essential last check before taking the machine on the road.

47.1a Replenish the crankcase with the correct oil...

47.1b ...and check the oil level with the dipstick provided

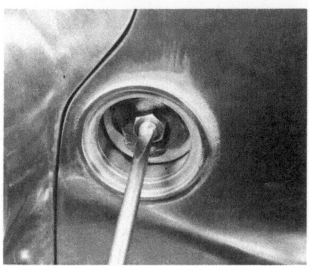
47.2a Adjust the clutch lifting mechanism by turning its screw

47.2b Carry out clutch cable adjustment by rotating the coarse adjuster...

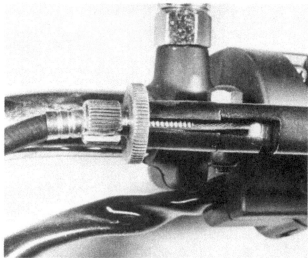
47.2c ...before using the adjuster ring at the handlebar end for fine adjustment

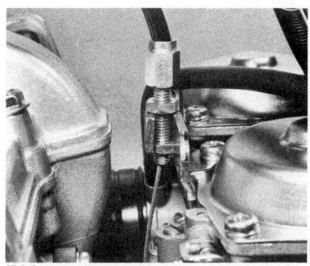
47.3 Carry out adjustment of the throttle cable by rotating the adjuster in its retaining clip

49 Taking the rebuilt machine on the road

1   Any rebuilt machine will need time to settle down, even if parts have been replaced in their original order. For this reason it is highly advisable to treat the machine gently for the first few miles to ensure oil has circulated throughout the lubrication system and that any new parts fitted have begun to bed down.
2   Even greater care is necessary if the engine has been rebored or if a new crankshaft has been fitted. In the case of a rebore, the engine will have to be run-in again, as if the machine were new. This means greater use of the gearbox and a restraining hand on the throttle until at least 1000 miles (1600 km) have been covered.
3   Suzuki recommend that for the initial 500 miles (800 km) running-in, an engine speed of 4000 rpm should not be exceeded. Between 500 and 1000 miles, the engine speed may be gradually increased up to 6000 rpm and over 1000 miles the full use of the engine rev range may be used, although this should be practised with some degree of care until an appreciably greater mileage is covered.
4   There is no point in keeping to any set speed limit; the main requirement is to keep a light loading on the engine and to gradually work up performance until the 1000 mile mark is reached. These recommendations can be lessened to an extent when only a new crankshaft is fitted. Experience is the best guide since it is easy to tell when an engine is running freely.
5   If at any time a lubrication failure is suspected, stop the engine immediately, and investigate the cause. If an engine is run without oil, even for a short period, irreparable engine damage is inevitable.
6   When the engine has cooled down completely after the initial run, recheck the various settings, especially the valve clearances. During the run most of the engine components will have settled into their normal working locations. Check the various oil levels, particularly that of the engine as it may have dropped slightly now that the various passages and recesses have filled. Although the ignition timing was set during engine reassembly, it is advisable to carry out a dynamic check at this point by connecting a stroboscopic lamp between an HT lead and sparking plug as instructed in Chapter 3. This will not only provide an accurate check of the ignition timing but will verify that the ATU is functioning correctly.

50 Fault diagnosis: engine

Symptom	Cause	Remedy
Engine will not start	Defective sparking plugs	Remove the plugs and lay them on cylinder heads. Check whether spark occurs when ignition is switched on and engine rotated.
	Ignition fault	Recheck connections and/or timing. If necessary, refer to Chapter 3.
Engine runs unevenly	Ignition and/or fuel system fault	Check each system independently, as though engine will not start.
	Blowing cylinder head gasket	Leak should be evident from oil leakage where gas escapes.
	Incorrect ignition timing	Check accuracy and if necessary reset.
Lack of power	Fault in fuel system or incorrect ignition timing	See above.
Heavy oil consumption	Cylinder block in need of rebore	Check for bore wear, rebore and fit oversize pistons if required.
	Damaged oil seals	Check engine for oil leaks.
Excessive mechanical noise	Worn cylinder bores (piston slap)	Rebore and fit oversize pistons.
	Worn camshaft drive chain (rattle)	Adjust tensioner or replace chain.
	Worn big-end bearings (knock)	Renew. See text.
	Worn main bearings (rumble)	Renew. See text.
Engine overheats and fades	Lubrication failure	Stop engine and check whether internal parts are receiving oil. Check oil level in crankcase.

51 Fault diagnosis: clutch

Symptom	Cause	Remedy
Engine speed increases as shown by tachometer but machine does not respond	Clutch slip	Check clutch adjustment for free play at handlebar lever. Check thickness of inserted plates.
Difficulty in engaging gears. Gear changes jerky and machine creeps forward when clutch is withdrawn. Difficulty in selecting neutral	Clutch drag	Check clutch adjustment for too much free play. Check clutch drums for indentations in slots and clutch plates for burrs on tongues. Dress with file if damage not too great.
Clutch operation stiff	Damaged, trapped or frayed control cable	Check cable and renew if necessary. Make sure cable is lubricated and has no sharp bends.

52 Fault diagnosis: gearbox

Symptom	Cause	Remedy
Difficulty in engaging gears	Selector forks bent	Renew.
	Gear clusters not assembled correctly	Check gear cluster arrangement and position of thrust washers.
Machine jumps out of gear	Worn dogs on ends of gear pinions	Renew worn pinions.
	Detent mechanism worn	Renew as required.
Gearchange lever does not return to original position	Broken return spring	Renew spring.

Chapter 2 Fuel system and lubrication

Refer to Chapter 7 for information relating to GS450 and GSX250/400 EZ models

Contents

Specifications

Fuel tank

	GS250	GSX250 and 400E	GSX400T
Overall capacity	11 lit (2.4/2.9 Imp/US gal)	14.5 lit (3.2/3.8 Imp/US gal)	12 lit (2.6/3.1 Imp/US gal)
Reserve capacity	2.0 lit (3.5/4.2 Imp/US pint)	3.0 lit (5.3/6.3 Imp/US pint)	4.0 lit (7.0/8.4 Imp/US pint)

Carburettors

	GS250 TT (UK) and GSX250 ET	GS250 TT (US) and GS250 TX (US)	GS250 TX (UK) and GS250 EX
Make	Mikuni	Mikuni	Mikuni
Type	BS30 SS	BS30 SS	BS30 SS
Venturi diameter	30 mm	30 mm	30 mm
Main jet	115	115	115
Main air jet	1.0	1.0	1.0
Needle jet:			
Left-hand	O-6	O-7	P-1
Right-hand	O-7	O-8	P-2
Jet needle	5CT 38	5CHT 40	5DFT 76
Needle position	3rd groove	3rd groove	3rd groove
Pilot jet	17.5	17.5	17.5
Pilot air jet	140	135	132.5
Pilot outlet	0.8	0.8	0.8
Float valve seat	2.0	2.0	2.0
Starter jet	35	35	35
Throttle valve	85	80	85

All models

Fuel level	4.0 ± 0.5 mm (0.16 ± 0.02 in)
Float height	21.4 ± 1.0 mm (0.84 ± 0.04 in)
Pilot screw:	
UK models	1½ turns out
US models	Preset

Carburettors

GSX400 models

Make	Mikuni
Type	BS34SS
Venturi diameter	34 mm
Main jet	117.5
Main air jet	0.6
Needle jet:	
GSX400ET and TT	O-9
GSX400EX and TX	P-1

Jet needle	5D69
Needle position	4th groove from top
Pilot jet	42.5
Pilot air jet	125
Pilot outlet	0.8
Valve seat	2.0
Starter jet	32.5
Throttle valve	135
Fuel level	5.0 ± 0.5 mm (0.2 ± 0.02 in)
Float height	22.4 ± 1.0 mm (0.88 ± 0.04 in)
Pilot screw	1½ turns out

Air cleaner

Element type	Oiled foam

Lubrication

	250 models	400 models
System type	Forced pressure, wet sump	
Oil capacity:		
Oil change	2.0 lit (3.6/4.2 Imp/US pint)	2.6 lit (4.6/5.4 Imp/US pint)
With filter change	2.6 lit (4.6/5.4 Imp/US pint)	2.9 lit (5.2/6.2 Imp/US pint)
Engine overhaul	2.6 lit (4.6/5.4 Imp/US pint)	3.0 lit (5.3/6.4 Imp/US pint)

Oil pump

Type	Trochoid
Outer rotor to housing clearance service limit	0.25 mm (0.010 in)
Inner rotor to outer rotor tip clearance service limit	0.20 mm (0.008 in)
Side clearance service limit	0.15 mm (0.006 in)
Pump reduction ratio:	
250 models	1.905 : 1 (75/24 x 25/41)
400 models	2.071 : 1 (76/28 x 29/38)

1 General description

The fuel system comprises a tank from which fuel is fed by gravity to the float chamber of each of the two carburettors. A single tap with a gauze filter is located beneath the tank, on the left-hand side. It contains provision for a reserve quantity of fuel, when the main supply is exhausted.

There are several types of tap fitted to the model types covered in this Manual; their differences and functions being fully described within the following Sections.

Two constant velocity (CV) Mikuni carburettors are mounted on angled brackets and fitted to the machine as a unit. The carburettors are controlled by a cable which connects the handlebar mounted twistgrip to the carburettor mounted cross shaft pulley. This shaft in turn operates the carburettor throttle butterfly valves, which in turn control the overall airflow through the instruments, and thus the engine speed. Each carburettor contains a diagraphm-type throttle valve which moves in response to changes in manifold depression and in this manner automatically controls the volume and strength of the mixture entering the combustion chamber. Because the carburettors react automatically the engine runs at the optimum setting at any given throttle twistgrip setting and engine load condition, and with compensation for variations in atmospheric pressure due to changes in altitude.

The CV carburettor is ideally suited to provide an accurately controlled mixture which conforms with the increasingly stringent emission laws in the US and Europe. This allows the overall mixture to be proportionally weaker than that of a conventional slide-type instrument, and in turn should give better fuel economy.

Engine lubrication is effected by the wet sump principle in which the reservoir of oil is contained within the engine sump. This oil is shared by the engine, primary drive and transmission components. The oil pump is of the Eaton trochoid type and is driven from a pinion engaged with and to the rear of the clutch.

Fig. 2.7 shows the layout of the lubrication system.

2 Fuel tank: removal and refitting

1 The fuel tank is retained by two guide channels which locate with a circular rubber block on each side of the steering head and a bolt passing through at the rear of the tank and into the frame. Machines fitted with a 'peardrop' style tank have a plastic trim moulding fitted around the forward edge of the tank; each half of this trim is secured to the tank by a single crosshead retaining screw.

2 Commence removal of the tank by positioning the tap lever in the 'On' position and releasing the fuel and vacuum hoses from the fuel tap stubs. The hoses are normally retained on the top stubs by spring clips, the ears of which should be pinched together to release them from the hose. Where necessary, remove the plastic trim from the front of the tank by unscrewing its retaining screws.

3 Release the dualseat from its mountings in order to gain access to the tank mounting bolt. Remove the bolt and lift the tank up at the rear. Maintaining a firm grip on the tank, pull it backwards whilst moving it gently from side to side in order to free the guide channels from the rubber mounting blocks.

4 Store the tank in a safe place whilst it is removed from the machine, well away from any naked lights or flames. It will otherwise represent a considerable fire or explosion hazard. Check that the tap is not leaking and that it cannot be accidentally knocked into the 'Prime' position. It is well worth taking simple precautions to protect the paint finish of the tank whilst in storage. Placing the tank on a soft protected surface and covering it with a protective cloth or mat may well avoid damage being caused to the finish by dirt, grit, dropped tools, etc.

5 To refit the tank, reverse the procedure adopted for its removal. Move it from side to side before it is fully home, so that

the rubber buffers engage with the guide channels correctly. If difficulty is encountered in engaging the front of the tank with the rubber buffers, apply a small amount of fuel to the buffers to ease location. Always carry out a leak check on the fuel pipe connections after fitting the tank and running the engine. Any leaks found must be cured; as well as the wastage involved, any fuel dropping onto hot engine castings may well result in a fire or explosion occurring. Take care not to trap any control cables or electrical leads between the tank and frame tubes.

3   Fuel tap: general description, removal, servicing and refitting

1   Any one of three types of fuel tap may be found to be fitted to the machines covered in this Manual. Two of these types are similar in appearance and incorporate 'On' and 'Res' positions for the tap lever. One of the two however, has a primary function in the form of a screw located near the base of the tap body, thus providing a means of allowing fuel to pass directly from the tank to the carburettors when the engine is not running. The third type of tap incorporates three lever positions; as well as the 'On' and 'Res' previously mentioned, there is also a 'Pri' position which provides the same priming function as for the screw.
2   With the tap lever turned to the 'On' position or 'Res' position, fuel can only flow to the carburettors when the engine is running. This is due to the tap diaphragm which is controlled by the induction pressure. If there is no fuel in the carburettor float chambers, as may be the case after carburettor dismantling, the primary function of the tap (where incorporated) should be utilised to allow an unrestricted flow of fuel from the tank to the float chambers.
3   Where the tap incorporates a 'Pri' position for the lever, the priming function can be utilised simply by turning the lever to that position. Where the tap incorporates a priming screw, the screw should be turned anti-clockwise so that the base of its head protrudes 2 – 3 mm (0.08-0.12 in) from the surface of the tap body. The screw should not be allowed to protrude more than 5 mm (0.20 in) otherwise fuel will leak from the screw housing. The screw must be retightened and the tap lever returned to the 'On' position as soon as the carburettors have been primed and the engine is running. Where the tap incorporates no priming function, the tap must be by-passed in order to allow fuel to pass to the float chambers whilst the engine is not running.
4   Before the fuel tap can be removed, it is first necessary to drain the tank. This is easily accomplished by removing the feed pipe from the carburettor float chamber and allowing the

contents of the tank to drain into a clean receptacle. This is only possible when a tap incorporating a priming function is fitted. Alternatively, the tank can be removed with fuel still in it and placed on one side, so that the fuel level is below the tap outlet. Care must be taken not to damage the tank paintwork whilst doing this.
5   The tap is held to the underside of the tank by two bolts with washers. There is an O-ring seal between the tap body and the tank, which must be renewed if it is damaged or if fuel leakage has occurred.
6   The filter screens, which are integral with the plastic level pipes, should be cleaned of any deposits using a soft brush and clean fuel. Because there is only a single tap to feed two carburettors, any restriction in fuel flow may lead to fuel starvation, causing missing and in extreme cases overheating due to a weak mixture.
7   Taps that do not incorporate a 'Pri' position for the lever are supplied as sealed units and must be renewed as such if leakage occurs around the lever joint. Taps which incorporate the three lever positions may be dismantled to the extent of removing the lever assembly in order to cure leakage around its joint. It is worth noting at this point that any leak around the lever joint may be due to the two plate securing screws having become loosened during machine use, in which case any leakage may be cured simply by retightening the screws. Although the tank must be drained before the lever assembly can be removed, there is no need to disturb the body of the tap.
8   To dismantle the lever assembly, remove the two crosshead screws passing through the plate on which the operating positions are inscribed. The plate can then be lifted away, followed by a spring washer, the lever itself and the seal behind the lever. The seal will have to be renewed if leakage has occurred. Check for any sediment build up within the tap body and if necessary clean it out using a similar method to that described for the filter screens.
9   Reassemble and refit the tap using a reversal of the above sequence. It is not necessary to use jointing compound or any other sealing medium to effect a seal.

4   Fuel feed hose: examination

1   The fuel feed hose is made from thin walled synthetic rubber and is of the push-on type. It is only necessary to replace the pipe if it becomes hard or splits. It is unlikely that the retaining clip should need replacing due to fatigue as the main seal between the pipe and union is effected by an interference fit.

2.2 Where necessary, remove the fuel tank trim...

2.3 ...before removing the tank mounting bolt

3.8 The fuel tap lever assembly is retained by two crosshead screws

5 Carburettors: removal and refitting

1 To improve access to the carburettor assembly, it is advisable to remove the fuel tank as described in Section 2 of this Chapter.

2 Commence removal of the carburettor assembly by detaching the throttle cable from the carburettor throttle operating pulley. To do this, loosen both the locknuts that retain the cable end adjuster to the carburettor mounted retaining clip. Rotate the throttle twistgrip so that the throttle is fully open and with the flat of a small screwdriver, hold the pulley in the fully open position whilst releasing the twistgrip to provide enough slack in the cable to allow the nipple to be released from its location in the pulley. Slide the cable end adjuster sideways to free it from the retaining bracket and tuck the cable out of the way on the frame top tube.

3 Loosen the clamps between the carburettors and air filter housing and release the housing from the three mounting points located on each inner side panel. Note when removing the retaining screws that one screw each side has electrical leads beneath its head; the position of these screws must be noted for reference when refitting. Note also that there are bushes located beneath all the screw heads and these must be removed before the housing can be slid rearwards to clear the carburettor mouths. it will also be necessary to ease each side panel outwards so that it clears the housing. Some difficulty was experienced in freeing the housing from the carburettor mouths and moving it back far enough to clear completely the carburettor assembly, but with the aid of an assistant and a certain amount of patience the job can soon be completed.

4 Finally, to remove the carburettor assembly from the machine, loosen the clamps between the carburettors and cylinder head and pull the carburettor assembly rearwards to clear the inlet manifolds and then to the right to clear the machine.

5 To refit the carburettor assembly, follow the above removal procedure in the reverse sequence whilst noting the following points. Take care to refit the electrical leads to their original locations beneath the heads of the filter housing retaining screws. Check that the throttle cable is correctly routed and adjusted and that all connections are correctly located and tightened so as to prevent any likelihood of fuel or air leakage.

6 Carburettors : dismantling, examination and reassembly

1 The two Mikuni CV carburettors are mounted as a unit on two angled metal brackets and are interconnected by the throttle linkage and the choke control rod. It is not normally necessary to separate the two carburettors in order to carry out a regular inspection of the internal components of each instrument. Instructions for separating the two units are, however, listed later in this Section.

2 Dismantling of each individual carburettor should be carried out as follows. Note that each carburettor should be dismantled separately, to prevent accidental interchange of parts. Before dismantling the carburettor, cover an area of the work surface with clean paper or rag. This will not only prevent any components that are placed upon it from becoming contaminated with dirt, moisture or grit but, by making them more visible, will also prevent the many small components removed from the carburettor body becoming lost.

3 Detach the float chamber from the carburettor body by unscrewing the four retaining screws and lock washers. Lift the float chamber away whilst taking care not to damage the gasket located beneath it. Make sure the float chamber gasket is in good condition. It should not be disturbed unless it shows sign of damage or has been leaking. It will, however, need to be removed if it is found necessary to check the float height.

4 Pull out the pivot pin from the twin float assembly and lift the floats away. The float needle can now be displaced from its seating and should be put aside in a safe place for examination at a later stage. It is very small and easily lost if care is not taken to store it in a safe place.

5 Unscrew and remove the single crosshead screw that serves to retain the float needle seat retaining plate in position between the two pivot pin columns of the carburettor body. Remove the plate and pull the needle seat from position, placing it together with the float needle.

6 Before attempting to remove either the pilot jet or main jet, check that the flat of the screwdriver to be used is a very good fit in the slot provided in each jet. Irreparable damage will occur to the jet if the slotted head is in any way malformed by an incorrectly fitting screwdriver. Unscrew and withdraw the pilot jet and follow this by unscrewing the main jet. Note the plain washer located beneath the head of the main jet and that the needle jet is prevented from dropping out of its location within the carburettor body by a small locating pin; this means that the jet can only be removed from the diaphragm housing.

7 Remove the diaphragm cover from the top of the carburettor body by unscrewing its four retaining screws and lockwashers. Lift out the piston spring, followed by the piston/diaphragm unit together with the jet needle. The needle jet should drop out of its location if the carburettor is now inverted into a cupped hand. If necessary, sharply tap the carburettor into the hand to jar the jet from position.

8 The jet needle is retained in the centre of the piston body by means of a plastic retainer and circlip. Removal of this circlip was found to be difficult. With no straight long-nose circlip pliers available, the only means found of dislodging the circlip was to ease it out of position by using the end of a thin-shanked parallel pin punch (as shown in the accompanying photograph) located in one of the circlip ends. A similar tool can of course be manufactured from any length of dowelled rod, although it must be rigid enough for the purpose. With the circlip removed, pull the plastic retainer out of the piston centre and dislodge the needle assembly.

9 Unscrew the pilot air jet from the side of the carburettor mouth. On UK models screw in the pilot adjuster screw until it seats lightly, counting the exact number of turns required to do so, so that it can be placed in the same position upon reassembly. Having recorded the screw position remove the screw from the carburettor. On US models the screw should not be disturbed because the adjustment is preset at the factory to comply with EPA regulations. Refer to Section 9 of this Chapter for further details.

10 The internal components of the carburettor have all now been removed. If, for any reason such as a worn or damaged linkage component, the two carburettors have to be separated, then proceed as follows. Loosen the two grub screws that retain the starter (choke) shaft to the two operating levers and withdraw the shaft from position. Take great care to retain the

small spring and steel ball located beneath the shaft in the left-hand carburettor as the shaft passes clear of its pivot. Remove each of the two mounting brackets by unscrewing their retaining screws. These screws may be found to be very tight and require careful use of an impact driver to free them. With both brackets removed, the carburettors are now connected only by the fuel cross-over pipe. Separate the two instruments by pulling them apart.

11 It is not recommended that the butterfly throttle valve assembly be removed from either carburettor as these components are not prone to wear. If wear occurs on the operating pivot a new carburettor will be required as air will find its way along the pivot bearings resulting in a weak mixture. If need be, the valve may be removed from its operating shaft by unscrewing its two retaining screws and rotating the shaft to allow it to drop clear of the carburettor. The shaft itself may be removed by removing the end cap to expose the E-ring at the shaft end and removing the E-ring to allow the shaft to be drawn out of the carburettor body.

12 Commence examination of each carburettor assembly by checking the condition of the floats. If they are damaged in any way, they should be renewed. The float needle and needle seating will wear after lengthy service and should be inspected carefully. Wear usually takes the form of a ridge or groove, which will cause the float needle to seat imperfectly. Always renew the seating and needle as a pair. An imperfection in one component will soon produce similar wear in the other.

13 After considerable service the piston needle and the needle jet in which it slides will wear, resulting in an increase in petrol consumption. Wear is caused by the passage of petrol and the two components rubbing together. It is advisable to renew the jet periodically in conjunction with the piston needle. Check the diaphragm for signs of perishing or for splits. If damage is evident the diaphragm must be renewed as a unit with the piston.

14 Check that all the mating surfaces on the carburettor body are flat by using a straight-edge laid across the mating surface. Ensure all O-rings and sealing gaskets are renewed when reassembling and refitting the carburettor and that, where applicable, they are correctly seated in their retaining grooves. The springs on the pilot screw (UK model only), the piston, the throttle linkage, the starter shaft and the jet needle should all be carefully inspected for signs of corrosion and fatigue and renewed if necessary.

15 On completion of the above listed examination procedure, the carburettor components to be reassembled should all be thoroughly cleaned in clean fuel whilst using a soft-bristled nylon brush to remove all traces of contamination. Compressed air may be used to dry the components and remove contamination from any recesses in the component casting. Great care must be taken to observe the necessary fire precautions when cleaning the components in fuel and to protect the eyes when using compressed air. Avoid using a piece of rag to clean any components since there is always a risk of particles of lint obstructing the internal passageways or the jet orifices.

16 Never use a piece of wire or any pointed metal object to clear a blocked jet. It is only too easy to enlarge the jet under these circumstances, and increase the rate of petrol consumption. If compressed air is not available, a blast of air from a tyre pump will usually suffice. As a last resort, a fine nylon bristle may be used.

17 Do not use excessive force when reassembling a carburettor because it is easy to shear a jet or some of the smaller screws. Furthermore, the carburettor is cast in a zinc based alloy which itself does not have a high tensile strength. If any of the castings are damaged during reassembly, they will almost certainly have to be renewed.

18 Reassembly is basically a reversal of the dismantling procedure, noting the following points. If the instruments have been separated and the throttle valve shaft removed from either carburettor body, then the oil seal through which the shaft passes should be removed from the carburettor body and replaced with a new item. Removal of this seal can be effected by careful use of the flat of a small screwdriver. Note that when a new seal is fitted, it must be lightly oiled and pressed into position with the groove facing outermost. Do not omit to inspect and if necessary renew the gasket washer which fits over the seal before refitting the E-ring. Ensure that the fuel cross-over pipe is a good fit in each carburettor bore before pushing the instruments together. Check that the two throttle linkage arms engage correctly so that the arm of the cable pulley lies between the spring loaded plunger and the throttle valve synchronisation screw.

19 Degrease the threads of each of the mounting bracket retaining screws and coat them with a thread locking compound before using the screws to retain the brackets in position. To ensure that the two instruments are in alignment with each other before the screws are finally tightened, it is a good idea to place both instruments side by side on a completely flat surface such as a sheet of plate glass, their mouths being in contact with the surface.

20 When refitting the starter shaft, lightly grease the length of the shaft before doing so and press the small spring and steel ball into its location in the carburettor body with the flat of a small screwdriver before passing the shaft over them. Take care to ensure that the indentation in each end of the shaft is in alignment with the hole in each operating lever before inserting and tightening the two grub screws.

21 Once the two instruments have been fitted together as a unit, check that both butterfly valves are set in identical positions and each has its top edge aligned with the innermost of the two bypass orifices in the carburettor bore. If this is not the case, bring the valves into alignment by rotating the balance screw located in the centre of the throttle linkage. Once the valves are aligned with each other, they may then be moved together to align with the by-pass orifice by rotating the throttle stop screw which is located between the two float chambers at the rear of the assembly.

22 If considered necessary, pay close attention to the figures accompanying this text whilst fitting the internal component parts into each carburettor. It should be noted that there are slight differences between the carburettors fitted to the 250 and 400 models covered in this Manual.

23 Take particular care when fitting the pistons to ensure that the needles pass cleanly into the needle jets. Note that the rubber diaphragm has a tag projecting from its outer edge; this tag must locate correctly in the corresponding slot in the carburettor body before the diaphragm cover is refitted.

24 Note when refitting the pilot screws, that each screw must first be screwed fully in and then set to its previously noted position, or, failing that, the recommended position. Finally, if an atomiser ring is fitted into the inlet side of the carburettor, ensure that it is pushed firmly into position before proceeding further.

25 On completion of assembly and installation on the machine the carburettors should be adjusted for mixture strength (where applicable) and synchronised. Refer to the following Section.

6.8 A parallel pin punch may be used to dislodge the jet needle retaining circlip

6.10a The starter (choke) shaft is retained to each of its operating levers by a single grub screw

6.10b Detach the two mounting brackets by removing their retaining screws

6.12a Inspect the float needle and its seating for wear...

6.12b ...whilst checking that the filter attached to the seating is clean

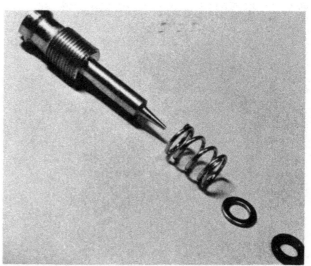

6.14 Renew all O-rings and carefully inspect all springs (pilot screw assembly shown)

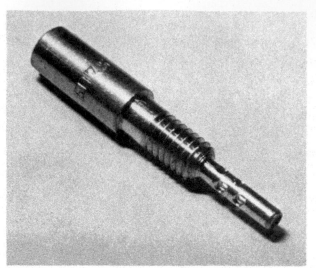

6.15a Thoroughly clean all components in fuel (pilot jet shown)...

6.15b ...and blow clear any blocked jets with compressed air (needle jet shown)

6.18a The complete carburettor assembly showing the throttle linkage and starter shaft assemblies

6.18b The fuel cross-over pipe must be correctly located

6.22a Commence assembly of each carburettor by inserting the needle jet through the diaphragm housing...

6.22b ...so that its end locates with the pin shown

6.22c Fit the washer beneath the head of the main jet...

6.22d ...before fitting the main jet into the needle jet

6.22e Insert the pilot jet into the float chamber column

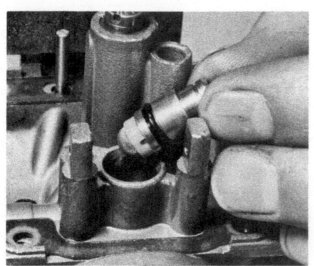

6.22f Fit a new O-ring to the float needle seat before pushing it into position...

6.22g ...and fitting the needle seat retaining plate and securing screw

6.22h Fit the float needle into its seat...

6.22i ...followed by the float assembly and pivot pin

6.22j With the base gasket fitted, locate the float chamber...

6.22k ...and secure it in position. Note the fitted pilot air jet (arrowed)

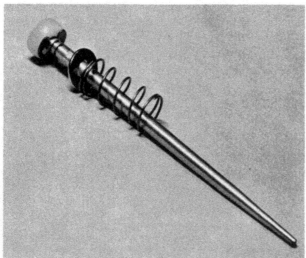
6.22l Assemble the jet needle components and insert it into the diaphragm piston...

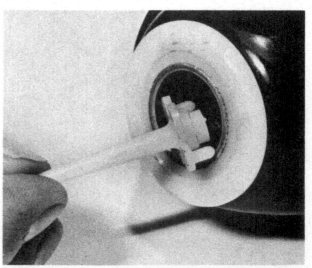
6.22m ....retain the needle assembly with the plastic retainer...

6.22n ...and hold the retainer in position with the circlip

6.23a Ensure the diaphragm tag is correctly located before fitting the spring...

6.23b ...and securing the cover in position with the four screws

6.24a Refit the pilot screw and set it to the previously noted position

6.24b Where fitted, the atomiser ring should be relocated in the carburettor mouth

7 Carburettors : synchronisation and adjustment

1   Synchronisatiom and adjustment of the carburettors should be carried out as a routine maintenance item. It may also be necessary when rough idling or poor performance is encountered and also after dismantling and reassembly has taken place. The procedure for these adjustments is made in two stages. The first stage is adjustment of the tick-over (idle) speed and mixture strength, by means of the shared throttle stop screw and the pilot screws respectively. It should be noted that the mixture strength (controlled by the pilot screw) is preset at the factory on US models so that the exhaust gases conform to the emission control regulations applicable to the area of original delivery. Subsequent adjustment of the pilot screw should not be made unless suitable equipment is available to check that on completion of adjustment the exhaust emissions do not exceed those indicated by the EPA. See Section 9 for further details. Adjustments to carburettors on US models should be confined to idle speed and synchronisation. The second stage, which is the synchronisation of the two carburettors, requires the use of two vacuum gauges together with the necessary adaptors and connection tubes. Vacuum gauges are somewhat expensive. For this reason, and because both stages of adjustment require some expertise, it is strongly recommend-

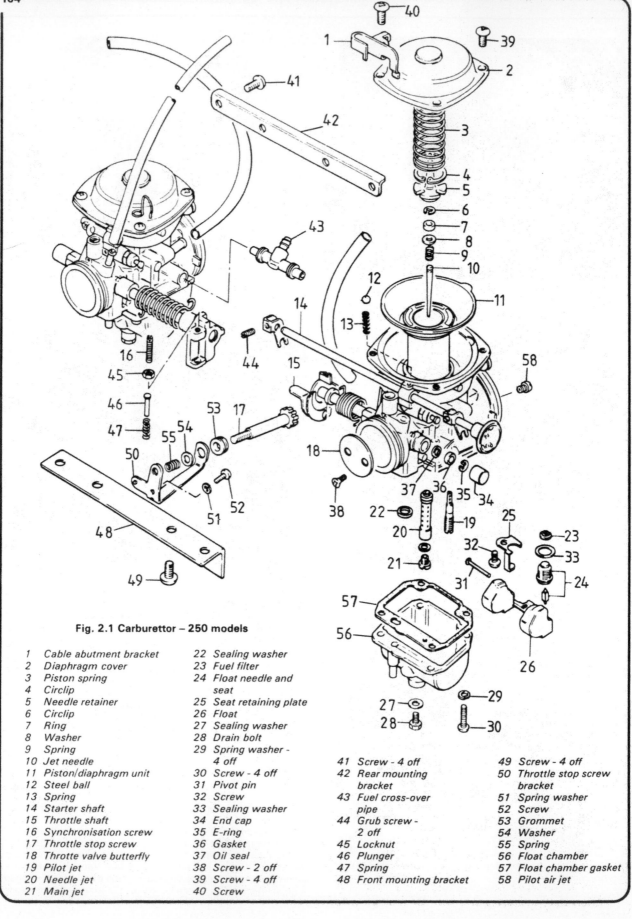

Fig. 2.1 Carburettor – 250 models

1	Cable abutment bracket	22	Sealing washer
2	Diaphragm cover	23	Fuel filter
3	Piston spring	24	Float needle and seat
4	Circlip		
5	Needle retainer	25	Seat retaining plate
6	Circlip	26	Float
7	Ring	27	Sealing washer
8	Washer	28	Drain bolt
9	Spring	29	Spring washer - 4 off
10	Jet needle		
11	Piston/diaphragm unit	30	Screw - 4 off
12	Steel ball	31	Pivot pin
13	Spring	32	Screw
14	Starter shaft	33	Sealing washer
15	Throttle shaft	34	End cap
16	Synchronisation screw	35	E-ring
17	Throttle stop screw	36	Gasket
18	Throttle valve butterfly	37	Oil seal
19	Pilot jet	38	Screw - 2 off
20	Needle jet	39	Screw - 4 off
21	Main jet	40	Screw

41	Screw - 4 off	49	Screw - 4 off
42	Rear mounting bracket	50	Throttle stop screw bracket
43	Fuel cross-over pipe	51	Spring washer
44	Grub screw - 2 off	52	Screw
45	Locknut	53	Grommet
46	Plunger	54	Washer
47	Spring	55	Spring
48	Front mounting bracket	56	Float chamber
		57	Float chamber gasket
		58	Pilot air jet

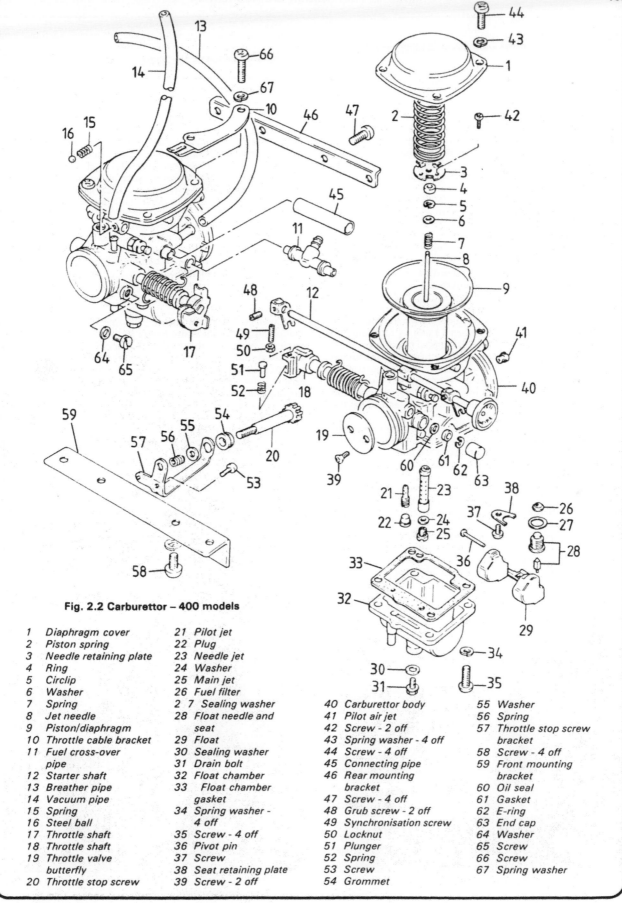

Fig. 2.2 Carburettor – 400 models

1	Diaphragm cover	
2	Piston spring	
3	Needle retaining plate	
4	Ring	
5	Circlip	
6	Washer	
7	Spring	
8	Jet needle	
9	Piston/diaphragm	
10	Throttle cable bracket	
11	Fuel cross-over pipe	
12	Starter shaft	
13	Breather pipe	
14	Vacuum pipe	
15	Spring	
16	Steel ball	
17	Throttle shaft	
18	Throttle shaft	
19	Throttle valve butterfly	
20	Throttle stop screw	

21	Pilot jet
22	Plug
23	Needle jet
24	Washer
25	Main jet
26	Fuel filter
27	Sealing washer
28	Float needle and seat
29	Float
30	Sealing washer
31	Drain bolt
32	Float chamber
33	Float chamber gasket
34	Spring washer – 4 off
35	Screw – 4 off
36	Pivot pin
37	Screw
38	Seat retaining plate
39	Screw – 2 off

40	Carburettor body
41	Pilot air jet
42	Screw - 2 off
43	Spring washer - 4 off
44	Screw - 4 off
45	Connecting pipe
46	Rear mounting bracket
47	Screw - 4 off
48	Grub screw - 2 off
49	Synchronisation screw
50	Locknut
51	Plunger
52	Spring
53	Screw
54	Grommet

55	Washer
56	Spring
57	Throttle stop screw bracket
58	Screw - 4 off
59	Front mounting bracket
60	Oil seal
61	Gasket
62	E-ring
63	End cap
64	Washer
65	Screw
66	Screw
67	Spring washer

ed that the machine is returned to a Suzuki Service Agent who will be able to carry out the work quickly and efficiently and for a reasonable charge.

2   Incorrect carburettor adjustment will affect performance and fuel economy adversely and may, in extreme cases, cause overheating problems. In addition, injudicious attention to the carburettors may prevent the machine from complying with certain emission regulations currently in force in some areas.

3   Before carrying out adjustments of the carburettors, it is important to check that the following items are adjusted correctly: Valve clearances, ignition timing and sparking plug gaps. Many engine faults which at first are thought to be due to carburettor maladjustments can often be traced to those components listed above.

4   Both stages of adjustment must be carried out with the engine at normal working temperature, preferably after the machine has been taken for a short run.

5   Start the engine and by means of the throttle stop screw located between the two instruments, set the engine tick-over speed to within the range 1150 - 1350 rpm (250 models) or 1050 - 1150 rpm (400 models). The mixture strength must now be adjusted on each carburettor in turn. Select one carburettor and screw in the pilot screw fully until it can be felt to seat lightly. **Do not overtighten** because the screw may break. Now unscrew the pilot screw until the engine speed is at its highest. This should occur between $1\frac{1}{4}$ and $1\frac{3}{4}$ turns out. the recommended datum setting for all UK model carburettors is $1\frac{1}{2}$ turns out. With the engine running at the highest speed within the specified range, the pilot adjustment for the carburettor in question is now correct. Repeat the procedure on the second carburettor and then readjust the tick-over to the specified speed. Refer to Routine Maintenance and check and adjust the amount of free play in the throttle cable.

6   Synchronisation of the two carburettors, as mentioned above, requires the use of a pair of vacuum gauges and the correct adaptors for interconnection with the inlet tracts. If the necessary equipment is available, it should be connected up and calibrated, following the manufacturer's recommendations. A blanking plug in the form of a crosshead screw is fitted to each side of the cylinder, just forward of the inlet stub flange. After removal of these screws and their sealing washers, the adaptors may be fitted.

7   Start the engine and set the speed to 1750 rpm with the throttle stop screw. If the gauge readings differ, rotate the throttle valve link adjuster screw either anti-clockwise or clockwise until the two readings are equal. Once the readings are equal, lock the adjuster screw in position with its locknut, reset the engine speed to 1750 rpm with the throttle stop screw and recheck the gauge readings. Finally, reset the engine speed to idle and remove the gauge assembly from the machine. Check that the sealing washer fitted to each blanking plug is in good condition and renewed if necessary before refitting and tightening the plugs.

8   Carburettors : checking the float chamber fuel level

1   If conditions of a continual weak mixture or flooding are encountered on one or other carburettor or if difficulty is experienced in tuning the carburettors, the fuel level in each carburettor float chamber should be checked and, if necessary, altered by adjusting the position of the float assembly.

2   The fuel level within each float chamber can be checked by using the recommended gauge set (Suzuki tool no 09913 - 14540) or by making up a gauge with a short length of clear plastic fuel line with an end adaptor threaded to fit into the hole vacated by the float chamber drain plug.

3   With the fuel tap set in the 'On' position, place a small clean container beneath the float chamber of one of the carburettors and remove the drain plug with its O-ring from the centre of the base of the chamber, allowing any fuel from the chamber to drain into the container. Whilst the fuel is draining, check the condition of the O-ring on the plug; it must be renewed if it is in any way damaged.

4   Connect the gauge adaptor into the hole vacated by the drain plug and position the gauge (or pipe) so that it is placed vertically alongside the float chamber and lower half of the carburettor body. Set the fuel tap to its priming function, so as to refill the float chamber, and start the machine. Return the tap lever to the 'On' or 'Res' position as soon as the engine is running and allow the engine to settle at its correct idling speed. If the fuel level within the float chamber is correct, then the level in the gauge should have settled at a point 4.0 ± 0.5 mm (0.16 $\pm$ 0.02 in) (250 models) or 5.0 ± 0.5 mm (0.2 $\pm$ 0.02 in) (400 models) below the float chamber to carburettor body mating surface. If this is not the case, then the gauge should be removed, the drain plug and O-ring refitted and the carburettor removed from the machine so that the float height may be checked and adjusted as follows.

5   The float level is correct when the distance between the uppermost edge of the floats (with the carburettor inverted) and the carburettor body mating surface is 21.4 ± 1.0 mm (0.84 $\pm$ 0.04 in) (250 models) or 22.4 ± 1.0 mm (0.88 $\pm$ 0.04 in) (400 models); the gasket must be removed from the mating surface before the measurement is taken and the needle valve kept just in contact with its seating surface whilst the measurement is made.

6   Adjustment of the float height is made by bending the small metal tang (tongue), situated on the float pivot and which abuts against the needle valve, in the direction required.

7.6 Remove the blanking plug to facilitate fitting of the vacuum gauge adaptor

7.7 Rotate the throttle valve linkage adjuster screws to equalise the vacuum gauge readings

8.6 Bend the float pivot tang to adjust the float height

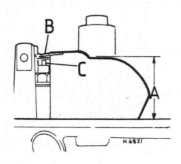

Fig. 2.3 Float height

A Float height   C Float valve
B Float tongue

9 Carburettor settings

1   Some of the carburettor settings, such as the sizes of the needle jets, main jets, and needle positions are pre-determined by the manufacturer. Under normal riding conditions it is unlikely that these settings will require modification. If a change appears necessary, it is often because of an engine fault, or an alteration in the exhaust system eg; a leaky pipe connection or silencer.
2   Apart from alterations of the pilot adjuster screws within the specified limits, some alterations to the mid-range mixture strength can be made by raising or lowering the jet needle (piston needle). This is accomplished by changing the position of the needle clip. Raising the needle will richen the mixture and lowering the needle will weaken it.
3   On US models the carburettors are manufactured to close tolerances to ensure that EPA exhaust emission regulations are complied with. The replaceable jets and needles are of the close tolerance type which are marked in a different letter and numeral style to aid identification when replacements become necessary. On renewal of these components ensure that the correct high-accuracy components are fitted. It should also be noted that on US models the pilot mixture is preset at the factory by means of the pilot screw, using exhaust analyser equipment. Subsequent adjustment of the pilot screw should not be made unless suitable equipment is available to check that the exhaust emissions remain within the limits dictated by the EPA. This applies equally to any other carburettor adjustments which might alter the exhaust emission content.

10 Air cleaner : removing, cleaning and refitting the element

1   The air cleaner housing is mounted immediately behind the carburettor assembly and contains the element which can be removed for cleaning or renewed when necessary.
2   To gain access to the element, detach the seat unit and release the housing lid by sliding its two retaining clips forward off their locations. it was found in practice that these clips were very difficult to dislodge, the best method of removal being to push one of the clips completely off its location with the flat of a large screwdriver, thus taking the tension off the remaining clip which may then be removed without much effort.
3   The element is retained within its housing by a plastic grid. This grid may be bent in two and lifted from its location within the housing to allow the element to be pulled clear. Inspect the element for tears; on no account should a torn element be refitted.
4   The element is oil impregnated and should be cleaned thoroughly in a non-flammable solvent such as white spirit (available as Stoddard solvent in the US) to remove all the old oil and dust. After cleaning squeeze out the sponge to remove the solvent and then allow a short time for any remaining solvent to evaporate. Do not wring out the sponge as this will cause damage and will lead to the need for early renewal. Reimpregnate the sponge with clean engine oil and gently squeeze out the excess.
5   Refit the element, its retaining grid and its housing lid by reversing the procedure used for removal.
6   Always clean the element at the intervals quoted in the Routine Maintenance Chapter at the front of the Manual. In dusty atmospheres, it will be advisable to increase the frequency of cleaning and re-impregnating.
7   Never run the engine without the element connected to the carburettor because the carburettor is specially jetted to compensate for the addition of this component. The resulting weak mixture will cause overheating of the engine with the risk of severe engine damage.

11 Exhaust system : general

1   The exhaust system on a four-stroke motorcycle will require very little attention, as, unlike two-stroke machines, it is not prone to the accumulation of carbon. The only points requiring attention are the general condition of the system, including mountings and protective finish, and ensuring that the system is kept airtight, particularly at the exhaust port.
2   Air leaks at the exhaust port, ie the joint between the exhaust pipe and the cylinder head, will cause mysterious backfiring when the machine is on overrun, as air will be drawn in causing residual gases to be ignited in the exhaust pipe. To this end, make sure that the sealing ring is renewed each time the system is removed.
3   Do not run the machine with the exhaust baffles removed, or with a quite different type of silencer fitted. The standard production silencers have been designed to give the best possible performance, whilst subduing the exhaust note to an acceptable level.
4   Whilst there are a number of good quality after-market exhaust systems available, there are others which may be of poor construction and fit and which may reduce performance rather than improve it. When purchasing such systems it is helpful to obtain recommendations from other owners who have had time to evaluate the system under consideration. Do not forget that there are noise limits which will be met by the more reputable manufacturers. It is not advised that a non-standard system is fitted during the warranty period, because this could result in subsequent claims being refuted.

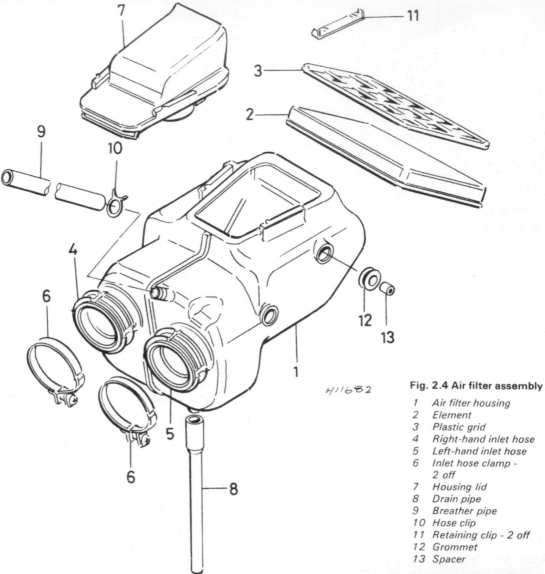

H11682

Fig. 2.4 Air filter assembly

1   Air filter housing
2   Element
3   Plastic grid
4   Right-hand inlet hose
5   Left-hand inlet hose
6   Inlet hose clamp -
    2 off
7   Housing lid
8   Drain pipe
9   Breather pipe
10 Hose clip
11 Retaining clip - 2 off
12 Grommet
13 Spacer

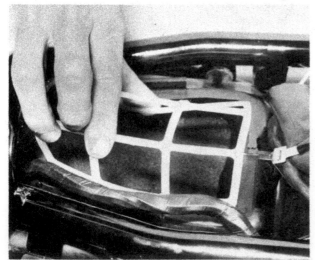

10.5a Refit the air filter element followed by its retainer

10.5b Locate the element retainer beneath the locating spigots

10.5c Secure the filter housing lid by sliding the clips into position

12 Exhaust system : removal and refitting

1   Commence removal of the exhaust system by first removing the bolts that secure the exhaust pipe clamps to the cylinder head; this must be done first as it prevents the weight of the complete system from being allowed to impose an unacceptable strain on the cylinder head stud threads once the rest of the mounting points are undone.

2   On systems where the silencers are detachable from the exhaust pipes, release the retaining clamps which secure the exhaust pipes to the silencers, remove the two balancer pipe to crankcase securing bolts and pull the pipe assembly forward and down to clear the machine. Each silencer is retained to its frame mounting by a single bolt; once this bolt is removed, the silencer may be lifted clear of the machine. Alternatively, the exhaust pipe assembly can be left attached to the machine until each silencer is removed.

3   On exhaust systems where the silencers are permanently attached to the pipes, the system must be removed from the machine in two halves. With the cylinder head connections already detached as previously described, fully loosen the balancer pipe retaining clamp(s) and with an assistant supporting one side of the system, support the opposite side. Release each silencer from its frame mounting by removing the securing bolt and move the complete system forward to clear the cylinder head before lowering it to clear the machine. At this point, the two halves of the system may be separated by pulling them apart whilst twisting slightly so that they separate at the balancer pipe connection.

4   The exhaust system may be refitted to the machine by using a reverse sequence to that given for removal whilst noting the torque loading figures given in the Specifications Section of Chapter 1. Always renew the sealing rings before inserting the exhaust pipe ends into the cylinder head.

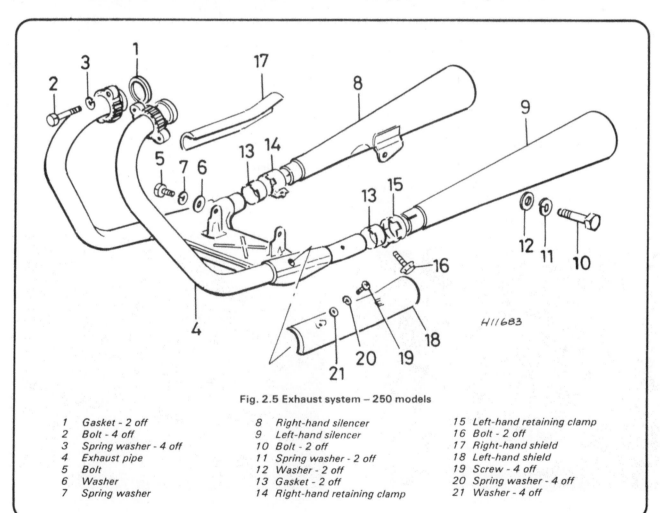

Fig. 2.5 Exhaust system – 250 models

1 Gasket - 2 off	8 Right-hand silencer	15 Left-hand retaining clamp
2 Bolt - 4 off	9 Left-hand silencer	16 Bolt - 2 off
3 Spring washer - 4 off	10 Bolt - 2 off	17 Right-hand shield
4 Exhaust pipe	11 Spring washer - 2 off	18 Left-hand shield
5 Bolt	12 Washer - 2 off	19 Screw - 4 off
6 Washer	13 Gasket - 2 off	20 Spring washer - 4 off
7 Spring washer	14 Right-hand retaining clamp	21 Washer - 4 off

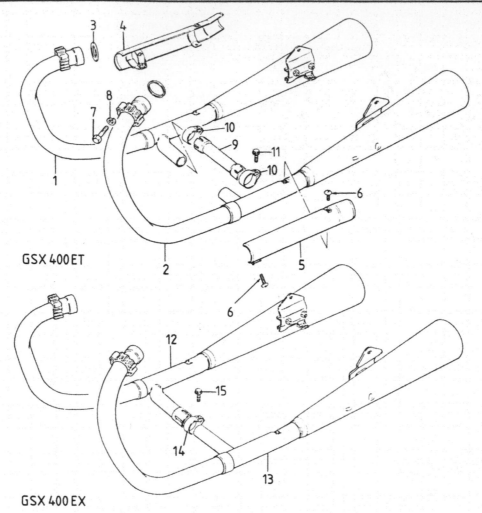

GSX 400 ET

GSX 400 EX

Fig. 2.6 Exhaust system – 400 models

1	Right-hand exhaust pipe	6	Screw - 6 off	11	Bolt - 2 off
2	Left-hand exhaust pipe	7	Bolt - 4 off	12	Right-hand exhaust pipe
3	Gasket - 2 off	8	Spring washer - 4 off	13	Left-hand exhaust pipe
4	Right-hand shield	9	Balancer pipe	14	Retaining clamp
5	Left-hand shield	10	Retaining clamp - 2 off	15	Bolt

12.4a Always renew each sealing ring...

12.4b ...before securing the exhaust pipe to the cylinder head

13 Engine and gearbox lubrication : general description

1   As previously described at the beginning of this Chapter, the lubrication system is of the wet sump type, with the oil being forcibly pumped from the sump to positions at the gearbox bearings, the main engine bearings, and the cambox bearings, all oil eventually draining back to the sump. The system incorporates a gear driven oil pump, a pressure relief valve, an oil filter with safety by-pass valve and an oil pressure switch. Oil vapours created in the crankcase are vented through a breather to the air cleaner box, where they are passed into the cylinder thus providing an oiltight system.

2   The oil pump is an Eaton trochoid twin rotor unit which is driven from a gear engaged with, and to the rear of, the clutch. A gauze filter plate is fitted to the intake side of the pump, which serves to protect the pump mechanism from impurities in the oil which might cause damage.

3   A corrugated paper oil filter is included in the system and is fitted within an enclosed chamber on the underside of the crankcase. Access to the filter is made through a finned cover. As the oil filter unit becomes clogged with impurities, its ability to function correctly is reduced, and if it becomes so clogged that it begins to impede the oil flow, a by-pass valve opens, and routes the oil flow through the filter core. This results in unfiltered oil being circulated throughout the engine, a condition which is avoided if the filter element is changed at regular intervals.

4   As previously mentioned, an oil breather is incorporated into the system. It is mounted in the top of the camshaft cover and is essential for an engine of this size with so many moving parts. It serves to minimise crankcase pressure variations due to piston and crankshaft movement, and also helps lower the oil temperature, by venting the crankcase. Furthermore, this system reduces the escape of unburnt oil into the atmosphere and so allows use of the machine in countries where stringent anti-pollution statutes are in operation. The breather tube carries the crankcase vapours to the air cleaner housing where they become mixed with the air drawn into the carburettors.

5   Excessive oil consumption is indicated by blue smoke being emitted from the exhaust pipes, coupled with a poor performance and fouling of sparking plugs. It is caused by either an excessive oil build-up in the oil breather chamber, or by oil getting past the piston rings. First check the oil breather chamber and air cleaner for oil sludge build up. If this is the fault, check the passageway from the air/oil separator in the oil breather chamber to the lower half of the crankcase. Blockage here will prevent oil flowing back into the crankcase resulting in oil build-up in the breather chamber and air cleaner tube.

6   Be sure to check the oil level in the sump before starting the engine. If the oil level is seen to be below the upper (F) mark on the filler cap dipstick, replenish with the correct amount of oil of the specified viscosity.

14 Oil pump : removal, examination and refitting

1   The oil pump is secured to the wall of the primary drive chamber behind the clutch unit. To gain access to the pump the engine oil should be drained, the primary drive cover detached and the clutch unit removed as described in the relevant Sections of Chapter 1.

2   To remove the pump, unscrew the three retaining screws and pull the unit from position. Displace the two O-rings in the casing wall. The oil pump pinion is retained on the pump shaft. Remove the circlip, lift the pinion off the shaft and push out the drive pin.

3   Remove the single screw from the reverse side of the pump body. The two halves of the pump body are located by two tight fitting dowel pins. Rather than levering the cases apart, which would damage the mating surfaces, the dowels should be driven out. Use a parallel shanked punch of a suitable size, whilst resting the pump across two strips of wood of a thickness sufficient to raise the pump off the workbench surface.

4   Separate the outer casing (reverse side) from the pump, leaving the driveshaft and rotors in place at this stage. Push out the drive shaft, together with the drivepin and then lift out the two rotors.

5   Rinse all the pump components in fuel whilst observing the necessary safety precautions and allow them to dry before carrying out a full examination. Before part reassembling the pump for the various measurements to be made, check the castings for cracks or other damage, especially the pump end covers.

6   Reassemble the pump rotors and measure the clearance between the outer rotor and the pump body, using a feeler gauge. If the clearance exceeds 0.25 mm (0.010 in) the rotor or the body must be renewed, whichever is worn. Measure the clearance between the outer rotor and the inner rotor with a feeler gauge. If this clearance is greater than 0.20 mm (0.008 in) the rotors must be renewed as a set.

7   Using a small sheet of plate glass or a straight edge placed across the pump housing, check the rotor endfloat. If the endfloat exceeds 0.15 mm (0.006 in) the complete pump must be renewed.

8   Examine the rotors and the pump body for signs of scoring, chipping or other surface damage which will occur if metallic particles find their way into the oil pump assembly. Renewal of the affected parts is the only remedy under these circumstances, bearing in mind that rotors must always be replaced as a matched set. Take note of the punch mark on each rotor; these marks must be aligned during reassembly.

9   Reassemble the pump by reversing the dismantling procedure. Make sure all parts of the pump are well lubricated with clean engine oil before the end cover is refitted and that there is plenty of oil between the inner and outer rotors. Degrease the threads of the single locking screw before coating them in thread locking compound and fitting and tightening the screw. Renew and lightly lubricate with clean engine oil the two O-rings before fitting the oil pump into the casing. Rotate the driveshaft as the screws are tightened down, to check that the oil pump revolves freely. A binding pump may be caused by dirt on the rotor faces or distortion of the cases, due to unequally tightened screws. It is recommended that a thread locking compound be used on these screws.

10   Check that the pump pinion locating pin and its retaining circlip are both correctly located before refitting the clutch assembly, the primary drive cover and replenishing the engine oil, whilst referring to the relevant Sections of Chapter 1.

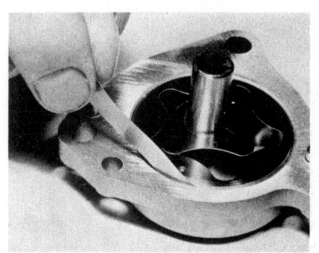

14.6a Measure the oil pump body to outer rotor clearance...

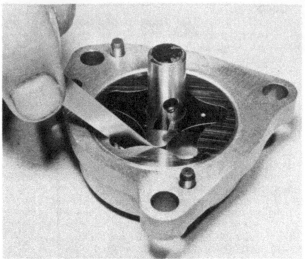

14.6b ...the outer rotor to inner rotor clearance...

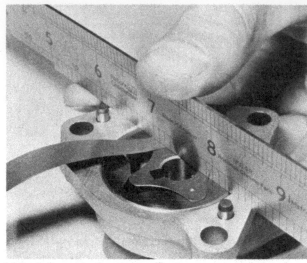

14.7 ...followed by the rotor endfloat

14.8 Check the rotor marks are aligned...

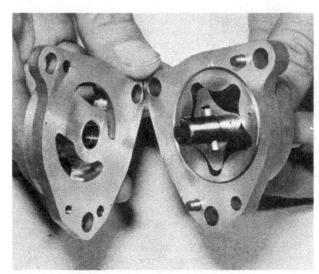

14.9a ...before refitting the pump end cover

14.9b Lubricate and fit new O-rings into the crankcase...

14.10 ...before fitting the oil pump assembly

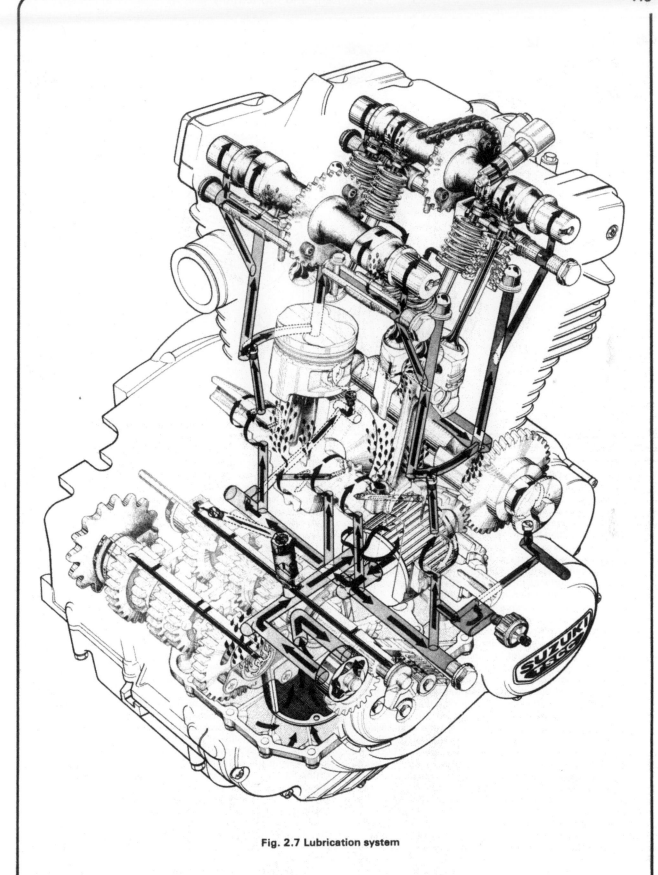

Fig. 2.7 Lubrication system

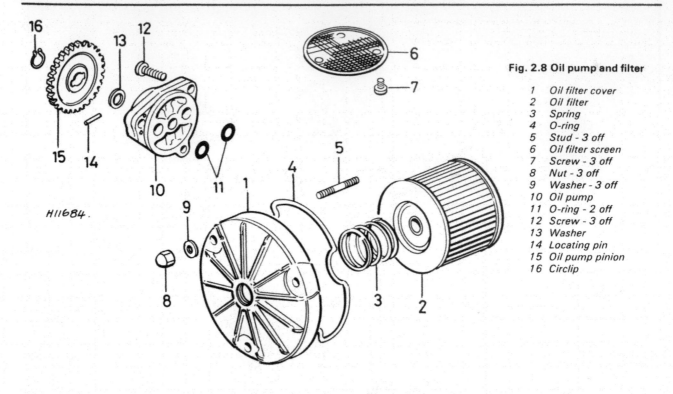

Fig. 2.8 Oil pump and filter

1   Oil filter cover
2   Oil filter
3   Spring
4   O-ring
5   Stud - 3 off
6   Oil filter screen
7   Screw - 3 off
8   Nut - 3 off
9   Washer - 3 off
10 Oil pump
11 O-ring - 2 off
12 Screw - 3 off
13 Washer
14 Locating pin
15 Oil pump pinion
16 Circlip

15 Checking the oil pressure

1   If it is found that the oil pressure warning light continues to operate after the machine has been started and run for a short time or whilst the machine is being ridden, then the engine should be switched off immediately, otherwise there is a risk of severe engine damage due to lubrication failure. The fault must be located and rectified before the engine is restarted and run, even for a brief moment. Machines fitted with plain shell bearings rely on high oil pressure to maintain a thin oil film between the bearing surfaces. Failure of the oil pressure will cause the working surfaces to come into direct contact, causing overheating and eventual seizure.

2   Before suspecting low oil pressure due to a defective pump, check the oil level with the filler cap dipstick and inspect the engine/gearbox unit for signs of serious leakage. Check also the electrical circuit to the warning light for continuity and inspect all the connections for signs of looseness and corrosion.

3   A blanking plug is fitted to the right-hand crankcase cover, directly below the ignition pick-up assembly housing. This plug should be substituted by a suitable adaptor piece to which the pressure gauge can be attached, via a flexible hose. The gauge should have a scale reading of 0 – 10 kg/cm^2 (0 – 140 psi).

4   With the pressure gauge connected and the crankcase oil level correct, start the engine and allow it to run for 10 minutes at 2000 rpm (summer conditions) or 20 minutes at 2000 rpm (winter conditions). On completion of this warming-up period raise the engine speed to 3000 rpm, when the pressure gauge should show a reading of 3.0 – 5.5 kg/sq cm (42.7 – 78.2 psi). A pressure reading lower than specified may be caused by a worn pump, a blocked oil strainer or oil filter element. A slow reduction in pressure, noted over a long period may well be caused by worn main or big-end bearings. Before dismantling the pump for inspection, clean the oil strainer and renew the filter, as described in the following Section of this Chapter.

5   Do not omit to check the condition of, and if necessary renew, the sealing washer located beneath the head of the blanking plug before refitting the plug to the crankcase cover.

16 Renewing the oil filter element and cleaning the oil strainer

1   The oil filter element is contained within a semi-isolated chamber in the base of the lower crankcase, closed by a finned cover retained by three domed nuts and plain washers. The oil strainer is contained within the rearmost of the two compartments covered by the finned sump cover and is in the form of a circular framed metal gauze plate which is retained to the crankcase by three screws.

2   It is recommended that the engine oil and oil filter element are changed at the same mileage interval. To effect efficient draining of the oil, start and run the engine until it reaches full operating temperature; this will heat and therefore thin the oil. Position a suitable receptacle beneath both the drain plug in the centre of the finned sump cover and the filter cover. Remove the drain plug and allow the oil to drain into the receptacle.

3   A coil spring is fitted between the cover and the filter element to keep the latter seated firmly in position. Be prepared for the cover to fly off after removal of the retaining nuts and their washers. Allow any oil contained within the filter housing to drain into the receptacle and withdraw the element from the housing.

4   No attempt should be made to clean the oil filter element; it must be renewed. When renewing the filter element it is wise to renew the filter cover O-ring at the same time. This will obviate the possibility of any oil leaks.

5   The by-pass valve, which allows a continued flow of lubrication if the element becomes clogged, is an integral part of the filter. For this reason routine cleaning of the valve is not required since it is renewed regularly.

6   With the filter element relocated in its housing, refit the retaining spring and push the cover into position over its retaining studs. Locate the retaining nuts and washers onto the stud threads and tighten them evenly and in small increments to prevent the spring slipping from its location in the cover and filter end.

7   Never run the engine without the filter element or increase the period between the recommended oil changes or oil filter changes. The recommended interval for both oil and filter

element changes is given in Routine Maintenance.

8 The oil strainer should be removed at every second oil and filter element change. To gain access to the strainer, remove the bolts that retain the finned sump cover in position. These bolts should be loosened a little at a time at first and in a diagonal sequence to prevent any risk of the cover becoming distorted. Before removing the bolts, note the position of the electrical cable retaining clips located beneath two of the bolt heads for reference when refitting. It will be found that even with all the retaining bolts removed, the cover will be stuck fast in position and will need to be tapped quite sharply around to its mating face with a soft-faced mallet in order to free it.

9 With the strainer removed from the crankcase, wash it and the sump cover in fuel whilst using a soft bristle brush to remove any stubborn traces of contamination. Be sure to observe the necessary fire precautions when carrying out this cleaning procedure.

10 Carefully inspect the strainer gauze for any signs of damage. If the gauze is split or holed, it must be renewed as it no longer forms an effective barrier between the sump and oil pump.

11 Fit the oil strainer back into the crankcase housing, taking care to fully tighten its retaining screws. Ensure that the mating surfaces of the sump cover and the lower crankcase half have been cleaned of all old gasket material and jointing compound and are properly degreased. Smear an even coating of Suzuki Bond No 4 (or equivalent) over the area of sump cover mating surface indicated in the accompanying photograph and over the corresponding area of crankcase mating surface. Fit the new gasket into position on the crankcase and place the sump cover in position on top of it. Refit the sump cover retaining bolts into their respective locations, remembering to position the electrical cable retaining clips in their previously noted positions beneath the bolt heads. Tighten the retaining bolts evenly and in a diagonal sequence, noting the final torque loading figures given in the Specificaitons Section of Chapter 1.

12 Finally, fit a new sealing washer beneath the head of the drain plug and refit and tighten the plug. Pour the correct amount of SAE 10W/40 engine oil through the crankcase filler point and check its level with the filler cap dipstick before refitting the filler cap and starting the engine.

16.2 The oil drain plug is located in the centre of the sump cover

16.3 A coil spring is fitted between the filter element and cover

16.11a Secure the oil strainer in position with the three screws

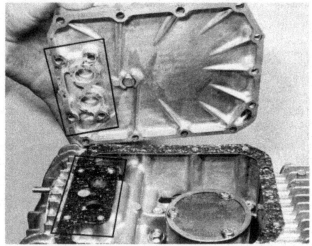

16.11b Smear an even coating of Suzuki Bond No 4 over the indicated crankcase and sump cover mating surfaces before fitting a new gasket

16.12a Pour the correct amount of recommended oil through the crankcase filler point...

16.12b ...and check the oil level with the dipstick

17 Fault diagnosis : fuel system and lubrication

Symptom	Cause	Remedy
Excessive fuel consumption	Air cleaner choked or restricted Fuel leaking from carburettor. Float stricking Badly worn or distorted carburettor Jet needle setting too high Main jet too large or loose Carburettor flooding	Clean or renew. Check all unions and gaskets. Float needle seat needs cleaning. Replace. Adjust as figure given in Specifications. Fit correct jet or tighten if necessary. Check float valve and replace if worn. Check float height.
Idling speed too high	Throttle cable out of adjustment or sticking	Adjust or remove and lubricate cable. Renew if necessary.
Engine gradually fades and stops	Fuel starvation	Check vent hole in filler cap. Sediment in filter bowl or float chamber. Dismantle and clean.
Engine runs badly. Black smoke from exhausts	Carburettor flooding	Dismantle and clean carburettor. Check for punctured float or sticking float needle.
Engine lacks response and overheats	Weak mixture Air cleaner disconnected or hose split Modified silencer has upset carburation	Check for partial block in carburettors. Reconnect or renew hose. Replace with original design.
Oil pressure warning light comes on	Lubrication system failure	Stop engine immediately. Trace and rectify fault before re-starting.
Engine gets noisy	Failure to change engine oil when recommended	Drain off old oil and refill with new oil of correct grade. Renew oil filter element and clean oil strainer.

Chapter 3 Ignition system

Refer to Chapter 7 for information relating to GS450 and GSX250/400 EZ models

Contents

Specifications

Ignition timing
Below 1650 ± 100 rpm	20° BTDC
Above 3500 ± 100 rpm	40° BTDC

Sparking plug
Make	NGK or ND
Type:	
250 UK models	DR8ES or. X27ESR-U
250 US models	D9EA or X27ES-U
400 models	DR8ES-L or X24ESR-U
Gap	0.6 – 0.7 mm (0.024 – 0.028 in)

Ignition coil
Primary winding resistance:	
GSX400TX	3.0 – 5.0 ohms
All others	3.5 – 4.5 ohms
Secondary winding resistance:	
GSX400TX	21 – 25 K ohms
All others	23 – 25 K ohms

Pulse generator coil
Resistance	60 – 80 ohms

1 General description

The ignition system fitted to the models covered in this Manual comprises a signal generating unit, a separate ignition control unit, two ignition (HT) coils, two sparking plugs and an automatic ignition timing unit (ATU).

The signal generating device is located on the right-hand end of the crankshaft and is contained within a housing formed by the crankcase cover. Direct access to the signal generator and to the ATU is gained through a circular cover which is attached to the crankcase cover by three crosshead screws. The device transmits signals in the form of electrical impulses, to the remotely mounted ignition control unit. This control unit, which contains a transistor capable of amplifying each signal received, is fitted to the electrical components mounting plate, behind the left-hand side panel. As each signal is received in the ignition control unit, a transistor contained therein operates 'on' or 'off', and opens and closes the correct circuit as required. Thus, the correct impulse for the crankshaft position, that is, whichever cylinder is firing, is received and passed on. By cutting off the primary current flowing on the primary side of the ignition coil, the sparking plugs are caused to spark. In other words, the primary current of the ignition coil is cut off by the transistor in this type of ignition system, whereas it is cut off by the contact breaker points in a conventional ignition system. The two ignition coils and the sparking plugs are the same items as fitted to a conventional type of ignition system.

There are several advantages that the transistorised ignition system holds over the traditional contact breaker points system. The most important advantage is that it removes the majority of mechanical components from the system. Because there are no contact breakers to wear, the owner is freed from the task of periodically adjusting or renewing them. Once the transistorised system has been set up, it need not be attended to unless it has been disturbed in the course of dismantling or failure occurs somewhere in the system. Other advantages of the transistorised ignition system include a greater resistance to the effects of vibration, dirt and moisture, and a constantly 'strong' spark at exactly the correct moment, with no wastage of electrical energy due to arcing etc.

The ATU serves to advance the ignition timing as the engine rpm rises. The mechanism is made up of two spring loaded weights which, under the action of centrifugal force created by the rotation of the crankshaft, fly apart and cause the signal generator to release an impulse earlier. If the mechanism does not operate smoothly, the timing will not advance smoothly, or it will tend to stick in one position. This will result in poor running in any but that one position. Sometimes the springs are prone to stretching which can cause the timing to advance too soon. It is best to check the ATU with the use of a stroboscopic timing lamp as described in Section 7 of this Chapter.

2 Testing the ignition system : general information

An electronic ignition system does not require maintenance in the generally accepted sense. With the exception of the ATU there are no mechanical parts, thus wear does not take place and the need for compensation by adjustment does not arise. Ignition problems in this type of system can be broken down as follows:

1   Loose, broken or corroded connections
2   Damaged or broken wiring
3   Wear or damage of the ATU
4   Faulty or inoperative electronic components.

The above are arranged in the order in which they are most likely to be found, and with the exception of number 4 should provide no undue problem in the event of fault finding or

rectification. Where part of the electronic side of the system fails, however, diagnosis becomes rather more difficult. The following sections provide details of the necessary test procedures, but it must be remembered that basic test equipment will be required. Most of the tests can be carried out with an inexpensive pocket multimeter. Many home mechanics will already have one and be conversant with its use. Failing this, they are easily obtainable from mail order companies or from electronics specialists.

When carrying out tests on the electronic ignition system, bear in mind that wrong connections could easily damage the component being tested. Adhere strictly to the test sequence described and be particularly careful to avoid reversed battery connections. Note that the system must **not** operate with one or more HT leads isolated, as the very high secondary winding voltage may destroy the ignition coil. Beware of shocks from the high tension leads or connections. Although not inherently dangerous, they can be rather unpleasant.

3 Ignition system : testing and maintenance

1   As stated earlier in this Chapter, the transistorised ignition system needs no regular maintenance once it has been set up and timed accurately. Occasional attention should however, be directed at the various connections in the system, and these must be kept clean and secure. A failure in the ignition system is comparatively rare, and usually results in a complete loss of ignition. Usually, this will be traced to the ignition control unit, and little can be done at the roadside to effect a repair. In the event of the control unit failing, it must be renewed as a repair is not practicable.

2   If the ignition control unit is thought to be at fault, it is recommended that it be removed and taken to a Suzuki dealer for testing. The dealer will have the use of a special control unit tester, and will be able to test the unit accurately and quickly. Testing at home is less practical and cannot be guaranteed to be accurate. At best, it will enable the owner to establish which part of the system is at fault, although replacement of the defective part remains the only effective cure.

3   For those owners possessing a multimeter and who are fully conversant with its use, a test sequence is given below for the ignition control unit and the signal generator. It is not recommended that the inexperienced attempt to test the system at home, as more damage could be sustained by the system.

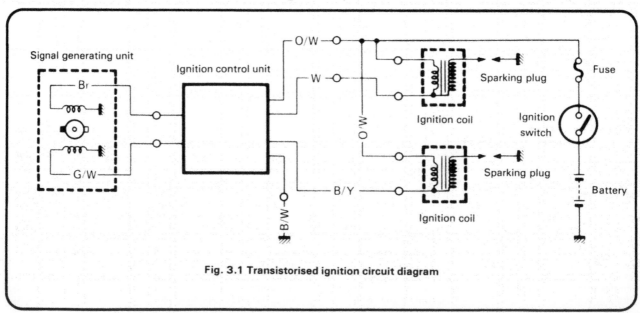

Fig. 3.1 Transistorised ignition circuit diagram

4 Ignition control unit : testing

1 Unscrew and remove both sparking plugs, reconnect each plug to its suppressor cap and position both plugs so that their bodies are in contact with the cylinder head casting and the electrodes are clearly visible from the right-hand side of the machine.

2 Remove the right-hand side panel from its frame connections and locate the small block connector which forms a connection between the signal generating unit and the ignition control unit. This connector should be found on the frame downtube, just behind the upper surface of the engine crankcase.

3 Set the multimeter to its resistance function (X1 ohms scale), pull apart the two halves of the block connector and turn the ignition switch to the 'On' position. Connect the negative (–) probe of the meter to the black/white (black/blue on GSX 400T models) lead connection on the half of the block nearest to the control unit. Connect the position (+) probe of the meter to the brown lead connection on the same half of the block and check to see if a spark appears between the electrodes of the right-hand cylinder sparking plug. Repeat the operation with the positive probe of the meter connected to the green/white lead connection and check for a spark between the electrodes of the left-hand cylinder sparking plug. The ignition control unit is fully serviceable if a spark occurs at both plugs during the test sequence.

4 It should be noted that this test is carried out on the assumption that the ignition coils are in correct working order. It follows therefore, that a faulty ignition coil will adversely affect the results of this test.

5 Signal generating unit : testing

1 Remove the right-hand side panel from its frame connections and locate the small block connector which forms a connection between the signal generating unit and the ignition control unit. This connector should be found on the frame downtube, just behind the upper surface of the engine crankcase.

2 Pull apart the two halves of the connector and with the multimeter set to its resistance function, place a probe on each of the two generating unit lead connections within the half of the block nearest to the unit, that is between the brown lead connection and the green/white lead connection. The resistance between the two lead connections should be between the range 60 – 80 ohms.

3 If the indicated resistance falls below the lower figure or the meter gives a reading of infinite resistance, then the signal generating unit is in need of replacement. Before discarding the unit, however, a second opinion should be sought from a Suzuki Service Agent to ensure your diagnosis is indeed correct. The signal generating unit can only be purchased and replaced as a complete unit; the individual pick-up coils cannot be renewed individually.

6 Ignition coils : testing

1 Each ignition coil is a sealed unit, designed to give long service without need for attention. They are located one each side of the frame top tube, directly above the engine cylinder head. If one of the ignition coils is suspected of partial or compete failure, its internal resistance and insulation can be checked by measuring the primary and secondary winding resistances. Note that it is very unlikely that both coils would fail simultaneously, and if this appears to be the case, be prepared to look elsewhere for the problem.

2 Set the multimeter to the ohms scale, and connect one probe lead to each of the thin low tension wires. Note that it does not matter which probe is connected to which lead, that is the orange/white lead to either the white or the black/yellow lead. A reading of 3.5 – 4.5 ohms (3.0 – 5.0 ohms on GSX400TX models) should be obtained if the primary windings are in good order.

3 Set the meter to the K ohms scale and repeat the test for the secondary windings; this time connecting one probe to the suppressor cap connection and the other to either the white or the black/yellow lead. A resistance of 23 – 25 K ohms (21 – 25 K ohms on GSX400TX models) should be obtained if the secondary windings are in good order.

4 If the coil has failed it is likely to have either an open or short circuit in the primary or secondary windings. This type of fault would be immediately obvious and would of course require the renewal of the coil concerned. Where the fault is less clear cut it is advisable to have the suspect coil tested on a spark gap tester by an official Suzuki Service Agent.

4.3 The ignition control unit (arrowed) is located beneath the left-hand side panel

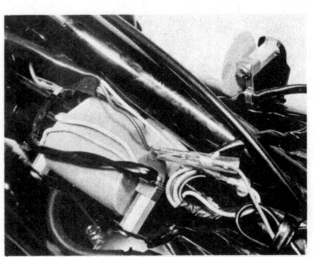

6.1 The ignition coils are located one either side of the frame top tube

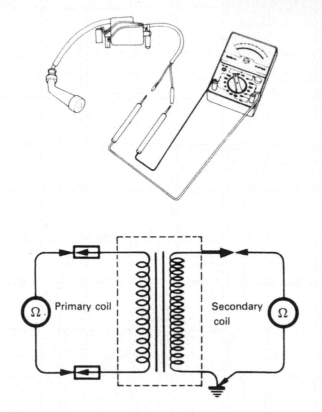

Ignition coil resistance		
Primary	O/W – W or B/Y	Approx 3.5 – 4.5Ω
Secondary	Plug cap – W or B/Y	Approx. 23 – 25kΩ

Fig. 3.2 Ignition coil test

7 Ignition timing : checking and resetting

1 Although the ignition system fitted to the models covered in this Manual appears to be of the fixed, non-adjustable, type, in practice a small amount of adjustment is possible. This adjustment facility may be of benefit in one or two ways. If the machine is to be fitted with, or has been recently fitted with, a new signal generating unit, the positioning of the device can be checked, and if necessary, adjusted. Also if the machine is malfunctioning, and the cause of the problem cannot be traced to any other source, an investigative check of the ignition timing may prove beneficial.

2 In order to check the accuracy of the ignition timing, it is first necessary to remove the small circular cover from the right-hand crankcase cover. The ignition timing on this system can only be checked using a stroboscopic timing lamp. In this way an additional task, that of checking the correct function of the ATU, may be accomplished simultaneously with checking the ignition timing.

3 Two basic types of 'strobe' are available, namely the neon and xenon tube types. Of the two, the neon type is much cheaper and will usually suffice if used in a shaded position, its light output being rather limited. The brighter but more expensive xenon types are preferable if funds permit, because they produce a much clearer image.

4 Connect the strobe to the left-hand cylinder high tension lead, following the maker's instructions. If an external 12 volt power source is required it is best **not** to use the machine's battery as spurious impulses can be picked up from the electrical system. A separate 12 volt car or motorcycle battery is preferable.

5 Start the engine and illuminate the ATU through the inspection aperture in the signal generating unit pick-up plate. With the engine running below 1500 rpm, the 'F' mark on the ATU should be in exact alignment with the index mark.

6 If the ignition timing is slightly incorrect, the two screws which hold the signal generator mounting plate should be slackened to facilitate movement of the plate. Move the plate a slight amount in the required direction, retighten the screws and recheck the timing. Repeat this process, if necessary, until the timing is correct.

7 To check the ATU for correct operation, increase the engine speed up to 3500 rpm whilst observing the timing marks with the aid of the strobe. If when increasing the engine speed from the commencement of advance at 1500 rpm, the timing marks are seen to move erratically, or if the advance range has altered appreciably, the ATU should be inspected for wear or malfunctioning as described in the following Section.

7.2 The ignition timing assembly is located beneath its circular cover

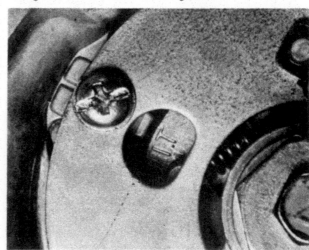

7.5 The timing marks are visible through the signal generating unit baseplate

Electrode gap check - use a wire type gauge for best results

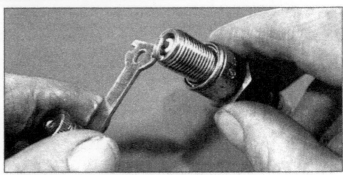

Electrode gap adjustment - bend the side electrode using the correct tool

Normal condition - A brown, tan or grey firing end indicates that the engine is in good condition and that the plug type is correct

Ash deposits - Light brown deposits encrusted on the electrodes and insulator, leading to misfire and hesitation. Caused by excessive amounts of oil in the combustion chamber or poor quality fuel/oil

Carbon fouling - Dry, black sooty deposits leading to misfire and weak spark. Caused by an over-rich fuel/air mixture, faulty choke operation or blocked air filter

Oil fouling - Wet oily deposits leading to misfire and weak spark. Caused by oil leakage past piston rings or valve guides (4-stroke engine), or excess lubricant (2-stroke engine)

Overheating - A blistered white insulator and glazed electrodes. Caused by ignition system fault, incorrect fuel, or cooling system fault

Worn plug - Worn electrodes will cause poor starting in damp or cold weather and will also waste fuel

8   Automatic timing unit (ATU) : examination and servicing

1   Fixed ignition timing is of little advantage as the engine speed increases, and it is necessary to incorporate a method of advancing the timing by centrifugal means. A balance weight assembly located behind the signal generating unit pick-up plate provides this method of advancing the timing. Full details for unit removal are given in Section 9 of Chapter 1.
2   The unit comprises spring loaded balance weights, which move outwards against the spring tension as centrifugal force increases. The balance weights must move freely on their pivots and be rust-free. The tension springs must also be in good condition. Keep the pivots lubricated and make sure the balance weights move easily, without binding. Most problems arise as a result of condensation within the engine, which causes the unit to rust and balance weight movement to be restricted.
3   If any malfunction or breakage has occurred, renew the complete unit, but if it appears to be in good condition, lightly oil it before refitting. Full details of refitting the unit are contained within Section 39 of Chapter 1.
4   The correct functioning of the ATU can be checked when carrying out ignition timing checks using a stroboscope as described in Section 7 of this Chapter.

8.2 Inspect the ATU spring and pivot assemblies

9   High tension (sparking plug) lead : examination

1   Erratic running faults and problems with the engine suddenly cutting out in wet weather can often be attributed to leakage from a high tension lead and sparking plug cap. If this fault is present, it will often be possible to see tiny sparks around the lead and cap at night. One cause of this problem is the accumulation of mud and road grime around the lead, and the first thing to check is that the lead and cap are clean. It is often possible to cure the problem by cleaning the components and sealing them with an aerosol ignition sealer, which will leave an insulating coating on both components.
2   Water dispersant sprays are also highly recommended where the system has become swamped with water. Both these products are easily obtainable at most garages and accessory shops. Occasionally, the suppressor cap of the lead itself may break down internally. If this is suspected, the components should be renewed.
3   Where the HT lead is permanently attached to the ignition coil, it is recommended that the renewal of the HT lead is entrusted to an auto-electrician who will have the expertise to solder on a new lead without damaging the coil windings.

10 Sparking plugs : checking and resetting the gap

1   The type of sparking plugs fitted to each model covered in this Manual must comply with that listed in the Specifications at the beginning of this Chapter. If in any doubt as to the type of sparking plugs that should be fitted to a particular type of machine, consult an official Suzuki Service Agent who will be able to recommend the correct sparking plug grade for the conditions in which the machine is to operate. The type of plug recommended by the manufacturer gives the best all round service.
2   Check the gap between the plug electrodes at the service interval recommended in the Routine Maintenance Chapter at the beginning of this Manual. To reset the gap, bend the outer electrode to bring it closer to the centre electrode and check that a 0.6 – 0.7 mm (0.024 – 0.028 in) feeler gauge can be inserted. Never bend the central electrode or the insulator will crack, causing engine damage if the particles fall in whilst the engine is running.
3   With some experience, the condition of the sparking plug electrodes and insulator can be used as a reliable guide to engine operating conditions.

4   Beware of overtightening the sparking plugs, otherwise there is risk of stripping the threads from the aluminium alloy cylinder heads. The plugs should be sufficiently tight to sit firmly on their copper sealing washers, and no more. Use a spanner which is a good fit to prevent the spanner from slipping and breaking the insulator.
5   If the threads in the cylinder head strip as a result of over tightening the sparking plugs, it is possible to reclaim the head by the use of a Helicoil thread insert. This is a cheap and convenient method of replacing the threads; most motorcycle dealers operate a service of this kind.
6   Make sure the plug insulating caps are a good fit and have their rubber seals. They should also be kept clean to prevent tracking. These caps contain the suppressors that eliminate both radio and TV interference.
7   Before fitting a sparking plug in the cylinder head, coat the threads sparingly with a graphited grease to aid future removal.

11 Fault diagnosis : ignition system

Symptom	Cause	Remedy
Engine will not start	Faulty ignition switch	Operate switch several times in case contacts are dirty. If lights and other electrics function, switch may need renewal.
	Starter motor not working	Discharged battery. Recharge the battery using an external charger.
	Short circuit in wiring	Check whether fuse is intact. Eliminate fault before switching on again.
	Completely discharged battery	If lights do not work, remove battery and recharge.
Engine misfires	Faulty connection	Check all ignition connections. Check wiring for breaks.
	Coil failure	Remove and test (see text).
	Sparking plug failure	Renew.
	Ignition system failure	Test to establish cause of problem.
	Incorrect grade of fuel	Drain and refill with correct grade.
	Faulty ATU	Remove and check operation.
Engine lacks power and overheats	Retarded ignition timing	Check timing. Check whether auto-timing unit has jammed.
Engine 'fades' when under load	Pre-ignition	Check grade of plugs fitted; use recommended grades only.

Chapter 4 Frame and forks

Refer to Chapter 7 for information relating to GS450 and GSX250/400 EZ models

Contents

Specifications

Frame

Type	Cradle, welded tubular construction

Front forks

Type	Telescopic, hydraulically damped
Travel	140 mm (5.50 in)
Spring free length service limit:	
GS250 TT (UK) models	103 mm (4.10 in) and 391 mm (15.40 in)
All other models	496 mm (19.50 in)
Oil capacity per leg	150 cc (5.28/5.07 Imp/US fl oz)
Oil type:	
GS250 (US) models	Fork oil
All other models	50/50 mixture of 10W/30 motor oil and automatic transmission fluid (ATF)

Rear suspension

Type	Swinging arm fork, controlled by two suspension units
Travel:	
GSX 250 models	85.0 mm (3.40 in)
GSX 400T models	95.0 mm (3.70 in)
All other models	100.0 mm (3.90 in)
Swinging arm pivot shaft runout service limit ...	0.3 mm (0.012 in)
Rear suspension unit type	Coil spring and hydraulic damper, 5-way adjustable

Torque wrench settings

	lbf ft	kgf m
Front fork damper rod retaining bolt	11.0 – 18.0	1.5 – 2.5
Lower fork yoke clamp bolts	18.0 – 29.0	2.5 – 4.0
Upper fork yoke clamp bolts	14.5 – 21.5	2.0 – 3.0
Front fork leg cap bolts	11.0 – 21.5	1.5 – 3.0
Steering stem clamp bolt	11.0 – 18.0	1.5 – 2.5

Steering stem adjusting ring	29.0 – 36.0	4.0 – 5.0
Steering stem top bolt	26.0 – 37.5	3.6 – 5.2
Handlebar clamp retaining bolts	8.5 – 14.5	1.2 – 2.0
Swinging arm pivot shaft nut	36.0 – 42.0	5.0 – 5.8
Rear suspension unit retaining nuts	14.5 – 21.5	2.0 – 3.0
Front footrest bolts:		
10 mm	19.5 – 31.0	2.7 – 4.3
8 mm	11.0 – 18.0	1.5 – 2.5
Rear footrest bolts	19.5 – 31.0	2.7 – 4.3
Exhaust silencer bracket mounting bolt	19.5 – 31.0	2.7 – 4.3

1 General description

Two types of frame are utilised on the models covered in this Manual. Both are of conventional welded tubular steel construction; the only difference being that whereas the frame utilised on the 400 models is of the full duplex cradle design, the frame utilised on the 250 models has a single front downtube which splits into a duplex cradle at its base.

The front forks fitted to all models are of the conventional telescopic type, having internal, oil-filled dampers. The fork springs are contained within the fork stanchions and each fork leg can be detached from the machine as a complete unit, without dismantling the steering head assembly.

Rear suspension is of the swinging arm type, using oil filled suspension units to provide the necessary damping action. The units are adjustable so that the spring ratings can be varied within certain limits to match the load carried.

2 Front fork legs : removal and fitting

1 Place the machine securely on its centre stand, leaving plenty of working area at the front and sides. Arrange wooden blocks beneath the crankcase so that the front wheel is raised clear of the ground.
2 Remove the front wheel as detailed in Section 3 of Chapter 5.
3 Unscrew and remove the two bolts which pass through the brake caliper support bracket securing it to the fork leg. Swing the caliper unit back and suspend it from the frame by means of a length of wire or string. The hydraulic hose need not be disconnected from the caliper unit.
4 Remove the speedometer cable guide from the base of the left-hand fork leg by unscrewing its single retaining bolt. Thread the speedometer cable through its guide on the front mudguard and secure it to a point on the frame, clear of the fork assembly.
5 Detach the mudguard and remove it from between the fork legs. The mudguard is secured by two bolts with lock washers passing into each fork leg.
6 If it is intended to dismantle the fork legs for inspection and servicing, then it is a good idea to slacken the top cap bolts before removing the legs from their yokes.
7 Loosen the clamp bolts which retain the fork legs in the upper and lower yokes. The fork legs can now be eased downwards, out of position. If the clamps prove to be excessively tight, they may be gently sprung, using a large screwdriver. This must be done with great care, in order to prevent breakage of the clamps, necessitating renewal of the complete yoke. The fork legs can now be dismantled for inspection and renovation as described in Section 4 of this Chapter.
8 Fitting of the fork legs is a straightforward reversal of the removal sequence, noting that the tops of the fork stanchions should be just flush with the top of the upper yoke. When tightening the forks and related components, start with the wheel spindle and work upwards, noting the torque settings given in the Specifications of this Chapter and of Chapter 5. If the fork legs were dismantled for inspection, do not omit to tighten fully the top cap bolts to the specified torque loading.

3 Steering head assembly : removal and fitting

1 The steering head will rarely require attention unless it becomes necessary to renew the bearings or if accident damage has been sustained. It is theoretically possible to remove the lower yoke together with the fork legs, but as this entails a considerable amount of unwieldy manoeuvring this approach is not recommended. A possible exception may arise if the fork stanchions have been damaged in an accident and are jammed in the lower yoke, and in this case a combination of this Section and Section 2 must be applied.
2 Remove the front wheel brake caliper, front mudguard and fork legs as described in Section 2 of this Chapter.
3 Before disconnecting any electrical leads from the multitude of electrical components mounted on the headstock assembly, it is advisable to isolate the battery from the machine's electrical system by removing one of the leads from its battery terminal. This will safeguard against any danger of components within the electrical system becoming damaged by short circuiting of the exposed connector ends.
4 It is now necessary to remove the dualseat in order to allow removal of the fuel tank from its frame mountings. Full instructions for removal of these components are contained within Section 15 of this Chapter and Section 2 of Chapter 2. This is so access can be gained to the connector terminal for the electrical wires leading to the handlebar mounted instrument console. Separate these terminal halves and thread the wires forward until they are free to hang below the console.
5 Disconnect the speedometer and tachometer cables from the base of the instrument console. Remove the speedometer cable from the machine and thread the tachometer cable clear of the steering head assembly. The instrument console may now be freed from its mountings on the upper fork yoke by removing the two mounting bolts and domed nuts with their associated washers. Carefully note the positions of these washers for reference when refitting.
6 The procedure from this point onwards must depend on individual circumstances. For obvious reasons, the full dismantling sequence is described here, but it is quite in order to avoid as much of the dismantling as possible by careful manoeuvring of the ancillary components. Obviously, much depends on a commonsense approach and a measure of ingenuity on the part of the owner.
7 Remove the screws which secure the headlamp lens and reflector assembly to the headlamp shell. Disconnect the headlamp bulb connector and the parking lamp bulb (UK and European models) and place the unit to one side. Trace the various multi-pin connections which enter the headlamp shell, making a quick sketch to show their relative positions. Separate the connectors and thread the leads out through the back of the shell, securing them to a point on the machine clear of the steering head assembly. The complete headlamp shell and indicator assembly may now be pulled forward out of its location between the fork yokes.
8 Trace the electrical lead from the ignition switch to its black connector and separate the two halves of the connector. Thread the lead clear of any frame components so that it is free to hang below the switch. If considered necessary, the ignition switch

may now be removed from the upper fork yoke by unscrewing and removing the two retaining bolts and lock washers.

9 Cover the forward length of the frame top tubes with an old blanket, or similar, to protect the paintwork from damage. Remove the single bolt that retains the hydraulic brake hose/electric loom guide plate to the underside of the lower yoke.

10 Detach the handlebars from their mounting points on the fork upper yoke. The handlebars are held by two U-clamps retained by two bolts and spring washers each. On 400 models plastic plugs are fitted to the Allen bolt heads: these should be removed to allow fitting of the Allen key. Lift the complete handlebar assembly clear of the fork yoke and whilst taking care to keep the brake reservoir upright, carefully manoeuvre the assembly towards the rear of the machine until it can be rested on the padding placed across the frame top tubes. Note that, if spilt, the brake hydraulic fluid will act as an efficient paint remover and will also damage any plastic components.

11 Loosen the clamp bolt located at the rear of the upper yoke and from the top of the yoke remove the large chrome bolt together with the washer. From the underside tap the upper yoke upwards until it frees the steering column. Support the weight of the lower yoke and, using a C-spanner, remove the steering head bearing adjuster ring. If a C-spanner is not available a soft brass drift and hammer may be used to slacken the nut.

12 Remove the dust excluder and the bearing cone (250 models) or inner race (400 models) once the adjuster nut has been detached. The bottom yoke, complete with steering column, can now be lowered from position. On 250 models make provision to catch the ball bearings as they are released: only the lower bearings will drop free since the upper bearings will most probably remain seated in the cup race retaining them. 400 models are fitted with caged roller bearings the rollers of which are securely restrained by the cages, and thus there is no danger of their falling out.

13 Full details of steering head bearing examination and renovation are given in Section 5 of this Chapter.

14 Fitting of the steering head assembly is a direct reversal of the removal procedure, taking into account the following points.

250 models

15 Position the eighteen steel balls on the lower bearing cone, holding them in position with grease. Insert the steering stem into the head lug until the bearing cone and balls locate with the bearing cup. Fill the upper bearing cup with grease, insert the eighteen steel balls and fit the upper bearing cone, dust excluder and adjuster ring. Tighten the adjuster ring until resistance is felt and then loosen the ring $\frac{1}{8}$ to $\frac{1}{4}$ of a turn. Adjustment is correct when all play is taken up but the yoke will move freely from lock-to-lock. It is possible to place several tons pressure on the steering head bearings if they are over-tightened. The usual symptom of overtight bearings is a tendency for the machine to roll at low speeds, even though the handlebars may appear to turn quite freely.

400 models

16 Pack the lower taper roller bearing with grease and insert the steering stem through the head lug. Grease the upper bearing and fit it together with the dust excluder and the adjusting ring. Adjustment of the steering head bearings requires the use of a torque wrench and a peg spanner adaptor which fits the adjusting ring and to which may be fitted the torque wrench drive. Suzuki recommend the use of tool number 09940-14910; if, however, this is not available a home-made tool can be fabricated from a length of thick-walled tube the end of which has been filed away to form pegs. The tube should be welded to an old socket to provide the drive. Tighten the adjuster ring to a torque setting of 4.0 – 5.0 kgf m (29.0 – 36.0 lbf ft) and then turn the yoke through its full range five or six times to seat the bearing. Slacken the adjuster ring by $\frac{1}{2}$ a turn. Adjustment is now complete.

All models

17 Ensure that after the front mudguard, speedometer/tachometer, indicator lamp/headlamp and handlebar assemblies have been fitted, all control cables, electrical wires, etc are correctly routed, refitted and reconnected. The headlamp beam height should be adjusted; see the relevant Section in Chapter 6. The handlebar must be refitted in the top yoke clamps with the punch marks in the correct positions; that is with the punch mark on the handlebar in line with the rear mating face of the lower clamp. Ensure that the clamp retaining bolts are tightened evenly so that the gaps between the front and rear mating surfaces of each clamp are equal. Take note, throughout the fitting procedure, of the torque loading figures given in the Specifications Section of this Chapter.

2.5 Remove the two bolts to free the front mudguard from each fork leg

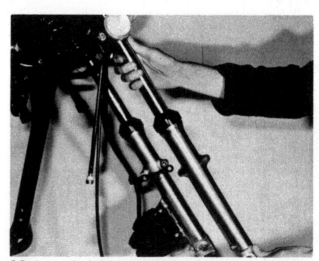

2.8a Locate the fork leg assembly into the steering head yokes...

2.8b ...and tighten the lower yoke pinchbolt...

2.8c ...followed by the upper yoke pinchbolt...

2.8d ...whilst noting the fitted position of the stanchion in relation to the yoke

4 Front fork legs : dismantling, examination, renovation and reassembly

1   It is advisable to dismantle each fork leg separately using an identical procedure. There is less chance of unwittingly exchanging parts if this approach is adopted. Commence by draining each fork leg of damping oil; there is a drain plug in each lower leg above and to the rear of each wheel spindle housing.

2   Clamp the fork lower leg between the jaws of a vice. Take care to prevent damage to the alloy finish of the leg by placing some form of protection between the vice jaws and the leg; thin wooden blocks or soft aluminium alloy angle pieces are ideal for this purpose. It is not advisable to use rag or similar as a form of protection as this will not prevent the fork leg slipping out of the vice jaws. Unscrew and remove the socket screw from the recess in the front wheel spindle housing.

3   Remove the top cap bolt from the end of the fork stanchion and withdraw the spring(s). Some fork springs have variable pitch coils. Note carefully whether the close coils are at the top or bottom of the fork on removal. The spring must be refitted in the same manner on subsequent reassembly. Where there are two springs fitted within each fork leg, take note of the seating ring fitted between the two springs. If this shows signs of excessive wear or damage then it must be renewed. Using the flat of a screwdriver, carefully prise the dust excluder from position and slide it up the fork stanchion. The stanchion can now be pulled out of the fork lower leg. The damper rod assembly will be withdrawn with the stanchion. To separate the two items, pull the damper rod seat off the end of the rod and invert the stanchion to allow the rod assembly to slide from position.

4   The oil seal fitted to the top of the lower leg should be removed only if it is to be renewed, because damage will almost certainly be inflicted when it is prised from position. The spring clip which retains the seal may be displaced from its location within the fork leg by inserting the flat of a small screwdriver into one of the clip indentations provided. With the clip thus removed, the seal may be levered out of position by placing the flat of a screwdriver beneath its lower edge. Take great care when removing both of these items not to damage the alloy edge of the fork leg with the screwdriver. If easily removed, the spacer ring located beneath the seal may be drawn from position for inspection and cleaning.

5   The front forks do not contain bushes. The fork legs slide directly against the outer hard chrome surface of the fork stanchions. If wear occurs, indicated by slackness, the fork leg complete will have to be renewed, possibly also the fork stanchion. Wear of the fork stanchion is indicated by scuffing and penetration of the hard chrome surface.

6   After an extended period of service the fork springs may take a permanent set. If the spring lengths are suspect they should be measured and the readings obtained compared with the service limits given in the Specifications Section of this Chapter. It is advisable to fit new fork springs if their overall length has decreased beyond the service limit given. Always renew the springs as a set, never separately.

7   Check the outer surface of the stanchion for scratches or roughness, it is only too easy to damage the oil seal during the re-assembly if these high spots are not eased down. The stanchions are unlikely to bend unless the machine is damaged in an accident. Any significant bend will be detected by eye, but if there is any doubt about straightness, roll the stanchion tubes on a flat surface such as a sheet of plate glass. If the stanchions are bent they must be renewed. Unless specialised repair equipment is available it is rarely practicable to effect a satisfactory repair to a damaged stanchion.

8   The piston ring fitted to the damper rod may wear if oil changes at the specified intervals are neglected. If damping has become weakened and does not improve as a result of an oil change, the piston ring should be renewed. Check also that the oilways in the damper rod have not become obstructed. Suzuki

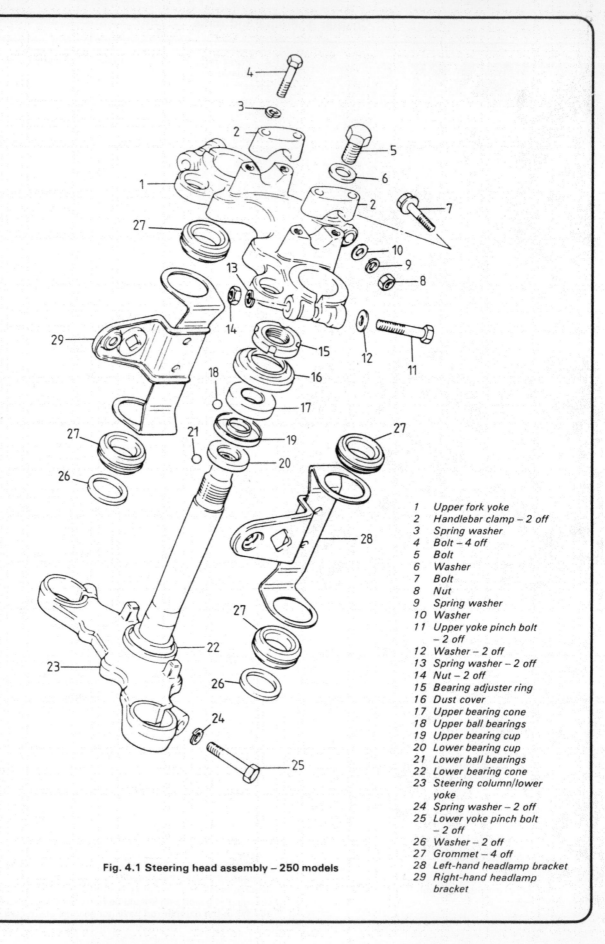

Fig. 4.1 Steering head assembly – 250 models

1 Upper fork yoke
2 Handlebar clamp – 2 off
3 Spring washer
4 Bolt – 4 off
5 Bolt
6 Washer
7 Bolt
8 Nut
9 Spring washer
10 Washer
11 Upper yoke pinch bolt
   – 2 off
12 Washer – 2 off
13 Spring washer – 2 off
14 Nut – 2 off
15 Bearing adjuster ring
16 Dust cover
17 Upper bearing cone
18 Upper ball bearings
19 Upper bearing cup
20 Lower bearing cup
21 Lower ball bearings
22 Lower bearing cone
23 Steering column/lower
   yoke
24 Spring washer – 2 off
25 Lower yoke pinch bolt
   – 2 off
26 Washer – 2 off
27 Grommet – 4 off
28 Left-hand headlamp bracket
29 Right-hand headlamp
   bracket

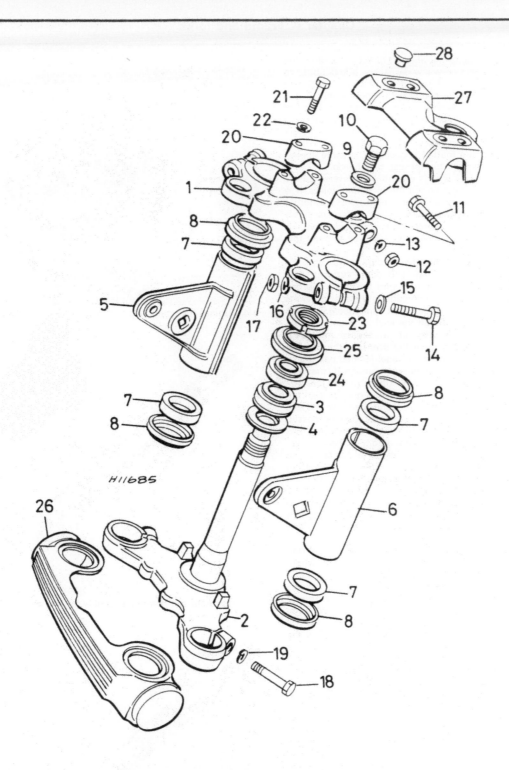

Fig. 4.2 Steering head assembly – 400 models

1	Upper fork yoke	11	Bolt	19	Spring washer – 2 off
2	Steering column/lower yoke	12	Nut	20	Handlebar clamp – 2 off
3	Lower bearing race	13	Spring washer	21	Bolt – 4 off
4	Washer	14	Upper yoke pinch bolt	22	Spring washer – 4 off
5	Right-hand headlamp bracket		– 2 off	23	Bearing adjuster ring
6	Left-hand headlamp bracket	15	Washer – 2 off	24	Upper bearing race
7	Cap – 4 off	16	Spring washer – 2 off	25	Dust cover
8	Grommet – 4 off	17	Nut – 2 off	26	Lower yoke cover
9	Washer	18	Lower yoke pinch bolt	27	Handlebar clamp cover
10	Bolt		– 2 off	28	Plug – 4 off

provide no information as to the amount of set allowed on the damper rod spring before renewal is necessary. If in doubt as to the condition of the spring, ask the advice of an official Suzuki service agent or compare the spring with a new item.

9   Closely examine the dust excluder for splits or signs of deterioration. If defective, it must be renewed, as any ingress of road grit will rapidly accelerate wear of the oil seal and fork stanchion. It is advisable to renew any gasket washers fitted beneath screw heads as a matter of course. The same applies to the O-rings fitted beneath the top cap bolt heads.

10   Reassembly of the fork legs is essentially a reversal of the dismantling procedure, whilst noting the following points. It is essential that all fork components are thoroughly washed in solvent and wiped clean with a lint-free cloth before assembly takes place. Any trace of dirt inside the fork leg assembly will quickly destroy the oil seal or score the stanchion to outer fork leg bearing surfaces.

11   Before fitting the new oil seal, it should be coated with the recommended fork oil on its inner and outer surfaces. This serves to make the fitting of the seal into the lower fork leg easier and also reduces the risk of damage to the seal when the fork stanchion is passed through it. Great care should be taken when fitting the stanchion through the seal.

12   Suzuki recommend the use of a service tool (No 09940-50111) with which to drive the seal into the fork leg recess. With the spacer ring fitted, the seal located partially in its recess and the fork stanchion passed through it, the tool, which takes the form of a short length of metal tube approximately 3 in long, with an inner diameter just greater than the outer diameter of the stanchion and an outer diameter just less than that of the outer diameter of the oil seal, may be passed over the stanchion and used to tap the seal home by using it as a form of slide hammer. If this tool is not readily available it can, of course, be fabricated from a piece of metal tubing of the appropriate dimensions. Care should be taken however, to ensure that the end of the tube that makes contact with the seal is properly chamfered, free of burrs and absolutely square to the fork stanchion. Alternatively, a suitable socket may be used to drive the seal into position before the stanchion is fitted. Ensure that the seal is driven home squarely and is properly retained in position by the spring clip.

13   Refit the damper rod assembly into the stanchion and refit the damper rod seat. Fit the fork spring(s) (and spacer, where fitted), taking care to insert them with the coils in the same position as noted during the dismantling sequence. On models with a single spring fitted to each fork leg, the smaller diameter end of the spring should face the bottom of the fork leg. Refit the cap bolt with its O-ring and tighten it finger-tight. This will serve to keep the damper rod in position during refitting of the socket screw into the wheel spindle housing recess. Insert the stanchion into the fork lower leg and prepare the socket screw for insertion into the wheel spindle housing recess by degreasing its thread and fitting a new gasket washer. Refer to the figure accompanying this text and coat the thread of the screw with a thread locking compound and a sealing compound as shown before fitting it into its recess and tightening it to the specified torque loading.

14   Relocate the dust excluder over the fork lower leg. Refit the drain plug and gasket washer and unscrew the cap bolt. Replenish the fork leg with the correct quantity of fork oil. Refer to the Specifications at the beginning of this Chapter for both the oil type and quantity for each model type. Fit the cap bolt and O-ring to the fork stanchion and nip it tight. The bolts should be fully tightened to the specified torque loading once the fork legs have been refitted to the machine.

15   The fork legs are now ready to be fitted to the machine but before doing so, it is worth pausing at this point to consider the advantages of fitting fork gaiters. It is a fact that the life of the oil seal can be lengthened considerably by doing this, with the additional advantage that the lower part of the fork stanchion is also protected. Several manufacturers provide gaiters which will fit the type of forks fitted to these machines.

4.4 Displace the spring clip to allow removal of the fork leg seal

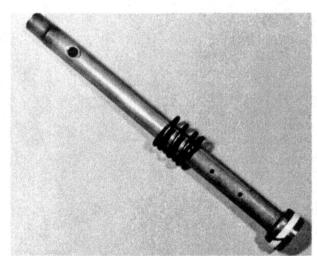
4.8 Inspect the damper rod assembly for ring and spring wear

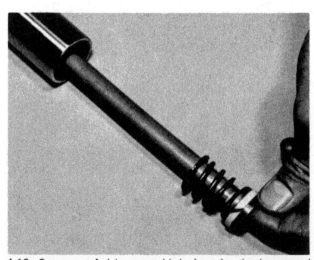
4.13a Commence fork leg assembly by inserting the damper rod into the fork stanchion...

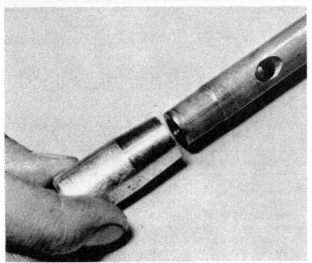
4.13b ...refitting the damper rod seat...

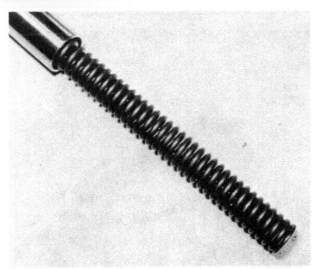
4.13c ...refitting the fork spring...

4.13d ...and loosely refitting the cap bolt and O-ring to prevent rotation of the damper rod

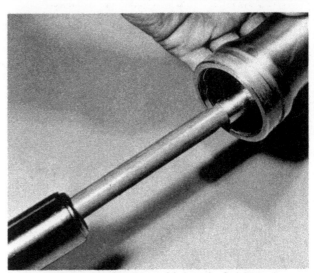
4.13e Insert the stanchion assembly into the fork lower leg...

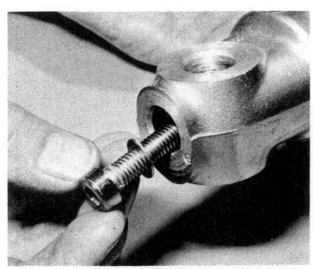
4.13f ...and lock it in position with the socket screw and washer

4.14a Fit a serviceable dust excluder

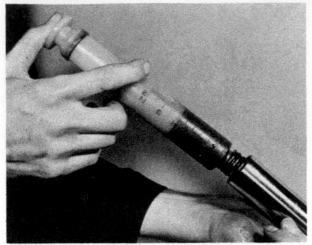

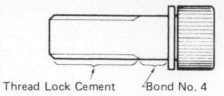

Thread Lock Cement   Bond No. 4

Fig. 4.4 Recommended compounds used on damper rod socket screw

4.14b Remove the cap bolt to allow replenishment of the fork leg oil

5 Steering head bearings : examination and renovation

250 models

1   Before reassembly of the front forks is commenced, examine the steering head races. The ball bearing tracks of the respective cup and cone bearings should be polished and free from indentations or cracks. If wear or damage is evident, the cups and cones must be renewed as a complete set.

2   Clean and examine the balls in each bearing assembly. These should also be polished and show no signs of surface cracks or blemishes. If any are defective then the complete set

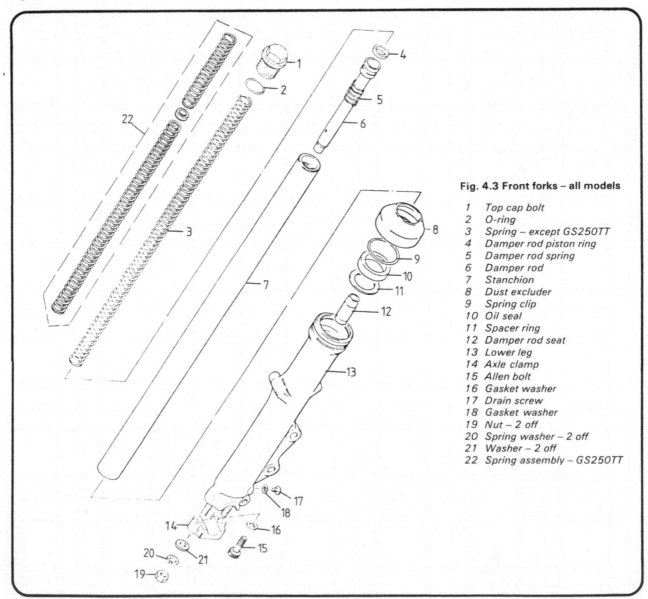

Fig. 4.3 Front forks – all models

1   Top cap bolt
2   O-ring
3   Spring – except GS250TT
4   Damper rod piston ring
5   Damper rod spring
6   Damper rod
7   Stanchion
8   Dust excluder
9   Spring clip
10 Oil seal
11 Spacer ring
12 Damper rod seat
13 Lower leg
14 Axle clamp
15 Allen bolt
16 Gasket washer
17 Drain screw
18 Gasket washer
19 Nut – 2 off
20 Spring washer – 2 off
21 Washer – 2 off
22 Spring assembly – GS250TT

must be renewed. Remember that a set of balls for each bearing is relatively cheap and that it is therefore not worth the risk of refitting items that are in doubtful condition.

3   Eighteen balls are fitted both in the top and bottom races. This arrangement will leave a gap but an extra ball must not be fitted otherwise the balls will press against each other, accelerating wear and making the steering stiff.

4   The bearing outer races are a drive fit in the steering head and may be removed by passing a long drift through the inner bore of the steering head and drifting out the defective item from the opposite end. The drift must be moved progressively around the race to ensure that it leaves the steering head evenly and squarely.

5   The lower of the two inner races fits over the steering stem and may be removed by carefully drifting it up the length of the stem with a flat-ended chisel, or similar. Again, take care to ensure that the race is kept square to the stem.

6   Fitting of the new bearing races is a straightforward procedure, taking note of the following points. Ensure that the race locations within the steering head are clean and free of rust; the same applies to the steering stem. Lightly grease the stem and head locations to aid fitting of the races and drift each race into position whilst keeping it square to its location. Fitting of the outer races into the steering head will be made easier if the opposite end of the head to which the race is being fitted has a wooden block placed against it to absorb some of the shock as the drift strikes the race.

400 models

7   Examination and renewal of the taper roller bearings fitted to the 400 models is essentially the same as for the 250 models. If pitting or damage to the rollers is found the complete bearing must be renewed. This applies equally if damage to either outer race is evident. Provided that correct adjustment and lubrication is maintained there is no reason why a set of steering head bearings should not last the life of the machine.

6   Front yokes : examination

1   To check the top yoke for accident damage, push the fork stanchions through the bottom yoke and fit the top yoke. If it lines up, it can be assumed the yokes are not bent. Both yokes must also be checked for cracks. If they are damaged or cracked, fit serviceable replacements.

7   Steering link : general description and renewal

1   A steering head lock is attached to the underside of the lower yoke of the forks by two screws and lock washers. When in a locked position, a tongue extends from the body of the lock when the handlebars are on full lock in either direction and abuts against a plate welded to the base of the steering head. In consequence, the handlebars cannot be turned until the lock is released.

2   If the lock malfunctions it must be renewed. A repair is impracticable. When the lock is changed the key must be changed too, to match the new lock.

8   Frame assembly : examination and renovation

1   If the machine is stripped for a complete overhaul, this affords a good opportunity to inspect the frame for cracks of other damage which may have occurred in service. Check the points at which the front downtube(s) and the frame top tubes join the steering head: these are the points where fractures are most likely to occur. The straightness of the tubes concerned will show whether the machine has been involved in a previous accident.

2   Check carefully areas where corrosion has occurred on the frame. Corrosion can cause a reduction in the material thickness and should be removed by use of a wire brush and derusting agents. After the machine has covered a considerable mileage, it is advisable to examine the frame closely for signs of cracking or splitting at the welded joints.

3   If the frame is broken or bent, professional attention is required. Repairs of this nature should be entrusted to a competent repair specialist, who will have available all the necessary jigs and mandrels to preserve correct alignment. Repair work of this nature can prove expensive and it is always worhwhile checking whether a good replacement frame of identical type can be obtained at a reasonable cost.

4   Remember that a frame which is in any way damaged or out of alignment will cause, at the very least, handling problems. Complete failure of a main frame component could well lead to a serious accident.

9   Swinging arm fork : removal, examination, renovation and refitting

1   The rear fork of the frame is of the swinging arm type. This unit is of tubular steel construction and pivots on a shaft which passes through its tubular crossmember and both sides of the main frame assembly. Two needle roller bearings and two inner bushes provide an efficient bearing for the unit, these two assemblies being kept separate by means of a tubular spacer placed in the centre of the crossmember. The pivot shaft is retained in position by a plain washer and nut.

2   Any wear in the swinging arm pivot bearings will cause imprecise handling of the machine, with a tendency for the rear end of the machine to twitch or hop. Worn swinging arm bearings can be detected by placing the machine on its centre stand and pulling and pushing vigorously on the rear wheel in a horizontal direction. Any play will be magnified by the leverage effect.

3   When wear develops in the swinging arm, necessitating renewal of the bushes and possibly the bearings, the renovation procedure is quite straightforward. Commence by removing the rear wheel as described in Section 11 of Chapter 5.

4   The final drive chainguard is secured by two bolts. Removal is not strictly necessary, although it will facilitate swinging arm detachment.

5   Remove the lower cap nut and plain washer from each of the rear suspension units and pull each unit away from its swinging arm fork mounting stud so that the fork swings down. Leave the suspension units hanging from the frame, but slacken the top nut so that they are free to move. This facilitates reassembly.

6   Remove the pivot shaft retaining nut and washer and draw the shaft out of position whilst supporting the swinging arm. The shaft may need a gentle tap with a soft-faced mallet and drift to displace it. Pull the final drive chain across so that it clears the swinging arm fork left-hand end. The swinging arm is now free and can be lifted out to the rear.

7   Remove the dust cap and thrust washer from each side of the swinging arm crossmember and then pull out the two short bushes. Push out the long central spacer, using a long shanked screwdriver.

8   The caged needle roller bearings should now be cleaned of all grease by wiping them with a fuel soaked rag, taking the necessary fire precautions whilst doing so. If the bearings are seen to be breaking up or if they are corroded through ingress of water due to lack of lubrication, they may be drifted out of position, using a suitable length of steel rod. Do not remove the bearings merely for inspection as the cages will be damaged by the drift. Lubricate the outside of the new bearing cages before driving them into place and ensure that they are fitted with the punch marked face outwards. A socket or short length of thick-walled metal tube with an external diameter which is slightly less than that of the bearing should be utilised to drive the

bearing into place. This tool will serve to help keep the bearing square to the end of the swinging arm crossmember as it is fitted.

9 Check the pivot shaft for straightness by rolling it on a flat surface such as a sheet of plate glass. Alternatively, place the shaft on two V-blocks and measure the amount of runout on the shaft with a dial gauge. If the amount of runout found exceeds 0.3 mm (0.012 in), replace the shaft with a straight item.

10 Inspect the condition of the two dust caps and thrust washers. if the caps show the slightest sign of damage or deterioration then they must be renewed as they not only serve to keep grease in the bearings but water and road dirt out. Worn bushes should also now be replaced with new items.

11 Reassemble and refit the swinging arm by reversing the removal and dismantling procedures. Apply a molybdenum disulphide grease to the central spacer, the bearings, the bushes and the inner edges of the dust caps as the unit is assembled. Grease the length of the pivot shaft shank before inserting it and retaining it in position with the plain washer and nut. Tighten the nut to the torque figure given in the Specifications of this Chapter. Note also the torque figure given for the rear suspension unit retaining nuts.

12 Where a grease nipple is fitted to the crossmember of the swinging arm, ensure that the lubrication procedure is now completed by pumping grease in through this nipple.

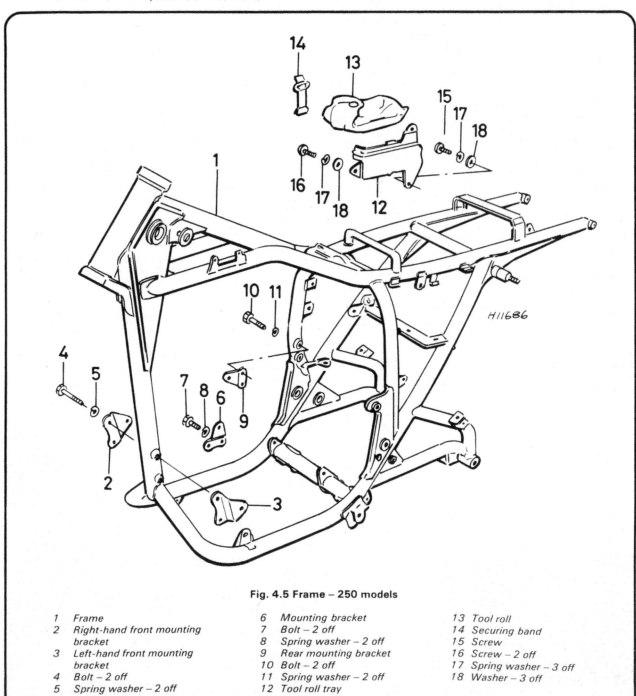

Fig. 4.5 Frame – 250 models

1	Frame	6	Mounting bracket	13	Tool roll
2	Right-hand front mounting bracket	7	Bolt – 2 off	14	Securing band
3	Left-hand front mounting bracket	8	Spring washer – 2 off	15	Screw
4	Bolt – 2 off	9	Rear mounting bracket	16	Screw – 2 off
5	Spring washer – 2 off	10	Bolt – 2 off	17	Spring washer – 3 off
		11	Spring washer – 2 off	18	Washer – 3 off
		12	Tool roll tray		

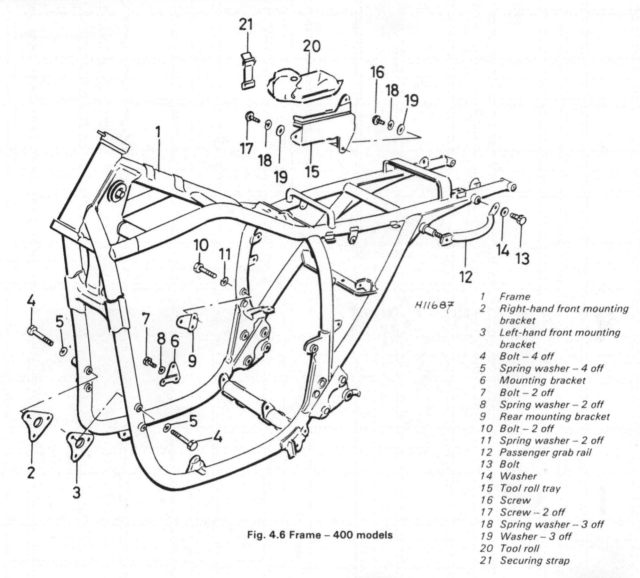

H11687

1   Frame
2   Right-hand front mounting
    bracket
3   Left-hand front mounting
    bracket
4   Bolt – 4 off
5   Spring washer – 4 off
6   Mounting bracket
7   Bolt – 2 off
8   Spring washer – 2 off
9   Rear mounting bracket
10 Bolt – 2 off
11 Spring washer – 2 off
12 Passenger grab rail
13 Bolt
14 Washer
15 Tool roll tray
16 Screw
17 Screw – 2 off
18 Spring washer – 3 off
19 Washer – 3 off
20 Tool roll
21 Securing strap

Fig. 4.6 Frame – 400 models

9.8a Clean and examine the caged needle roller bearings

9.8b Use a length of steel rod to drift worn or damaged bearings from position...

9.8c ...and a length of thick-walled metal tube to drift new bearings into position

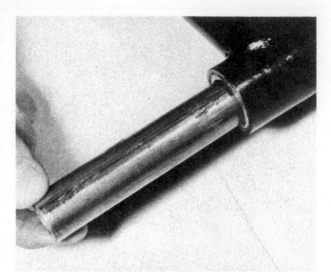

9.11a Commence reassembly of the swinging arm by greasing and inserting the central spacer

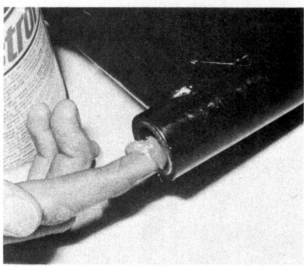

9.11b Generously grease the bearings...

9.11c ...before inserting each bush into positon

9.11d Grease and fit serviceable thrust washers and dust caps

9.11e Check the torque arm to swinging arm connection for security...

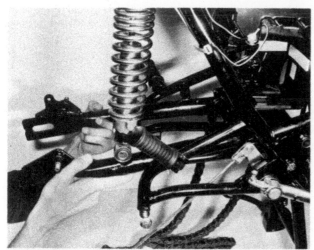

9.11f ...before lifting the swinging arm fork into position...

9.11g ...and securing it with the pivot shaft, plain washer and nut

9.11h The rear suspension unit retaining nuts should be tightened to the correct torque loading

must be renewed. This applies equally if the damper rod has become bent or its chromed surface badly corroded or damaged. Check also for deterioration of the rubber mounting bushes and of the rubber buffer within the spring. The piston housing must be free of damage if it is to function correctly.

3   If necessary, the suspension units can be removed from the frame and swinging arm attachment studs simply by removing the upper and lower cap nut and plain washers and pulling the unit out away from the machine.

4   Refitting of the units is a reversal of the removal procedure whilst noting the following points. If refitting the original or used units, take the opportunity, whilst they are removed from the machine, to give them a thorough clean. Do not, under any circumstances, grease the chromed surface of the damper rod. With the units fitted to the machine, torque load the retaining nuts to the figure given in the Specifications Section of this Chapter.

5   Note that in the interests of good roadholding it is essential that both suspension units have the same load setting. If renewal is necessary, the units must be replaced as a matched pair.

10 Rear suspension units – examination, removal and refitting

1   Rear suspension units of the hydraulically damped type are fitted to the Suzuki GS and GSX models covered in this Manual. Each unit comprises a hydraulic damper, effective primarily on rebound, and a concentric spring. It is secured to the frame and swinging arm by means of rubber-bushed lugs at each end of the unit. The units are provided with an adjustment of the spring tension, giving five settings. The settings can be easily altered without removing the units from the machine by using a tommy bar in the hole directly below the springs. Turning clockwise will increase the spring tension and stiffen the rear suspension, turning anti-clockwise will lessen the spring tension and therefore soften the ride. As a general guide the softest setting is recommended when no pillion passenger is carried. The hardest setting should be used when a heavy load is carried, and during high-speed riding. The intermediate positions may be used as conditions dictate.

2   There is no means of draining the units or topping up, because the dampers are built as a sealed unit. If the damping fails or if the units start to leak, the complete damper assembly

11 Centre stand : examination and servicing

1   The centre stand is retained to the underside of the frame by two bolts which serve as pivot shafts. Each of these bolts has a spring washer located beneath its head and is prevented from leaving its retaining nut by a spring clip which passes through the end of its threaded section. A bush is fitted between each pivot shaft and its centre stand retaining tube.

2   The pivot assembly on centre stands is often neglected with regard to lubrication and this will eventually lead to wear. It is prudent to remove the pivot bushes from time to time and grease them thoroughly. This will prolong the effective life of the stand. Check before refitting the pivot shafts, that the spring washers are not flattened and are in good condition; they must be renewed if they no longer serve their locking function. Do not omit to refit the spring clips.

3   Check that the return spring is in good condition. A broken or weak spring may cause the stand to fall whilst the machine is in motion, with the resulting danger that once it catches in some obstacle the balance of the machine will be drastically affected.

Fig. 4.7 Swinging arm – all models

1 Swinging arm fork
2 Brake torque arm
3 Spacer
4 Needle roller bearing
   – 2 off
5 Bush – 2 off
6 Thrust washer – 2 off
7 Dust cap – 2 off
8 Washer
9 Nut
10 Pivot bolt
11 Bolt – 2 off
12 Spring washer – 2 off
13 Nut – 2 off
14 R-pin – 2 off
15 Right-hand suspension unit
16 Left-hand suspension unit
17 Passenger grab rail
18 Nut – 4 off
19 Washer – 2 off
20 Washer – 5 off
21 Washer – 4 off
22 Washer – 4 off

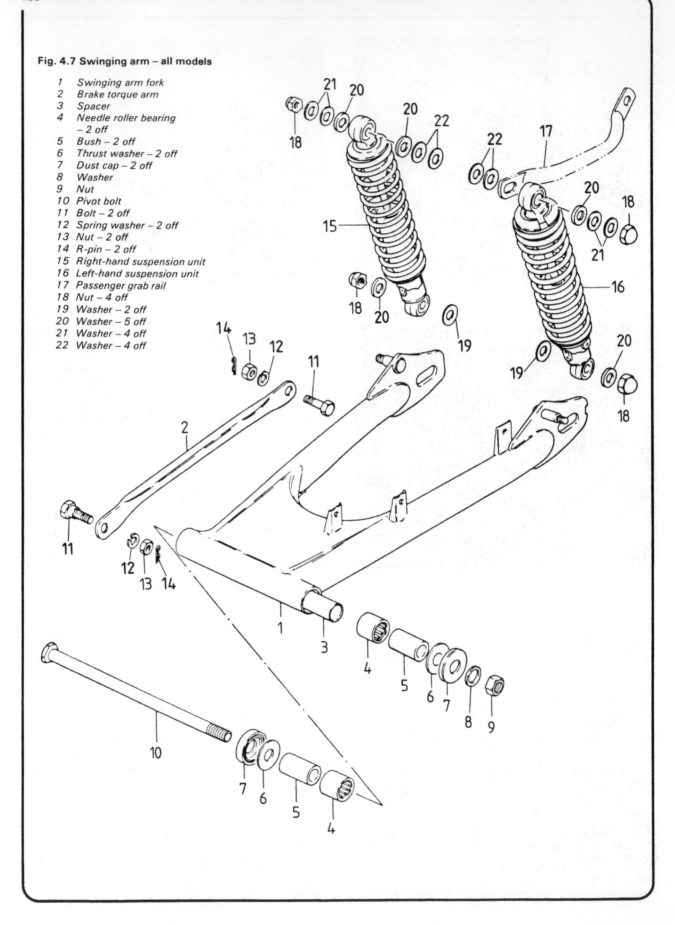

10.1 Rotate the adjuster to alter the spring setting

10.4 Note the position of the pivot washers and grab rail end

11.1 Inspect the centre stand spring and pivot points

12 Prop stand : examination and servicing

1   The prop stand bolts to a lug attached to the rear of the left-hand lower frame tube. An extension spring ensures that the stand is retracted when the weight of the machine is taken off the stand. When properly retracted, the stand should be tight against its stop and well out of the way.

2   Check that the pivot bolt is secure and that the pivot surfaces are well lubricated with either grease or motor oil. The extension spring must be free from fatigue and in good condition. A serious accident is almost inevitable if the stand extends whilst the machine is on the move.

12.1 Inspect the prop stand spring and pivot

13   Footrests : examination and renovation

1   The front footrests fitted to the 250 models are of the bolt-on metal bracket type with renewable rubber pads. If they are bent in a spill or through the machine falling over, they can be removed and straightened in a vice whilst heated to a dull red with a blow lamp, or welding torch.

2   The front footrests fitted to the 400 models and the pillion footrests fitted to all models are pivoted on clevis pins, the front footrests only being spring loaded in the down position. If an accident occurs, it is probable that the footrest peg will move against the spring loading and remain undamaged. A bent peg may be detached from the mounting, after removing the clevis pin securing split pin and the clevis pin itself. The damaged peg can be straightened in a vice, using a blowlamp flame to apply heat at the area where the bend occurs. The footrest rubber will, of course, have to be removed as the heat will render it unfit for service.

3   Note that if there is evidence of failure of the metal either before or after straightening, it is advised that the damaged component is renewed. If a footrests breaks in service, loss of machine control is almost inevitable.

4   When refitting the clevis pin to a pivoting type of footrest assembly, ensure that, where fitted, the return spring is correctly located and that a new split-pin is used to retain the clevis pin in position.

14 Rear brake pedal and shaft assembly : examination and renovation

1   The rear brake pedal is secured by a single pinch bolt to the splined brake pivot shaft. In the event of damage, the pedal may be removed and treated similarly to a bent footrest as described in the previous Section.

2   The pinch bolt should be removed completely before pulling the pedal off its shaft. Note the spring washer located beneath the bolt head and detach the pedal return spring from its attachment points. Check that the washer has not flattened and is in good condition; it must be renewed if it no longer serves its locking function. Inspect the return spring for signs of fatigue or failure and renew it if necessary.

3   Once straightened, the pedal should be refitted to its shaft in a similar position to that noted before removal. Suzuki provide an aid to correct pedal positioning by punch marking the rear edge of the pedal shaft end; the punch mark should align with the gap in the pedal boss through which the pinch bolt passes.

4   The warning relating to footrest breakage applies equally to the brake pedal because it follows that failure is most likely to occur when the brake is applied firmly, which is when it is required most.

5   It is advisable, if the brake pedal has been in any way damaged, to check the condition of its pivot shaft. This is also the case if operation of the pedal is noticed to be stiff and the pedal fails to return immediately when released.

6   With the pedal removed, the shaft arm must be disconnected from the brake operating rod and the stop lamp switch operating spring. Detach the rod by removing the split-pin and plain washer from the end of the clevis pin. Remove the clevis pin and detach the rod end from the shaft arm. With the rod thus released it is a simple operation to unhook the end of the switch spring from its location on the shaft arm.

7   Push the shaft out of its location in the frame. If the shaft has become seized in position it can be removed by placing the end of a large tommy bar against its end and tapping the end of the bar with a hammer. If excessive force is required to free the shaft, take great care to support the frame members around the shaft before tapping it free and be very careful not to damage the shaft splines. It may be found necessary to apply penetrating oil around the shaft to frame boss mating surfaces in order to help free a seized shaft.

8   With the shaft removed, inspect its length for straightness using a straight edge placed alongside it for comparison. If bent, the shaft must be renewed. If the shaft was found to have seized through lack of lubrication or if the bearing surfaces are found to be corroded, clean the bearing face of the shaft and of the frame boss with fine grade emery paper before refitting the shaft into the boss and checking for excessive play between the two components. If the shaft is thought to be excessively worn then it must be renewed.

9   Grease the length of the shaft with a high quality lithium based lubricant and insert the shaft into its frame location. Reconnect the brake operating rod and the stop lamp switch operating spring; use a new split-pin on the end of the rod clevis pin. Lightly grease the brake pedal return spring to lessen the risk of its being corroded by road salts and fit it and the brake pedal to the splined end of the shaft. Before riding the machine, carry out a final check to ensure that both the rear brake and the stop lamp are in correct adjustment and are functioning correctly.

15 Dualseat : removal and refitting

1   Removal of the dualseat is achieved by releasing the seat from its rear mounting points and pulling the seat up to clear the mountings and then rearwards to release it from the forward frame mounting.

2   On machines equipped with a seat fairing, the seat may be released simply by turning the key in the centrally positioned lock which is located just to the rear of the pillion passenger grab rail.

3   On machines equipped with a conventional type of seat, the seat may be released by unscrewing and removing the two dome-headed bolts which are located one each side of the rear section of the seat.

14.3 Note the shaft end punch mark when positioning the brake pedal

16 Speedometer and tachometer heads : removal, inspection and fitting

1   The speedometer and tachometer are mounted together in a single console which is sited on top of the steering head assembly. They are secured in position by two crosshead screws which pass through the base of the console casing and into the base of each instrument.

2   The instruments may be detached from the machine by removing the above mentioned screws with their associated washers, detaching the top of the console by removing its four securing screws and disconnecting the drive cables to allow the instruments to be pulled far enough out of the console to detach any electrical connections from its base.

3   If either instrument fails to record, check the drive cable first before suspecting the head. If the instrument gives a jerky response it is probably due to a dry cable, or one that is trapped or kinked.

4   The speedometer and tachometer heads cannot be repaired by the private owner, and if a defect occurs a new instrument has to be fitted. Remember that a speedometer in correct working order is required by law on a machine in the UK and also in many other countries.

16.1 Each instrument head is secured by two crosshead screws

5 Refer to the following Sections for details of servicing the instrument drive assemblies. On completion of servicing the instruments and their drive assemblies, fit them using the reverse procedure to that given for removal. Check that all disturbed electrical connections are properly remade and that the drive cables are correctly routed.

17 Speedometer and tachometer drive cables : examination and renovation

1 It is advisable to detach the speedometer and tachometer drive cables from time to time in order to check whether they are adequately lubricated and whether the outer cables are compressed or damaged at any point along their run. A jerky or sluggish movement at the instrument head can often be attributed to a cable fault.

2 To grease the cable, uncouple both ends and withdraw the inner cable. (On some model types this may not be possible in which case a badly seized cable will have to be renewed as a complete assembly). After removing any old grease, clean the inner cable with a petrol soaked rag and examine the cable for broken strands or other damage. Do not check the cable for broken strands by passing it through the fingers or palm of the hand, this may well cause a painful injury if a broken strand snags the skin. It is best to wrap a piece of rag around the cable and pull the cable through it, any broken strands will snag on the rag.

3 Regrease the cable with high melting point grease, taking care not to grease the last six inches closest to the instrument head. If this precaution is not observed, grease will work into the instrument and immobilise the sensitive movement.

4 If the cable breaks, it is usually possible to renew the inner cable alone, provided the outer cable is not damaged or compressed at any point along its run. Before inserting the new inner cable, it should be greased in accordance with the instructions given in the preceding paragraph. Try to avoid tight bends in the run of the cable because this will accelerate wear and make the instrument movement sluggish.

18 Speedometer and tachometer drives : location and examination

1 The speedometer is driven from a gearbox fitted over the front wheel spindle on the left-hand side of the wheel hub. Drive is transmitted through a dog plate fixed to the hub which engages with the drive gear in the gearbox Provided that the gearbox is repacked with grease from time to time, very little wear should be experienced. In the event of failure, the complete gearbox should be renewed.

2 The tachometer drive is taken from the cylinder head cover at a point between the two cylinders. The drive is taken from the overhead exhaust camshaft by means of skew-cut pinions and then by a flexible cable to the tachometer head. It is unlikely that the internal drive will give trouble during the normal service life of the machine, particularly since it is fully enclosed and effectively lubricated.

19 Cleaning the machine

1 After removing all surface dirt with a rag or sponge which is washed frequently in clean water, the machine should be allowed to dry thoroughly. Application of car polish or wax to the cycle parts will give a good finish, particularly if the machine receives this attention at regular intervals.

2 The plated parts should require only a wipe with damp rag, but if they are badly corroded, as may occur during the winter when the roads are salted, it is permissible to use one of the proprietary chrome cleaners. These often have an oily base which will help to prevent corrosion from recurring.

3 If the engine parts are particularly oily, use a cleaning compound such as Gunk or Jizer. Apply the compound whilst the parts are dry and work it in with a brush so that it has an opportunity to penetrate and soak into the film of oil and grease. Finish off by washing down liberally, taking care that water does not enter the carburettors, air cleaners or the electrics.

4 If possible, the machine should be wiped down immediately after it has been used in the wet, so that it is not garaged under damp conditions, which will promote rusting. Make sure that the chain is wiped and re-oiled to prevent water from entering the rollers and causing harshness with an accompanying rapid rate of wear. Remember there is less chance of water entering the control cables and causing stiffness if they are lubricated regularly as described in the Routine Maintenance Section.

20 Cleaning the plastic mouldings

1 The moulded plastic cycle parts, which include the side-panels, seat fairing and headlamp fairing (where fitted), need treating in a different manner from normal metal cycle parts.

2 These plastic parts will not respond to cleaning in the same way as painted metal parts; their construction may be adversely affected by traditional cleaning and polishing techniques, and lead as a result, to the surface finish deteriorating. It is best to wash these parts with a household detergent solution, which will remove oil and grease in a most effective manner.

3 Avoid the use of scouring powder or other abrasive cleaning agents because this will score the surface of the mouldings making them more receptive to dirt, and permanently damaging the surface finish.

21 Fault diagnosis : frame and forks

Symptom	Cause	Remedy
Machine veers to the left or right with hands off handlebars	Wheels out of alignment Forks twisted Frame bent	Check and realign. Strip and repair. Strip and repair or renew.
Machine tends to roll at low speeds	Steering head bearings not adjusted correctly or worn	Check adjustment and renew bearings if necessary.
Machine tends to wander	Worn swinging arm bearings	Check and renew bearings.
Forks judder when front brake applied	Steering head bearings slack Fork components worn	Adjust bearings. Strip forks, and renew all worn parts.
Forks bottom	Short of oil	Replenish with correct viscosity oil.
Fork action stiff	Fork legs out of alignment Bent shafts, or twisted ie. yokes	Slacken clamp bolts, front wheel spindle and top bolts. Pump forks several times and tighten from bottom upwards. Strip and renew parts, if damaged.
Machine pitches badly	Defective rear suspension units or ineffective fork damping	Check damping action. Check grade and quantity of oil in front forks

Chapter 5 Wheels, brakes and tyres

Refer to Chapter 7 for information relating to GS450 and GSX250/400 EZ models

Contents

Specifications

Wheels

Type:
GSX250 and GSX400 E models .... Cast aluminium alloy
GS250 and GSX400 T models .... Chromed steel rims and steel spokes
Rim runout service limit:
Axial .... 2.0 mm (0.08 in)
Radial .... 2.0 mm (0.08 in)
Spindle runout (front and rear) .... 0.25 mm (0.010 in)

Brakes

Hydraulic fluid type .... SAE J1703 (UK) or DOT 3/4 (USA)
Front brake type .... Single hydraulically operated disc
Disc thickness .... 5.0 ± 0.010 mm (0.20 ± 0.004 in)
Service limit .... 4.5 mm (0.18 in)
Disc runout service limit .... 0.30 mm (0.012 in)
Master cylinder piston diameter .... 13.957 – 13.984 mm (0.549 – 0.550 in)
Master cylinder bore .... 14.000 – 14.043 mm (0.551 – 0.553 in)
Caliper cylinder bore .... 42.850 – 42.926 mm (1.687 – 1.690 in)
Rear brake type .... Internally expanding, single leading shoe, drum
Drum 1D service limit .... 160.70 mm (6.330 in)
Lining thickness service limit .... 1.50 mm (0.06 in)

Tyres

Size:		Front	Rear
GS 250 models		3.00 – 18 4PR	3.50 – 17 4PR
GSX 400T models		90/90 – 19 52S	110/90 – 17 60S
All other models		3.00S18 4PR	3.50S18 4PR

Pressures – front	GS models	GSX 400T models	All other models
Normal riding:			
Solo	21 psi (1.50 kg/cm²)	21 psi (1.50 kg/cm²)	25 psi (1.75 kg/cm²)
Dual	25 psi (1.75 kg/cm²)	25 psi (1.75 kg/cm²)	25 psi (1.75 kg/cm²)
Continuous high speed riding:			
Solo	25 psi (1.75 kg/cm²)	25 psi (1.75 kg/cm²)	28 psi (2.00 kg/cm²)
Dual	28 psi (2.00 kg/cm²)	28 psi (2.00 kg/cm²)	32 psi (2.25 kg/cm²)

Pressures – rear
    Normal riding:
        Solo .... 25 psi (1.75 kg/cm²)   25 psi (1.75 kg/cm²)   28 psi (2.00 kg/cm²)
        Dual .... 32 psi (2.25 kg/cm²)   32 psi (2.25 kg/cm²)   32 psi (2.25 kg/cm²
    Continuous high speed riding:
        Solo .... 28 psi (2.00 kg/cm²)   25 psi (1.75 kg/cm²)   32 psi (2.25 kg/cm²)
        Dual .... 32 psi (2.25 kg/cm²)   36 psi (2.50 kg/cm²)   36 psi (2.50 kg/cm²)

Recommended minimum tread depth:
    Front .... 1.6 mm (0.06 in)
    Rear .... 2.0 mm (0.08 in)

Torque wrench settings

	lbf ft	kgf m
Front wheel spindle retaining nut	26.0 – 37.5	3.6 – 5.2
Front wheel spindle/fork leg clamp nuts	11.0 – 18.0	1.5 – 2.5
Rear wheel spindle retaining nut	36.0 – 58.0	5.0 – 8.0
Rear wheel sprocket retaining nuts	18.0 – 29.0	2.5 – 4.0
Wheel spoke nipples	3.0 – 3.5	0.4 – 0.5
Brake disc retaining bolts	11.0 – 18.0	1.5 – 2.5
Brake caliper bracket/fork leg retaining bolt	18.0 – 29.0	2.5 – 4.0
Brake caliper spindle bolts	11.0 – 14.5	1.5 – 2.0
Brake caliper bleed screw	5.0 – 6.5	0.7 – 0.9
Brake hose union bolts	14.5 – 18.0	2.0 – 2.5
Master cylinder clamp bolts	3.5 – 6.0	0.5 – 0.8
Rear brake pedal arm bolt	7.0 – 11.0	1.0 – 1.5
Rear brake cam lever bolt	3.5 – 6.0	0.5 – 0.8
Rear torque arm retaining nut	14.5 – 21.5	2.0 – 3.0

1 General description

The design of wheel fitted to the machines covered in this Manual varies between the model types. There are two basic designs; the traditional type of wheel, that is a chromed steel rim laced to an alloy hub by steel spokes, and a ten-spoke cast alloy wheel. Both types of wheel utilise conventional tubed tyres, the sizes of which are given in the Specifications Section of this Chapter.

All models are equipped with a hydraulically operated, single disc, front brake which utilises a single piston type of caliper. Braking on the rear wheel is provided by a rod operated, drum brake, incorporating a single leading shoe.

2 Front wheel : examination and renovation

1 Place the machine on its centre stand and position blocks underneath the engine crankcase so that the front wheel is raised clear of the ground. Spin the wheel and check the rim alignment; this should be no more than 2.0 mm (0.08 in) out of true.
2 On machines fitted with conventional wire-spoked wheels, small irregularities in alignment can be corrected by tightening the spokes in the affected area although a certain amount of experience is necessary to prevent over-correction. Any flats in the wheel rim will be evident at the same time. These are more difficult to remove and in most cases it will be necessary to have the wheel rebuilt on a new rim. Apart from the effect on stability, a flat will expose the tyre bead and walls to greater risk of damage if the machine is run with a deformed wheel.
3 Check also for loose and broken spokes. Tapping the spokes is the best guide to tension. A loose spoke will produce a quite different sound and should be tightened by turning the nipple in an anti-clockwise direction. Always check for runout by spinning the wheel again. If the spokes have to be tightened by an excessive amount, it is advisable to remove the tyre and tube as detailed in Section 16 of this Chapter. This will enable the protruding ends of the spokes to be ground off, thus preventing them from chafing the inner tube and causing punctures.
4 On machines fitted with cast alloy wheels, Suzuki recommend that a wheel which is more that the specified 2.0 mm (0.8 in) out of alignment should be renewed. This is, however,

a counsel of perfection; a runout somewhat greater than this can probably be accommodated without noticeable effect on steering. No means is available for straightening a warped wheel without resorting to the expense of having the wheel skimmed on all faces. If warpage was caused by impact during an accident, the safest measures is to renew the wheel complete.
5 When inspecting cast alloy wheels, carefully check the complete wheel for cracks and chipping, particularly at the spoke roots and the edge of the rim. As a general rule a damaged wheel must be renewed as cracks will cause stress points which may lead to sudden failure under heavy load. Small nicks may be radiused carefully with a fine file and emery paper (No 600 – No 1000) to relieve the stress. If there is any doubt as to the condition of a wheel, advice should be sought from a Suzuki repair specialist.
6 Each cast alloy wheel is covered with a coating of lacquer, to prevent corrosion. If damage occurs to the wheel and the lacquer finish is penetrated, the bared aluminium alloy will soon start to corrode. A whitish grey oxide will form over the damaged area, which in itself is a protective coating. This deposit however, should be removed carefully as soon as possible and a new protective coating of lacquer applied.
7 Note that on both wheel types, it is possible that worn wheel bearings may cause rim runout. These should be renewed as described in Section 4 of this Chapter.

3 Front wheel : removal and refitting

1 Place the machine on the centre stand so that it is resting securely on firm ground with the front wheel well clear of the ground. If necessary, place wooden blocks below the crankcase to raise the wheel.
2 Displace the split-pin from the wheel spindle retaining nut and remove the nut with its plain washer. Unscrew the two nuts, spring washers and plain washers that secure the spindle clamp to the base of the left-hand fork leg and pull the clamp off its two retaining studs. Draw the wheel spindle out of the right-hand fork leg and allow the wheel to drop clear of the brake caliper assembly. Manoeuvre the wheel clear of the fork legs having first fully withdrawn the spindle to free the speedometer gearbox.

3   Take great care not to operate the front brake lever once the brake disc is removed from between the brake pads since fluid pressure may displace the piston and cause fluid leakage. Additionally, the distance between the pads will be reduced, making refitting of the wheel extremely difficult. To prevent any chance of this happening, it is a good idea to place a hardwood wedge between the two pads directly the wheel is removed.

4   The speedometer gearbox may be left attached to the drive cable unless it is necessary to remove it from the machine for inspection or renewal, in which case it may be detached from the cable by unscrewing the knurled retaining ring.

5   Refit the wheel by reversing the removal procedure, noting the following points. Ensure that the brake disc is free of grease or oil contamination before commencing the fitting operation. Ensure that the speedometer gearbox is well greased before relocating it in the wheel hub. Note that the drive tangs of the gearbox must fit in the grooves within the wheel hub and that the embossed arrow-mark on the gearbox casing must point upwards.

6   Take care, when lifting the wheel into position between the fork legs, to ensure that the brake disc enters between the brake pads cleanly otherwise damage to the pads will result. Once the wheel is aligned between the fork legs, insert the spindle and loosely refit the spindle clamp, nuts and washers, tightening the nuts finger-tight.

7   Tighten the wheel spindle retaining nut to the specified torque loading before locking it in position with a new split-pin. The two spindle clamp retaining nuts should now be tightened down evenly to the specified torque loading so that the gap between the clamp and the base of the fork leg is seen to be equal either side of the wheel spindle.

8   Carry out a final check to ensure that the wheel spins freely. Operate the brake several times to ensure its correct operation and recheck all disturbed connections for security.

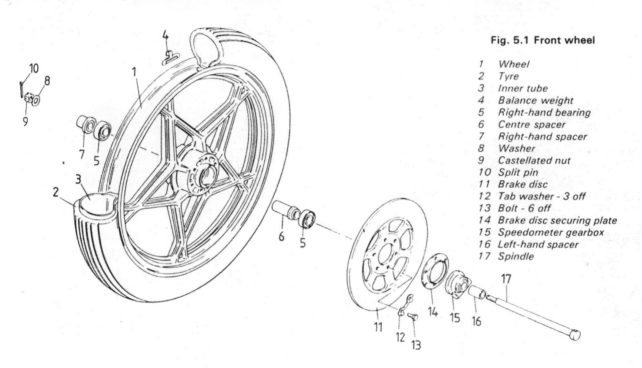

Fig. 5.1 Front wheel

1   Wheel
2   Tyre
3   Inner tube
4   Balance weight
5   Right-hand bearing
6   Centre spacer
7   Right-hand spacer
8   Washer
9   Castellated nut
10   Split pin
11   Brake disc
12   Tab washer - 3 off
13   Bolt - 6 off
14   Brake disc securing plate
15   Speedometer gearbox
16   Left-hand spacer
17   Spindle

3.5a The speedometer gearbox may be dismantled for inspection and lubrication

3.5b Note the location of the gearbox drive plate...

3.5c ...before fitting the retaining washer and circlip

3.5d The drive plate tangs must locate in the wheel hub grooves

3.5e The embossed arrow-mark on the gearbox casing must point upwards

3.6 Note the fitted position of the plastic dust cover on the wheel hub

3.7 Always lock the wheel spindle nut with a new split-pin

4 Front wheel bearings : removal, examination and fitting

1   Access to the front wheel bearings may be made after removal of the wheel from the forks. The speedometer gearbox together with the spacer should be detached from the left-hand side of the wheel hub whilst the plastic dust cover should be prised from its location on the right-hand side of the wheel hub to allow withdrawal of the spacer located beneath it. Note that this cover is not fitted to all the models covered in this Manual.
2   In order to avoid damage to the disc during bearing removal, it is advisable to remove the disc in accordance with the instructions given in Section 10 of this Chapter.
3   Position the wheel on a work surface with its hub supported by wooden blocks so that enough clearance is left beneath the wheel to drive the bearing out. Ensure the blocks are placed as close to the bearing as possible, to lessen the risk of distortion occurring to the hub casting whilst the bearings are being removed or fitted.
4   Place the end of a long-handled drift against the upper face of the lower bearing and tap the bearing downwards out of the wheel hub. The spacer located between the two bearings may be moved sideways slightly in order to allow the drift to be positioned against the face of the bearing. Move the drift

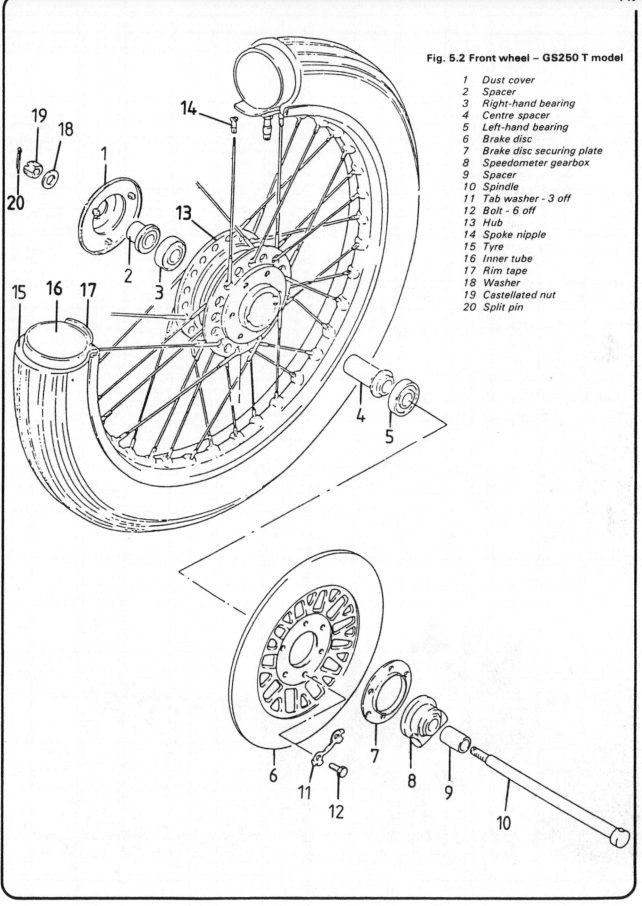

Fig. 5.2 Front wheel – GS250 T model

1 Dust cover
2 Spacer
3 Right-hand bearing
4 Centre spacer
5 Left-hand bearing
6 Brake disc
7 Brake disc securing plate
8 Speedometer gearbox
9 Spacer
10 Spindle
11 Tab washer - 3 off
12 Bolt - 6 off
13 Hub
14 Spoke nipple
15 Tyre
16 Inner tube
17 Rim tape
18 Washer
19 Castellated nut
20 Split pin

around the face of the bearing whilst drifting it out of position, so that the bearing leaves the hub squarely.

5 With the one bearing removed, the wheel may be lifted and the spacer withdrawn from the hub. Invert the wheel and remove the second bearing, using a similar procedure to that used for the first.

6 Remove all the old grease from the hub and bearings, giving the latter a final wash in petrol Check the bearings for signs of play or roughness when they are turned. If there is any doubt about the condition of a bearing, it should be renewed.

7 If the original bearings are to be refitted, then they should be repacked with the recommended grease before being fitted into the hub. New bearings must also be packed with the recommended grease. Ensure that the bearing recesses in the hub are clean and both bearing and recess mating surfaces lightly greased. The two bearings and central spacer may now be fitted. With the hub well supported by the wooden blocks, drift the first bearing into position using a length of metal tube, or a socket, of suitable diameter and a soft-faced mallet. Invert the wheel, insert the spacer and fit the second bearing using the same procedure as that given for the first. Take great care to ensure that the bearings enter the housings perfectly squarely otherwise the housing surface may be broached.

8 Refit the disc to the wheel hub and clean and lightly grease the spacers before reassembling the hub components and refitting the wheel between the fork legs. Take care to refit all the hub components in their original locations; if in doubt, refer to the figure accompanying this text.

5 Front disc brake assembly : examination and brake pad renewal

1 Check the front brake master cylinder, hoses and caliper units for signs of leakage. Pay particular attention to the condition of the hoses, which should be renewed without question if there are signs of cracking, splitting or other exterior damage. With the machine positioned on its centre stand on an area of flat ground and with the handlebars positioned so that the front wheel is pointing forward, check the hydraulic fluid level by referring to the upper and lower level lines on the exterior of the transparent reservoir body.

2 Replenish the reservoir after removing the cap on the brake fluid reservoir and lifting out the diaphragm plate. The cap is either a screw-on/off fitting, or is retained by four screws, depending on the model. The condition of the fluid can be checked at the same time. Checking the fluid level is one of the maintenance tasks which should **never be neglected**. If the fluid is below the lower level mark, brake fluid of the correct specification must be added. **Never** use engine oil or any fluid other than that recommended. Other fluids have unsatisfactory characteristics and will rapidly destroy the seals. The fluid level is unlikely to fall other than a small amount, unless leakage has occurred somewhere in the system. If a rapid change of level is noted, a careful check for leaks should be made before the machine is used again. It is also worth noting that Suzuki recommend that the brake hoses should be renewed every two years in the interests of safety.

3 The brake pads should be inspected for wear. Each has a red groove, which marks the wear limit of the friction material. When this limit is reached, both pads must be renewed, even if only one has reached the wear mark. In normal use both pads will wear at the same rate and therefore both must be renewed. To facilitate the checking of brake pad wear the caliper is provided with an inspection window, closed by a small cover. Prise the cover from position in order to inspect the pads.

4 If the brake action becomes spongy, or if any part of the hydraulic system is dismantled (such as when a hose has been renewed) it is necessary to bleed the system in order to remove all traces of air. Follow the procedure in Section 9 of this Chapter.

5 To gain access to the pads for renewal, detach the brake caliper from its support bracket by removing the two securing bolts and pull the caliper rearwards to expose both pads. Withdraw the pads together with the shim plate.

6 Fit the new pads together with the new shim plate and refit the caliper by reversing the removal procedure. The caliper piston should be pushed fully into the caliper so that there is sufficient clearance between the brake pads to allow the caliper to fit over the disc. Note the torque figure given in the Specifications Section of this Chapter before refitting and tightening the two caliper to support bracket securing bolts.

7 Ensure that during the pad removal and refitting operation, the brake lever is not operated at any time. This will operate and displace the caliper piston; reassembly is then considerably more difficult.

8 In the interests of safety, always check the function of the brakes; pump the brake lever several times to restore full braking power, before taking the machine on the road.

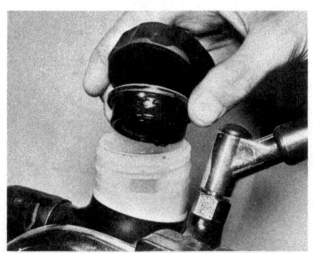

5.2 Remove the reservoir cap and diaphragm assembly

5.3 Inspect the brake pads for wear through the caliper window

5.5a Detach the brake caliper from its support bracket by removing the two securing bolts

5.5b Withdraw each brake pad from the support bracket

6 Front disc brake caliper : dismantling, examination, renovation and reassembly

1 Select a suitable receptacle into which to drain the brake fluid. Place the receptacle, together with some clean rag, beneath the caliper unit. Remove the bolt which retains the hose union to the caliper and allow the fluid to drain from the hose into the receptacle. Take great care not to allow hydraulic fluid to spill onto paintwork; it is a very effective paint stripper. Hydraulic fluid will also damage rubber and plastic components. If any small spillage occurs, wipe up the fluid immediately using the rag mentioned above.
2 Loosen the caliper to support bracket securing bolts. Remove the support bracket to fork leg securing bolts and lift the complete unit away from the machine. With the unit placed on a clean work surface, detach the caliper from its support bracket by removing the two already loosened securing bolts.
3 The caliper piston boot should now be prised out of position by using the flat of a small screwdriver whilst taking care not to scratch the surface of the caliper bore. The piston can be displaced by applying an air jet to the hydraulic fluid feed orifice of the caliper. Take great care not to use too high an air pressure when doing this and place a thick wad of clean rag over the end of the caliper bore; this will serve to both catch the piston and prevent any injury through hydraulic fluid being ejected from the caliper under pressure. It is still, however, advisable to wear some form of eye protection during this operation.
4 Clean the caliper components thoroughly, only in hydraulic brake fluid. **Never** use petrol or cleaning solvent for cleaning hydraulic brake parts otherwise the rubber components will be damaged. Discard all the rubber components as a matter of course. The replacement cost is relatively small and does not warrant re-use of components vital to safety. Check the piston and caliper cylinder bore for scoring, rusting or pitting. If any of these defects are evident it is unlikely that a good fluid seal can be maintained and for this reason the components should be renewed.
5 Renew the two spindles with their rubber boots from the brake caliper support bracket by pulling them from position. Clean the spindles and their retaining bores within the support bracket and examine the rubber boots for any signs of damage or deterioration. If necessary, renew any items that appear to be in a doubtful condition, before applying Suzuki brake grease to the length of each spindle and refitting them to the bracket. Note that the spindles differ insomuch as the lower of the two has a slightly stepped shank.

6 Assemble the caliper unit by reversing the dismantling sequence. When assembling, pay particular attention to the following points. Ensure that during assembly all the components are kept absolutely clean. Apply a generous amount of brake fluid to the inner surface of the cylinder and to the periphery of the piston, then assemble. Do not assemble the piston with it inclined or twisted. When installing the piston push it slowly into the cylinder while taking care not to damage the piston seal.
7 With both the caliper and support bracket serviced and reassembled, join the two components together by refitting the two retaining bolts, finger-tight. Refer to the figure accompanying this text and note the position of the shim, guide and spring plates. Refit the complete assembly to the fork leg and tighten both the fork to support bracket and the bracket to caliper unit securing bolts to the specified torque loading.
8 Before reconnecting the brake hose union to the caliper, check the condition of the two gasket washers located one either side of the union. Renew these washers if necessary and fit and tighten the union bolt to the recommended torque loading.
9 Bleed the brake after refilling the reservoir with new hydraulic brake fluid, then check for leakage while applying the brake lever. Push the machine forward and bring it to a halt by applying the brake. Do this several times to ensure that the brake is operating correctly before taking the machine for a test run. During the run, use the brakes as often as possible and on completion, recheck for signs of fluid loss.

7 Front disc brake master cylinder : removal, examination, renovation and refitting

1 The master cylinder and hydraulic reservoir take the form of a combined unit mounted on the right-hand side of the handlebars, to which the front brake lever and stop lamp switch are attached. The master cylinder is actuated by the front brake lever, and applies hydraulic pressure through the system to operate the front brake when the handlebar lever is manipulated. The master cylinder pressurises the hydraulic fluid in the brake pipe which, being incompressible, causes the piston to move in the caliper unit and apply the friction pads to the brake. If the master cylinder seals leak, hydraulic pressure will be lost and the braking action rendered much less effective.
2 Before the master cylinder can be removed, the system must be drained. Place a clean container below the caliper unit

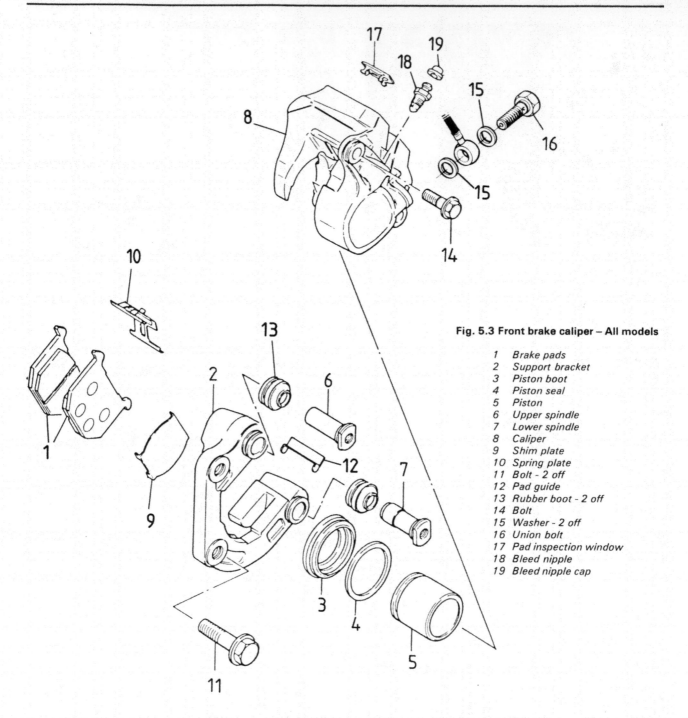

Fig. 5.3 Front brake caliper – All models

1   Brake pads
2   Support bracket
3   Piston boot
4   Piston seal
5   Piston
6   Upper spindle
7   Lower spindle
8   Caliper
9   Shim plate
10 Spring plate
11 Bolt - 2 off
12 Pad guide
13 Rubber boot - 2 off
14 Bolt
15 Washer - 2 off
16 Union bolt
17 Pad inspection window
18 Bleed nipple
19 Bleed nipple cap

and attach a plastic tube from the bleed screw on top of the caliper unit to the container. Open the bleed screw one complete turn and drain the system by operating the brake lever until the master cylinder reservoir is empty. Close the bleed screw and remove the pipe.

3   Place a pad of clean rag beneath the point where the brake hose adjoins the master cylinder. This simple precaution is essential to prevent brake fluid dripping onto, and therefore damaging, plastic and painted components located beneath the hose union once the securing bolt is removed. Detach the union bolt rubber cover and remove the bolt. Once any excess fluid has drained from the union connection, wrap the end of the hose in rag or polythene and attach it to a point on the handlebars.

4   Remove the stop lamp switch from the master cylinder unit casting and whilst keeping the unit upright, remove the two bolts that hold the handlebar clamp to the unit body. Lift the unit away from the machine and empty any surplus fluid contained within it into the container used during the draining procedure. Remove the brake lever from the unit body and withdraw the piston components located beneath it starting with the rubber boot. This boot should be carefully eased out of the unit body with the flat of a small screwdriver. Using a pair of straight-nose circlip pliers, remove the piston retaining circlip located beneath the boot and pull out the piston, piston cap and the spring.

5   Remove the reservoir from the unit body by unscrewing the four retaining screws. Separate the reservoir body from the cap,

the diaphragm plate and the diaphragm. Remove the O-ring from the recess in the unit body.

6   Place all the master cylinder component parts in a clean container and wash them thoroughly in new brake fluid. Lay the parts out on a sheet of clean rag or paper and examine each component as follows.

7   Inspect the unit body for signs of stress fracture around both the brake lever pivot and handlebar mounting points. Carry out a similar inspection around the hose union boss. Examine the cylinder bore for signs of scoring. If any of these faults are found then the unit body must be renewed.

8   Inspect the piston surface for scoring and renew it if necessary. It is advisable to discard all the rubber components of the piston assembly as a matter of course as the replacement cost is relatively small and does not warrant re-use of components vital to safety. The same advice applies to the O-ring between the unit and reservoir bodies. Finally, inspect the brake hose union bolt threads for signs of failure and renew the bolt if in the slightest doubt. Renew each of the gasket washers, located one either side of the hose union.

9   Check before reassembly that any traces of contamination within the reservoir body have been removed. Check also that the diaphragm is not perished or split before relocating it, its plate and the reservoir cap on the reservoir body. It must be noted at this point that any reassembly work must be undertaken in ultra-clean conditions. Particles of dirt will only serve to score the working parts of the cylinder and thereby cause early failure of the system.

10   When reassembling and refitting the master cylinder, follow the removal and dismantling procedures in the reverse order whilst paying particular attention to the following points. Make sure that the piston components are fitted the correct way round and in the correct order. Immerse all of these components in new brake fluid prior to assembly.

11   When refitting the unit to the handlebars, note that the upper clamp bolt must be tightened fully before the lower clamp bolt, so that there is no gap at the top of the clamp. A small clearance should be present at the bottom of the clamp; say 2 mm (0.08 in). Note also the torque figures given in the Specifications Section for both the clamp bolts and the hose union bolt.

12   Bleed the system after refilling the reservoir with new hydraulic brake fluid, then check for leakage while applying the brake lever. Push the machine forward and bring it to a halt by applying the brake. Do this several times to ensure that the brake is operating correctly before taking the machine for a test run. During the run, use the brakes as often as possible and on completion, recheck for signs of fluid loss.

13   The component parts of the master cylinder assembly (and the caliper assembly) may wear or deteriorate in function over a long period of use. It is however, generally difficult to foresee how long each component will work with proper efficiency and from a safety point of view it is best to change all the expendable parts every two years on a machine that has covered a normal mileage.

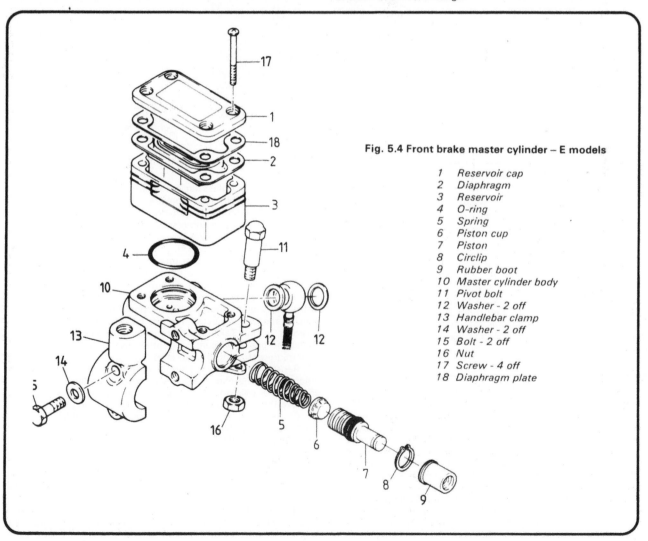

Fig. 5.4 Front brake master cylinder – E models

1   Reservoir cap
2   Diaphragm
3   Reservoir
4   O-ring
5   Spring
6   Piston cup
7   Piston
8   Circlip
9   Rubber boot
10   Master cylinder body
11   Pivot bolt
12   Washer - 2 off
13   Handlebar clamp
14   Washer - 2 off
15   Bolt - 2 off
16   Nut
17   Screw - 4 off
18   Diaphragm plate

Fig. 5.5 Front brake master cylinder – T models

1   Reservoir cap
2   Diaphragm
3   Reservoir
4   O-ring
5   Spring
6   Piston cup
7   Piston
8   Washer
9   Circlip
10 Rubber boot

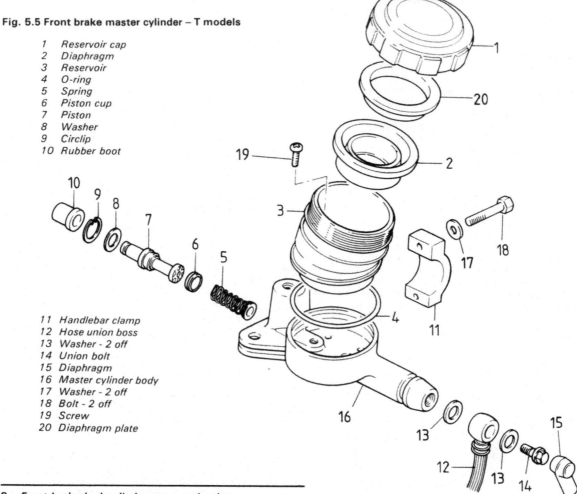

11 Handlebar clamp
12 Hose union boss
13 Washer - 2 off
14 Union bolt
15 Diaphragm
16 Master cylinder body
17 Washer - 2 off
18 Bolt - 2 off
19 Screw
20 Diaphragm plate

8   Front brake hydraulic hose : examination

1   An external brake hose is used to transmit hydraulic pressure to the caliper unit once the front brake lever is applied. This hose is of the flexible type, its lower half being fitted with an armoured surround.

2   When the brake assembly is being overhauled, or at any time during a routine maintenance or cleaning procedure, check the condition of the hose for signs of leakage, damage, deterioration or scuffing against any cycle components. The union connections at either end of the hose must also be in good condition, with no stripped threads or damaged sealing washers. Adhere strictly to the torque figure quoted when tightening these bolts as they are easily sheared if over-tightened.

9   Bleeding the front brake hydraulic system

1   As mentioned earlier, brake action is impaired or even rendered inoperative if air is introduced into the hydraulic system. This can occur if the seals leak, the reservoir is allowed to run dry or if the system is drained prior to the dismantling of any component part of the system. Even when the system is refilled with hydraulic fluid, air pockets will remain and because air will compress, the hydraulic action is lost.

2   Check the fluid content of the reservoir and fill almost to the top. Remember that hydraulic brake fluid is an excellent paint stripper, so beware of spillage, especially near the petrol tank. Do not omit to refit the reservoir cap to prevent dirt entering the

system.

3   Place a clean glass jar below the brake caliper unit and attach a clear plastic tube from the caliper bleed screw to the container. Place some clean hydraulic fluid in the container so that the pipe is always immersed below the surface of the fluid.

4   Pump the brake lever several times in rapid succession, finally holding the lever in the 'fully on' position. Loosen the bleed screw one quarter of a turn so that the brake fluid is seen to run down the tube into the container. This will cause the pressure within the system to be released thereby causing the brake lever to move from its 'fully on' position to touch the throttle twistgrip. Directly this happens, nip the bleed screw shut. As the fluid is ejected from the bleed screw the level in the reservoir will fall. Take care that the level does not drop too low whilst the operation continues, otherwise air will re-enter the system, necessitating a fresh start.

5   Repeat the above procedure until no further air bubbles emerge from the end of the plastic pipe. Hold the brake lever against the twistgrip and tighten the caliper bleed screw. Remove the plastic tube **after** the bleed screw is closed.

6   Check the brake action for sponginess, which usually denotes there is still air in the system. If the action is spongy, continue the bleeding operation in the same manner, until all traces of air are removed.

7   Bring the reservoir up to the correct level of fluid and refit the diaphragm, sealing gasket and cap. Check the entire system for leaks. Recheck the brake action and check the stop lamp for correct operation.

8   Note that fluid from the container placed below the brake caliper unit whilst the system is bled, should not be reused, as it will have become aerated and may have absorbed moisture.

10 Front brake disc : examination, removal and fitting

1   The brake disc can be checked for wear and for warpage whilst the front wheel is still in the machine. Using a micrometer, measure the thickness of the disc at the point of greatest wear. If the measurement is much less than the recommended service limit given in the Specifications section of this Chapter, the disc should be renewed. Check the warpage (run out) of the disc by setting up a suitable pointer close to the outer periphery of the disc and spinning the front wheel slowly. If the total warpage is more than 0.30 mm (0.012 in), the disc should be renewed. A warped disc, apart from reducing the braking efficiency, is likely to cause juddering during braking and will also cause the brake to bind when it is not in use.
2   The disc should also be checked for bad scoring on its contact area with the brake pads. If any of the above mentioned faults are found, then the disc should be removed from the wheel for renewal.
3   To detach the disc, first remove the wheel as described in Section 3 of this Chapter. The disc is retained by six bolts screwed into the hub, which are linked in pairs by tab washers. Bend down the ears of the tab washers and remove the bolts. The disc can then be eased off the hub boss.
4   Fit the disc by reversing the removal procedure. Ensure that the six bolts are tightened evenly and in a diagonal sequence to the recommended torque loading. Do not reuse the same tabs at the tab washer ends, renew the washers if necessary.

11 Rear wheel : examination, removal, renovation and refitting

1   Place the machine on its centre stand so that the rear wheel is raised clear of the ground. Check the condition of the wheel as described in Section 2 of this Chapter.
2   Before attempting wheel removal, it is first necessary to gain access to the ends of the wheel spindle. This can be achieved by two methods, the first being to detach both silencers, or in the case of machines fitted with a system where the silencers are welded to the pipes, the complete exhaust system. Alternatively, position a jack beneath the rear wheel and extend it just enough to take the weight of the wheel. Remove the cap nut and plain washer from the lower attachment point of each rear suspension unit and pull each unit outwards to release it from its mounting stud. Once the torque arm and brake operating rod are detached from the brake assembly, the wheel may be steadily jacked up so that the ends of the wheel spindle appear above the tops of the silencers.
3   Detach the torque arm by removing the split-pin which passes through the end of the bolt retaining the arm to the brake backplate. Remove the retaining nut and spring washer and withdraw the bolt to release the arm. Detach the brake rod by removing the adjuster nut and depressing the brake pedal so that the rod leaves the trunnion in the brake operating arm. Refit the nut to secure the rod spring.
4   If a final drive chain which incorporates a spring link is fitted, place a length of clean rag or paper beneath the machine, in line with the chain run, to prevent the chain coming into direct contact with the ground once it is released from the wheel sprocket. Turn the rear wheel until the joining link in the drive chain is accessible, then, using pointed-nose pliers, prise off the spring link, and separate the chain. Loosely reassemble the joining link and store it safely to avoid loss. Lift the chain off the wheel sprocket and lay the ends on the length of rag or paper.
5   Remove the wheel spindle nut after displacing the split-pin. Withdraw the wheel spindle from the left-hand side and catch the right-hand spacer as it falls free. Where an endless chain is fitted, the sprocket, still meshed with the chain, can be pulled out, away from the hub to free the wheel. The cush drive hub will come away with the sprocket, leaving the cush drive rubber inserts in position. Tilt the wheel slightly and remove it from between the fork arms.
6   Once the faults found in the wheel are rectified, the wheel may be refitted by reversing the removal procedure. Ensure that the wheel spacers are correctly positioned before inserting the wheel spindle; if in doubt as to the position of any of the wheel components, refer to the figure accompanying this text.
7   Fit the spindle retaining nut finger-tight. Where a chain incorporating a spring link is fitted, move the wheel fully forward, refit the chain over the rear wheel sprocket and connect the ends by inserting the joining link. Note that the link securing clip must be fitted with the closed end facing the direction of the chain.
8   Feed the rear brake operating rod through the brake lever trunnion and fit the adjusting nut onto the end of the rod to retain it in position. Ensure that the spring is correctly positioned between the butt flange on the operating rod and the forward facing face of the lever trunnion.
9   Relocate the torque arm retaining bolt in the brake backplate and refit the arm over its end. Retain the arm in position by fitting and tightening the retaining nut to the recommended torque and locking it in position by fitting a new

9.3 Attach a bleed tube to the caliper bleed screw

10.4 The brake disc securing bolts must be locked in position with the tab washer ends as shown

split-pin. The torque arm must be properly attached to the brake backplate. Failure to ensure this will mean that on the first application of the rear brake, the wheel will lock, with disastrous consequences.

10 If the rear wheel has been raised to accommodate removal of the spindle, the rear suspension units should be refitted to their swinging arm attachment points and the cap nuts

tightened to the correct torque loading before adjusting the final drive chain tension as described in Section 17 of this Chapter. 11 Tighten the wheel spindle retaining nut to the torque figure given in the Specifications Section of this Chapter and fit the new split-pin to retain it in position. Finally, adjust the rear brake pedal free play as described in Section 14 of this Chapter.

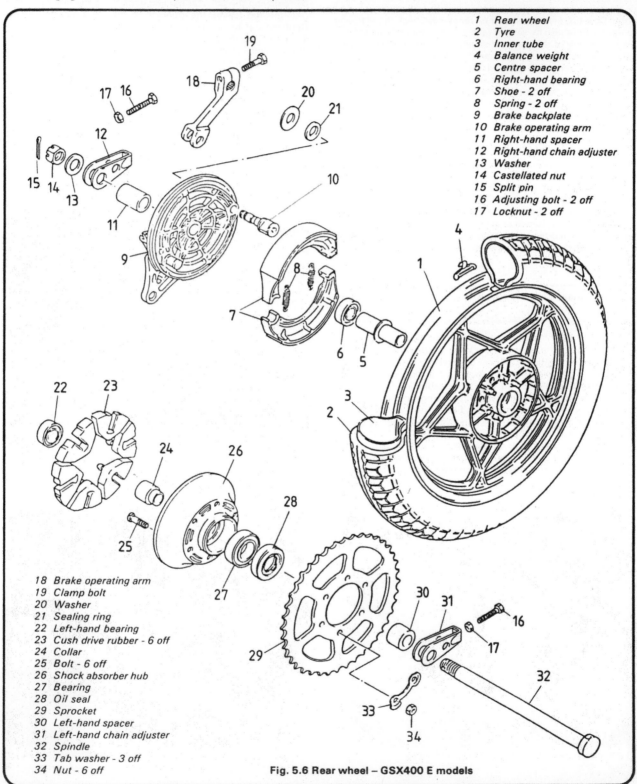

1   Rear wheel
2   Tyre
3   Inner tube
4   Balance weight
5   Centre spacer
6   Right-hand bearing
7   Shoe - 2 off
8   Spring - 2 off
9   Brake backplate
10 Brake operating arm
11 Right-hand spacer
12 Right-hand chain adjuster
13 Washer
14 Castellated nut
15 Split pin
16 Adjusting bolt - 2 off
17 Locknut - 2 off

18 Brake operating arm
19 Clamp bolt
20 Washer
21 Sealing ring
22 Left-hand bearing
23 Cush drive rubber - 6 off
24 Collar
25 Bolt - 6 off
26 Shock absorber hub
27 Bearing
28 Oil seal
29 Sprocket
30 Left-hand spacer
31 Left-hand chain adjuster
32 Spindle
33 Tab washer - 3 off
34 Nut - 6 off

Fig. 5.6 Rear wheel – GSX400 E models

11.6a Lift the rear wheel assembly into the swinging arm fork

11.6b Check the position of the rear wheel spacers...

11.7 ...before fitting the wheel spindle retaining nut, finger-tight

11.8 Reconnect the brake operating rod to the brake lever

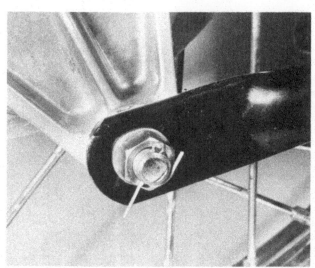

11.9 Lock the torque arm retaining nut in position by fitting a new split pin

11.11 Fit a new split-pin through the tightened wheel spindle retaining nut to lock it in position

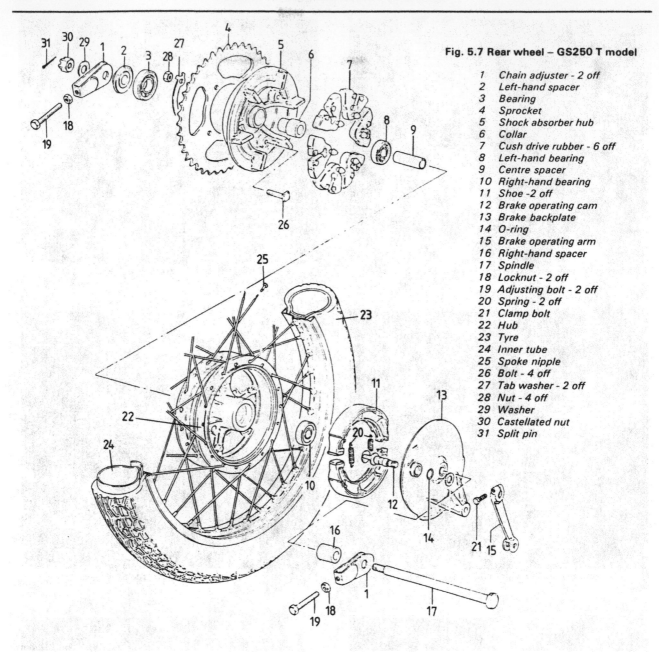

Fig. 5.7 Rear wheel – GS250 T model

1   Chain adjuster - 2 off
2   Left-hand spacer
3   Bearing
4   Sprocket
5   Shock absorber hub
6   Collar
7   Cush drive rubber - 6 off
8   Left-hand bearing
9   Centre spacer
10 Right-hand bearing
11 Shoe -2 off
12 Brake operating cam
13 Brake backplate
14 O-ring
15 Brake operating arm
16 Right-hand spacer
17 Spindle
18 Locknut - 2 off
19 Adjusting bolt - 2 off
20 Spring - 2 off
21 Clamp bolt
22 Hub
23 Tyre
24 Inner tube
25 Spoke nipple
26 Bolt - 4 off
27 Tab washer - 2 off
28 Nut - 4 off
29 Washer
30 Castellated nut
31 Split pin

12 Rear wheel bearings : removal, examination and fitting

1   The rear wheel assembly has three journal ball bearings. One bearing lies each side of the wheel hub and the third bearing is fitted in the cush drive assembly to which is attached the sprocket.
2   The cush drive sprocket unit can be removed from engagement with the final drive chain after the rear wheel spindle has been withdrawn and the wheel lifted over to the right-hand side of the swinging arm fork.
3   The two bearings contained within the rear wheel hub may be drifted out of position by using a procedure similar to that given for the front wheel bearings. It will of course, be necessary to detach the brake backplate assembly from the wheel hub before commencing this operation.
4   Before the cush drive bearing is drifted out of position, it is necessary to remove the stepped hollow spindle from its centre.

Suzuki recommend that the sprocket is removed from the cush drive drum but in practice it was found that this was not necessary provided that care was taken in supporting the assembly during bearing removal. Note that where an oil seal is fitted, this will be drifted out along with the bearing. It is good practice to renew this seal along with any defective bearings.
5   The procedure for examination and fitting of the bearings and the hub component parts should be adopted from the relevant paragraphs in Section 4 of this Chapter whilst making any necessary reference to the figure accompanying this text. Note that the two bearings fitted in the wheel hub differ in that the right-hand item is of the rubber sealed type whereas the left-hand item is of the metal plate sealed type. Do not omit to apply a light smear of grease to the lip of the cush hub bearing oil seal and apply a solution of soapy water to the rubber pads within the wheel hub so as to make easier insertion of the cush drum vanes into the pads.

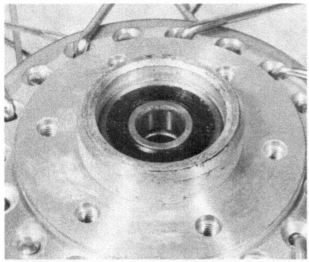
12.5a With the right-hand bearing fitted in position...

12.5b ...grease and insert the bearing spacer...

12.5c ...followed by the left-hand bearing

12.5d Drift the left-hand bearing into position...

12.5e ...until it is located correctly in the wheel hub

12.5f With the bearing inserted into the cush drive hub...

12.5g ...lightly grease and fit the stepped hollow spindle

12.5h Insertion of the cush drive vanes will be made easier by applying a soap solution to the rubber pads

13 Rear brake assembly : dismantling, examination, renovation and reassembly

1   The rear brake assembly complete with the brake backplate can be withdrawn from the rear wheel hub after removal of the wheel from the swinging arm fork. With the wheel laid on a work surface, brake backplate uppermost, the brake backplate may be lifted away from the hub. It will come away quite easily, with the brake shoe assembly attached to its back.

2   Examine the condition of the brake linings. If they are worn beyond the specified limit or uneven, then the brake shoes should be renewed. The linings are bonded on and cannot be supplied separately.

3   If oil or grease from the wheel bearings has badly contaminated the linings, the brake shoes should be renewed. There is no satisfactory way of degreasing the lining material. Any surface dirt on the linings can be removed with a stiff-bristled brush. High spots on the linings should be carefully eased down with emery cloth.

4   Examine the drum surface for signs of scoring, wear beyond the service limit or oil contamination. All of these conditions will impair braking efficiency. Remove all traces of dust, preferably using a brass wire brush, taking care not to inhale any of it, as it is of an asbestos nature, and consequently harmful. Remove oil or grease deposits, using a petrol soaked rag.

5   If deep scoring is evident, due to the linings having worn through to the shoe at some time, the drum must be skimmed on a lathe, or renewed. Whilst there are firms who will undertake to skim a drum whilst it is fitted to the wheel, it should be borne in mind that excessive skimming will change the radius of the drum in relation to the brake shoes, therefore reducing the friction area until extensive bedding in has taken place. Also full adjustment of the shoes may not be possible. If in doubt about this point, the advice of one of the specialist engineering firms who undertake this work should be sought.

6   Note that it is a false economy to try to cut corners with brake components; the whole safety of both machine and rider being dependent on their good condition.

7   Removal of the brake shoes is accomplished by folding the shoes together so that they form a 'V'. With the spring tension relaxed, both shoes and springs may be removed from the brake backplate as an assembly. Detach the springs from the shoes and carefully inspect them for any signs of fatigue or failure. If in doubt, compare them with a new set of springs.

8   Before fitting the brake shoes, check that the brake operating cam is working smoothly and it is not binding in its pivot. The cam can be removed by withdrawing the retaining bolt on the operating arm and pulling the arm off the shaft. Before removing the arm, it is advisable to mark its position in

relation to the shaft, so that it can be relocated correctly.

9   Remove any deposits of hardened grease or corrosion from the bearing surface of the brake cam shaft by rubbing it lightly with a strip of fine emery paper or applying solvent with a rag. Lightly grease the length of the shaft and the face of the operating cam prior to reassembly. Clean and grease the pivot stub set in the backplate.

10 Check the condition of the O-ring which prevents the escape of grease from the end of the cam shaft. If it is in any way damaged or perished then it must be renewed before the shaft is relocated in the backplate. Align and fit the operating arm with its plain washer. Note that the arm clamp bolt and nut should be torque loaded to the figure given in the Specifications Section of this Chapter.

11 Before refitting existing shoes, roughen the lining surface sufficiently to break the glaze which will have formed in use. Glasspaper or emery cloth is ideal for this purpose but take care not to inhale any of the asbestos dust that may come from the lining surface.

12 Fitting the brake shoes and springs to the brake backplate is a reversal of the removal procedure. Some patience will be needed to align the assembly with the pivot and operating cam whilst still retaining the springs in position; once they are correctly aligned though, they can be pushed back into position by pressing downwards in order to snap them into position. Do not use excessive force, or there is risk of distorting the brake shoes permanently.

13.9 Lightly grease the brake cam shaft before inserting it in position

13.10a Fit a serviceable O-ring over the shaft...

13.10b ...followed by the plain washer

13.10c Lock the operating arm in position by torque loading the clamp bolt

13.12 Align the brake shoes on the pivot and cam before pressing them into position

14 Adjusting the rear brake

1   Adjustment of the rear brake is correct when there is 20 – 30 mm (0.75 – 1.0 in) of up and down movement measured at the forward point of the brake pedal, between the fully off and on positions.

2   Suzuki recommend that the brake pedal is positioned so that the top surface of its foot piece is 10 mm (0.35 in) below the top edge of the footrest rubber; although this is only a recommendation and the initial position of the pedal is really dictated by rider preference. To position the pedal thus, loosen the locknut which retains the adjuster bolt at the rear end of the pedal and rotate the bolt in or out until the pedal is correctly positioned. Hold the adjuster bolt in position and retighten the locknut.

3   With the pedal set in relation to the footrest, check and if necessary, adjust its range of movement to comply with the information given in paragraph 1. This may be achieved by rotating the adjuster nut on the brake drum lever end of the brake operating rod.

4   Note that an indication of brake lining wear is provided by an indicator line cast into the brake backplate. If, when the brake is correctly adjusted and applied fully, the line on the end of the brake cam spindle is seen to align with a point outside the indicator line arc cast in the backplate, then the lining on the brake shoes can be assumed to have worn beyond limits, and should be renewed at the earliest opportunity.

5   On completion of brake adjustment, check that the stop lamp switch operates the stop lamp as soon as the brake pedal is depressed. If necessary, adjust the height setting of the switch in accordance with the instructions given in Chapter 6.

15 Cush drive : examination and renovation

1   The cush drive assembly is contained in the left-hand side of the wheel hub. It takes the form of six triangular rubber pads incorporating slots, that fit within the vanes of the hub. A heavily ribbed hub bolted to the rear sprocket engages with the slots to form a shock absorber which permits the sprocket to move within certain limits. This absorbs any surge or roughness in the transmission. The rubbers should be renewed when movement of the sprocket indicates bad compaction of the rubbers or if they commence to break up.

2   It should be noted that it will be a very difficult process to insert the vanes of the hub into new rubbers unless the rubbers are first lubricated with a solution of soapy water.

14.2 Alter the brake pedal position by rotating the adjuster bolt

14.4 An indication of brake lining wear is provided

15.1 Each cush drive rubber may be removed for renewal

16 Rear wheel sprocket : examination and renewal

1   The rear wheel sprocket is secured to the cush drive hub by four (250 models) or six (400 models) bolts whose nuts are locked in pairs by tab washers To remove the sprocket, bend back the locking tab at each end of the washers and undo the nuts.

2   The sprocket need only be renewed if the teeth are hooked, chipped, broken or badly worn. It is considered bad practice to renew one sprocket on its own; both drive sprockets should be renewed as a pair, preferably with a new final drive chain. If this recommendation is not observed, rapid wear resulting from the running of old and new parts together will necessitate even earlier replacement on the next occasion.

3   Fitting of the sprocket to the cush drive hub is simply a reversal of the removal procedure. Do not rebend the locking tab previously used; if necessary, renew the tab washers. Tighten the sprocket retaining nuts to the torque figure given in the Specifications Section of this Chapter before locking them in position.

17 Final drive chain : examination, adjustment and lubrication

1   As the final drive chain is fully exposed on all models it requires lubrication and adjustment at regular intervals. To adjust the chain place the machine on its centre stand and take out the split-pin from the rear wheel spindle and slacken the spindle nut. Slacken also the nuts securing the brake torque arm. Undo the locknut on the chain adjusters and turn the adjuster bolts inwards to tighten the chain. Marks on the adjusters must be in line with identical marks on the frame fork to align the rear wheel correctly. A final check can be made by laying a straight wooden plank alongside the wheels, each side in turn. Chain tension is correct if there is 20 – 30 mm (0.8 – 1.2 in) of slack in the centre of the lower chain run.

2   Note that a chain that is run exceptionally slack may strike the gear indicator switch which is retained on the gearbox wall. If the switch body breaks, lubricant loss from the gearbox may endanger the continued good health of the gearbox components. Do not run the chain overtight to compensate for uneven wear. A tight chain will place excessive stresses on the gearbox and rear wheel bearings leading to their early failure. It will also absorb a surprising amount of power.

3   After a period of running, the chain will require lubrication. Lack of oil will accelerate the rate of wear of both chain and sprockets and will lead to harsh transmission. The application of engine oil will act as a temporary expedient, but it is preferable to use one of the proprietary graphited greases contained within an aerosol can. This type of lubricant is thrown off the chain less easily than engine oil. Ideally the chain should be removed at regular intervals, and immersed in a molten lubrication such as Linklyfe or Chainguard after it has been cleaned in a paraffin bath. These latter lubricants achieve better penetration of the chain links and rollers and are less likely to be thrown off when the chain is in motion. Because the chain fitted as original equipment is of the endless type, that is it does not contain a spring master link, it should be appreciated that chain removal is a fairly complicated procedure which necessitates detachment of the swinging arm asssembly if the chain is to be removed without it being parted by use of a chain splitting tool. Because of this, it is obviously not desirable to keep removing the chain for lubrication, or to remove the chain so that it may be checked for wear. To carry out a check on the amount of chain wear with the chain fitted to the machine, wash the chain down with a petrol-soaked rag and stretch the chain to its full length by turning the adjuster bolts inwards. Select a length of chain in the middle of the chain run and count out 21 pins, that is, a 20 pitch length. Measure the distance between the 1st and the 21st pin. If this distance exceeds the service limit of 324.3 mm (12.77 in), the chain must be renewed. Note that this check

Tyre changing sequence - tubed tyres

 Deflate tyre. After pushing tyre beads away from rim flanges push tyre bead into well of rim at point opposite valve. Insert tyre lever adjacent to valve and work bead over edge of rim.

 Use two levers to work bead over edge of rim. Note use of rim protectors

Remove inner tube from tyre

When first bead is clear, remove tyre as shown

When fitting, partially inflate inner tube and insert in tyre

 Work first bead over rim and feed valve through hole in rim. Partially screw on retaining nut to hold valve in place.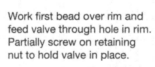

Check that inner tube is positioned correctly and work second bead over rim using tyre levers. Start at a point opposite valve.

 Work final area of bead over rim whilst pushing valve inwards to ensure that inner tube is not trapped

should ALWAYS be made after the chain has been washed out, but before any lubricant is applied, otherwise the lubricant may take up some of the play.

4   If it is found that the chain fitted as original equipment is stretched beyond limits the chain must be renewed, if necessary, with replacement sprockets. The original chain is of the endless type without a joining spring link and to allow removal of the chain and fitting of the new item the swinging arm must be detached. Refer to Chapter 4 for details. Some thought might be given to fitting a chain with a spring link. Although this introduces a weak point in the chain it does allow easy chain removal for periodic cleaning and efficient lubrication which will increase the chain's service life. Subsequent chain renewal is made easier too. If this course of action is chosen the old, endless type, chain can be removed by displacing a link pin using a chain-breaking tool. This will overcome the need for swinging arm removal.

5   When fitting a chain which incorporates a spring link, make sure that the spring link is seated correctly, with the closed end facing the direction of travel.

6   Note that replacement chains are now available in standard metric sizes from Renold Limited, the British chain manufacturer. When ordering a new chain, always quote the size, the number of chain links and the type of machine to which the chain is to be fitted.

7   Remember, on completion of examination and if necessary, adjustment or replacement of the chain, to ensure that both adjuster bolts are locked in position by tightening their locknuts. The wheel spindle retaining nut should be tightened to the torque loading specified at the beginning of this Chapter and locked in position by fitting a new split-pin. Finally, spin the rear wheel to ensure that it rotates freely, adjust the rear brake operating mechanism if necessary and retighten the torque arm retaining nuts.

18 Tyres : removal and fitting

1   At some time or other the need will arise to remove and replace the tyres, either as a result of a puncture or because replacements are necessary to offset wear. To the inexperienced, tyre changing represents a formidable task, yet if a few simple rules are observed and the technique learned, the whole operation is surprisingly simple.

2   To remove the tyre from either wheel, first detach the wheel from the machine. Deflate the tyre by removing the valve insert and when it is fully deflated, push the bead from the tyre away from the wheel rim on both sides so that the bead enters the centre well of the rim. Remove the locking cap and push the tyre valve into the tyre itself.

3   Insert a tyre lever close to the valve and lever the edge of the tyre over the outside of the wheel rim. Very little force should be necessary; if resistance is encountered it is probably due to the fact the tyre beads have not entered the well of the wheel rim all the way round the tyre. Note that where the machine is fitted with cast alloy wheels, the risk of damage to the wheel rims can be minimised by the use of proprietary plastic rim protectors placed over the rim flange at the point where the tyre levers are inserted. Suitable rim protectors may be fabricated very easily from short lengths (4 – 6 inches) of thick-walled nylon petrol pipe which have been split down one side using a sharp knife. The use of rim protectors should be adopted whenever levers are used and, therefore, when the risk of damage is likely.

4   Once the tyre has been edged over the wheel rim, it is easy to work around the wheel rim so that the tyre is completely free to one side. At this stage, the inner tube can be removed.

5   Working from the other side of the wheel, ease the other edge of the tyre over the outside of the wheel rim that is furthest away. Continue to work around the rim until the tyre is free completely from the rim.

6   If a puncture has necessitated the removal of the tyre, reinflate the inner tube and immerse it in a bowl of water to trace the source of the leak. Mark its position and deflate the tube. Dry the tube and clean the area around the puncture with a petrol soaked rag. When the surface has dried, apply rubber solution and allow this to dry before removing the backing from the patch and applying the patch to the surface.

7   It is best to use a patch of self-vulcanising type, which will form a very permanent repair. Note that it may be necessary to remove a protective covering from the top surface of the patch, after it has sealed into position. Inner tubes made from synthetic rubber may require a special type of patch and adhesive, if a satisfactory bond is to be achieved.

8   Before refitting the tyre, check the inside to make sure that the object which caused the puncture is not trapped. Check the outside of the tyre, particularly the tread area, to make sure nothing is trapped that may cause a further puncture.

9   If the inner tube has been patched on a number of past occasions, or if there is a tear or large hole, it is preferable to discard it and fit a new one. Sudden deflation may cause an accident, particularly if it occurs with the front wheel.

10 To fit the tyre, inflate the inner tube sufficiently for it to assume a circular shape but only just. Then push it into the tyre so that it is enclosed completely. Lay the tyre on the wheel at an angle and insert the valve through the rim tape and the hole in the wheel rim. Attach the locking cap on the first few threads, sufficient to hold the valve captive in its correct location.

11 Starting at the point furthest from the valve, push the tyre bead over the edge of the wheel rim until it is located in the central well. Continue to work around the tyre in this fashion until the whole of one side of the tyre is on the rim. It may be necessary to use a tyre lever during the final stages.

12 Make sure that there is no pull on the tyre valve and again commencing with the area furthest from the valve, ease the other bead of the tyre over the edge of the rim. Finish with the area close to the valve, pushing the valve up into the tyre until the locking cap touches the rim. This will ensure the inner tube is not trapped when the last section of the bead is edged over the rim with a tyre lever.

13 Check that the inner tube is not trapped at any point. Reinflate the inner tube, and check that the tyre is seating correctly around the wheel rim. There should be a thin rib moulded around the wall of the tyre on both sides, which should be equidistant from the wheel rim at all points. If the tyre is unevenly located on the rim, try bouncing the wheel when the tyre is at the recommended pressure. It is probable that one of the beads has not pulled clear of the centre well.

14 Always run the tyres at the recommended pressures and never under or over-inflate. The correct pressures are given in the Specifications Section of this Chapter.

15 Tyre replacement is aided by dusting the side walls particularly in the vicinity of the beads, with a liberal coating of french chalk. Washing-up liquid can also be used to good effect, but this has the disadvantage of causing the inner surfaces of the wheel rim to corrode. Do not be over generous in the application of lubricant or tyre creep may occur.

16 On machines equipped with wire-spoked wheels, never fit the inner tube and tyre without the rim tape in positon. If this precaution is overlooked there is good chance of the ends of the spoke nipples chafing the inner tube and causing a crop of punctures.

17 Never fit a tyre that has a damaged tread or side walls. Apart from the legal aspects, there is a very great risk of a blowout, which can have serious consequences on any two-wheel vehicle.

18 Tyre valves rarely give trouble, but it is always advisable to check whether the valve itself is leaking before removing the tyre. Do not forget to fit the dust cap which forms an effective second seal.

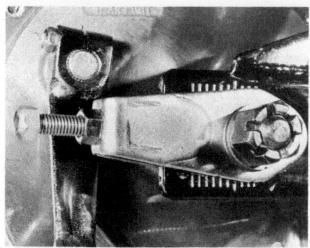

17.1 The marks in each chain adjuster must be in line with the same marks on each fork end

17.8 Retain each adjuster bolt in position by tightening its locknut

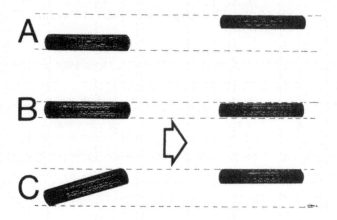

Fig. 5.8. Method of checking wheel alignment

A & C - Incorrect   B - Correct

19 Tyre valve dust caps

1   Tyre valve dust caps are often left off when a tyre has been replaced, despite the fact that they serve an important two-fold function. Firstly they prevent dirt or other foreign matter from entering the valve and causing the valve to stick open when the tyre pump is next applied. Secondly, they form an effective second seal so that in the event of the tyre valve sticking, air will not be lost.
2   Note that when a dust cap is fitted for the first time, the wheel may have to be rebalanced.

20 Wheel balancing

1   It is customary on all high performance machines to balance the front wheel complete with tyre and tube. The out of balance forces which exist are eliminated and the handling of the machine is improved in consequence. A wheel which is badly out of balance produces through the steering a most unpleasant hammering effect at high speeds.
2   Some tyres have a balance mark on the sidewall, usually in the form of a coloured dot. This mark must be in line with the tyre valve, when the tyre is fitted to the inner tube. Even then the wheel may require the addition of balance weights, to offset the weight of the tyre valve itself.
3   If the front wheel is raised clear of the ground and is spun, it will come to rest with the tyre valve or the heaviest part downward and will always settle in the same position. Balance weights must be added to a point diametrically opposite this heavy spot until the wheel will come to rest in ANY position after it is spun.
4   For machines fitted with the conventional type of wire-spoked wheels, balance weights which clip around the wheel spokes are normally available in 20 or 30 gram sizes. If they are not available, wire solder, wrapped around the spokes close to the spoke nipples, forms a good substitute.
5   For machines fitted with cast alloy wheels, Suzuki supply two weights of wheel balancing weight. These are both designed to clip to the wheel rim, the relevant Suzuki part numbers being 55411-47001 and 55412-47001.
6   Although the rear wheel is more tolerant to out-of-balance forces than is the front wheel, ideally this too should be balanced if a new tyre is fitted. Because of the drag of the final drive chain, it should be detached from the rear sprocket. If a chain of the endless type is fitted the rear wheel must be removed from the machine and placed on a suitable free running spindle. Balancing can then be carried out as for the front wheel.

21 Fault diagnosis : wheels, brakes and tyres

Symptom	Cause	Remedy
Handlebars oscillate at low speed	Buckle or flat in wheel rim, most probably front wheel	Check rim for damage by spinning wheel. Renew wheel if not true.
	Tyre pressure incorrect	Check, and if necessary adjust.
	Tyre not straight on rim	Check tyre fitting. If necessary, deflate tyre and reposition.
	Worn wheel or steering head bearings	Check and renew or adjust.
Machine tends to weave	Tyre pressures incorrect	Check, and if necessary adjust. If sudden, check for puncture.
	Suspension worn or damaged	Check action of front forks and rear suspension units. Check swinging arm for wear.
Machine lacks power and accelerates poorly	Front brake binding	Hot disc or caliper indicates binding. Overhaul caliper and master cylinder, fit new pads if required, check disc for scoring or warpage.
	Rear brake binding	Hot drum indicates binding. Check adjustment. Check drum for distortion.
Brake grab or judder when applied gently	Front: Pads badly worn or scored. Wrong type of pad fitted	Renew pads and check disc and caliper.
	Warped disc.	Renew.
	Rear: Ends of brake shoes not chamfered	Chamfer with a file.
	Elliptical brake drum	Lightly skim with a lathe by a specialist.
Brake squeal	Front: Glazed pads. Pads worn to backing metal	Sand pad surface to remove glaze then use brake gently for about 100 miles to permit bedding in. If worn to backing check that disc is not damaged and renew as necessary.
	Caliper and pads polluted with brake dust or foreign matter	Dismantle and clean. Overhaul caliper where necessary.
	Rear: Glazed brake shoes Shoes worn to backing metal Shoes and drum polluted as above	Remove glaze, inspect and clean.
Excessive front brake lever travel	Air in system	Find cause of air's presence. If due to leak, rectify, then bleed brake.
	Very badly worn pads	Renew, and overhaul system where required.
	Badly polluted caliper	Dismantle and clean.
Front brake lever feels spongy	Air in system	See above.
	Pads glazed	See above.
	Caliper jamming	Dismantle and overhaul.
Rear brake pull-off spongy	Brake cam binding in housing	Free and grease.
	Weak brake shoe springs	Renew if springs have not become displaced.
Harsh transmission	Worn or badly adjusted final drive chain	Adjust or renew.
	Hooked or badly worn sprockets	Renew as a pair.
	Loose rear sprocket	Check bolts.
	Worn damper rubbers	Renew rubber inserts.

Chapter 6 Electrical system

Refer to Chapter 7 for information relating to GS450 and GSX250/400 EZ models

Contents

Specifications

Battery

Capacity	12V, 12Ah
Electrolyte specific gravity	1.280 at 20°C (68°F)
Earth	Negative
Type	YB10L-A2

Alternator

No load voltage	75 volts or more at 5000 rpm
Regulated voltage	14 - 15 volts at 5000 rpm

Starter motor

Brush length service limit	9.0 mm (0.4 in)
Commutator undercut service limit	0.2 mm (0.008 in)

Starter relay

Resistance	3 - 4 ohms

Main fuse .... 15 amp

Bulbs

Headlamp:	
GSX250 and 400 models	12V, 35/35W
GS250 US models	12V, 45/40W
GS250 UK models	12V, 45/45W
Tail/stop lamp:	
GS250 US models	12V, 8/23W
All other models	12V, 5/21W

Directional indicators:	
GS250 US models	12V, 23W
All other models	12V, 21W
Pilot lamp	12V, 3.4W
Speedometer light	12V, 3.4W
Tachometer light	12V, 3.4W
High beam indicator light	12V, 3.4W
Indicator warning light	12V, 3.4W
Neutral indicator light	12V, 3.4W
Oil pressure indicator light	12V, 3.4W
Gear position indicator light:	
GSX400 models only	12V, 1.1W

1 General description

All of the Suzuki 250 and 400 models covered in this Manual incorporate a 12 volt electrical system which is powered by an alternator mounted on the extreme left-hand end of the crankshaft. This alternator, which as its name implies, produces alternating current, is of the 3-phase type, having a permanent magnet rotor and an eighteen-coil stator. During daylight running when no lights are in use, only two of the three output phases are utilised. Directly the lighting switch is operated to turn the lights on, the third phase is switched into circuit in order to provide the additional current demanded by the lighting circuit.

The ac current produced by the alternator is converted into a dc supply for the battery by means of a rectifier unit. This unit is contained within the same sealed casing as a regulator unit, which in turn serves to control the voltage output in order to prevent overcharging of the battery.

2 Testing the electrical system: general information

1 The electrical system incorporated in the Suzuki model types covered in this Manual lends itself easily to fairly comprehensive testing of the component parts. A certain amount of preliminary dismantling will be necessary to gain access to the components to be tested. Normally, removal of the seat and side panels will be required, with the possible addition of the fuel tank and headlamp unit to expose the remaining components.

2 Simple continuity checks may be made using a dry battery and bulb arrangement, but for most of the tests in this Chapter a pocket multimeter can be considered essential. Many owners will already possess one of these devices, but if necessary they can be obtained from electrical specialists, mail order companies or can be purchased from a Suzuki service agent as a 'pocket tester', part number 09900 - 25002.

3 Care must be taken when performing any electrical test, because some of the electronic assemblies can be destroyed if they are connected incorrectly or inadvertently shorted to earth. Instructions regarding meter probe connections are given for each test, and these should be read carefully to preclude any accidental damage during the test. Note that separate amp, volt and ohm meters may be used in place of the multimeter if necessary, noting that the appropriate test ranges will be required.

4 Where test equipment is not available, or the owner feels unsure of the procedure described, it is recommended that professional assistance is sought. Do not forget that a simple error can destroy a component such as the regulator/rectifier, resulting in expensive replacements being necessary.

3 Charging system: checking the output

1 Remove the right-hand sidepanel in order to gain acccess to the battery terminals. Set a multimeter to the 0–20 volts dc

scale and connect the red positive (+) probe to the positive battery terminal and the black negative (–) probe to the negative battery terminal.

2 Start the engine and raise the engine speed to 5000 rpm, or slightly more. Switch the headlamp on, to main beam, and note the meter reading. A reading of 14 - 15 volts dc will indicate that the system is functioning correctly.

3 If the voltage reading is significantly lower than that stated above, carry out the alternator output check described in the following Section. If this check is satisfactory, then it may be assumed that the regulator/rectifier unit is defective and needs to be renewed. Before committing oneself to the expense of purchasing a replacment unit, it is well worth carrying out continuity checks on all of the relevant electrical leads and physically inspecting any connections for faults. Confirmation of a defective unit may be proved by carrying out the checks stated in Section 6 of this Chapter. If the alternator output check is unsatisfactory, carry out the test described in Section 5 of this Chapter to decide whether the alternator stator or rotor is at fault.

4 On no account should the engine be run with the battery leads disconnected. This is because the resultant open voltage can destroy the regulator/rectifier unit diodes.

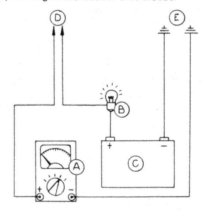

Fig. 6.1 Continuity test circuits

A Multimeter	D Test probes
B Bulb	E Earth (ground)
C Battery	

4 Alternator: checking the output

1 With the seat unit removed from the machine, trace the leads back from the alternator stator and disconnect them at the three push-connectors located within the rubber protective boot, which itself is located just to the rear of the air filter housing.

2 Set the multimeter to the 0 - 100 volts ac scale. Start the engine and raise the engine speed to 5000 rpm. Place the probes of the multimeter on the terminal connections to the alternator stator as indicated on the figure accompanying this text. The reading shown on the multimeter should be in excess of 75 volts ac for each of the three tests carried out.

3 If any one of the readings obtained is found to be below that stated above, suspect the alternator stator of being defective. This may be confirmed by carrying out the resistance test detailed in the following Section.

5 Alternator: stator resistance testing

1 With the leads to the alternator located and disconnected as described in the preceding Section the alternator stator may be checked for continuity or short-circuits. Set the multimeter to the resistance measuring function and check for continuity between the leads as shown in the accompanying figure. If non-continuity (open circuit) is found in any one test there is evidence that a coil has burned out.

2 Check also between each lead and a good earth point on the engine. The alternator coils must be isolated from the stator core to function correctly, thus, if continuity is found in any test short-circuiting should be suspected.

3 If any test indicates that damage to the stator has occurred the stator should be removed from the machine and given a close visual inspection. It is possible that lack of continuity has been caused by a lead breakage somewhere between the wiring connectors and the stator.

4 Before consigning a suspect stator to the scrap bin have the system checked by a Suzuki Service Agent, who will be able to confirm whether renewal is required or whether a repair is possible.

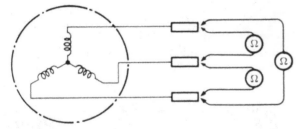

Fig. 6.2 Alternator output test voltmeter connections

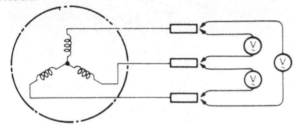

Fig. 6.3 Alternator resistance test ohmmeter connections

6 Voltage regulator/rectifier unit: testing

1 If, after carrying out the tests detailed in the preceding Sections, there is evidence to suggest that the regulator/rectifier unit is defective, then the following resistance test should be made to verify whether or not this is the case.

2 The unit concerned is a small integrated circuit (IC) housed in a finned alloy casing which is mounted on a plate located beneath the left-hand side panel. Its function is to convert the alternator output to direct current (dc) from alternating current (ac), this part of the function being executed by the rectifier stage. The regulator stage monitors the drain on the electrical system and controls the alternator current output to the battery. The unit is normally very reliable, there being no possibility of mechanical failure, but it can become damaged in the event of a short circuit in the electrical system or by poor or intermittent battery or earth connections.

3 With the sidepanel and the seat removed from the machine, trace the electrical leads from the unit and disconnect them at their nearest connections. With a multimeter set to its resistance function (X1 ohm range), refer to the following table and measure the resistance between the combinations of leads shown. Adhere strictly to the sequence given in the table and take great care not to allow the probes of the meter to short to earth or against each other. If, during the test sequence, the resistance readings shown on the meter scale differ from those given in the table, then the regulator/rectifier unit may be assumed to be defective and should be renewed.

prove of tester to	⊕ prove of tester to				
	R	W/Bl	W/R	Y	B/W
R		∞	∞	∞	∞
W/Bl	5-6Ω		∞	∞	∞
W/R	5-6Ω	∞		∞	∞
Y	5-6Ω	∞	∞		∞
B/W	35-45Ω	5-6Ω	5-6Ω	5-6Ω	

6.2 The finned regulator/rectifier unit is mounted on the left-hand side plate with the starter solenoid switch (lower left) and flasher relay unit (lower right)

7 Battery: examination and maintenance

1 A YB10L-A2 battery is fitted as standard to all the models covered in this Manual and is of the lead-acid type. The translucent plastic case of the battery permits the upper and lower levels of the electrolyte to be observed when the battery is exposed by removal of the right-hand sidepanel from the machine. Maintenance is normally limited to keeping the electrolyte level between the prescribed upper and lower limits and by making sure the vent pipe is not blocked and remains correctly routed.

2 Unless acid is spilt, as may occur if the machine falls over, the electrolyte should always be topped up with distilled water, to restore the correct level. If acid is spilt on any of the machine, it should be neutralised with an alkali such as washing soda and washed away with plenty of water, otherwise serious corrosion will occur. Top up with sulphuric acid of the correct specific gravity (1.280) only when spillage has occurred. Check that the vent pipe is well clear of the frame tubes or any of the other cycle parts, for obvious reasons. Note the instructions attached to the battery cover for the correct routing of the vent pipe.

3   If battery problems are experienced, the following checks will determine whether renewal is required. A battery can normally be expected to last for about 3 years, but this life can be shortened dramatically by neglect. In normal use, the capacity for storage will gradually diminish, and a point will be reached where the battery is adequate for all but the strenuous task of starting the engine. It follows that renewal will be necessary at this stage, as none of the models types covered in this Manual have an alternative kickstart mechanism fitted.

4   Remove the flat battery and examine the cell and plate condition near the bottom of the casing. An accumulation of white sludge around the bottom of the cells indicates sulphation, a condition which indicates the imminent demise of the battery. Little can be done to reverse this process, but it may help to have the electrolyte drained, the battery flushed and then refilled with new electrolyte. Most electrical wholesalers have facilities for this work.

5   Warping of the plates or separators is also indicative of an expiring battery, and will often be evident in only one or two of the cells. It can often be caused by old age, but a new battery which is overcharged will show the same failure. There is no cure for the problem and the need to avoid overcharging cannot be overstressed.

6   Try charging the suspect battery as described in the following Section. If the battery fails to accept a full charge, and in particular, if one or more cells show a low hydrometer reading, the battery is in need of renewal.

7   A hydrometer will be required to check the specific gravity of the electrolyte, and thus the state of charge. Any small hydrometer will do, but avoid the very large commercial types because there will be insufficient electrolyte to provide a reading. When fully charged, each cell should read 1.280, with little discrepancy between cells.

8   Note that it is seldom practicable to repair a cracked battery case because the acid present in the joint will prevent the formation of an effective seal. It is always best to renew a cracked battery, especially in view of the corrosion which will be caused if the acid continues to leak.

9   If the machine has remained unused for a period of time, it is advisable to remove the battery and give it a refresher charge every six weeks or so from a battery charger. If the battery is permitted to discharge completely, the plates will sulphate and render the battery useless.

10 Occasionally, check the condition of the battery terminals to ensure that corrosion is not taking place and that the electrical connections are tight. If corrosion has occurred, it should be cleaned away by scraping with a knife and then using emery cloth to remove the final traces. Remake the electrical connections whilst the joint is still clean, then smear the assembly with petroleum jelly (NOT grease) to prevent recurrence of the corrosion. Badly corroded connections can have a high electrical resistance and may give the impression of a complete battery failure.

---

8   Battery: charging procedure

---

1   The normal charging rate for batteries of up to 14 amp hour capacity is $1\frac{1}{2}$ amps. It is permissible to charge at a more rapid rate in an emergency but this shortens the life of the battery, and should be avoided. Because of this, go for the smallest charging rate available. Avoid 'quick charge' services offered by garages. This will indeed charge the battery rapidly, but will also overheat it and may halve its life expectancy. Never omit to remove the battery cell caps or neglect to check that the side vent is clear before recharging a battery, otherwise the gas created within the battery when charging takes place will burst the case with disastrous consequences. Do not attempt to charge the battery with it in-situ and with its leads still connected. This will only lead to failure of the regulator/rectifier unit.

2   Make sure that the battery charger connections are correct; red to positive and black to negative. When the battery is reconnected to the machine, the black lead must be connected to the negative terminal and the red lead to positive. This is most important, as the machine has a negative earth system. If the terminals are inadvertently reversed, the electrical system will be damaged permanently. The regulator/rectifier unit will be destroyed by a reversal of the current flow.

3   A word of caution concerning batteries. Sulphuric acid is extremely corrosive and must be handled with great respect. Do not forget that the outside of the battery is likely to retain traces of acid from previous spills, and the hands should always be washed promptly after checking the battery. Remember too that battery acid will quickly destroy clothing.

4   Note the following rules concerning battery maintenance:
Do not allow smoking or naked flames near batteries.
Do avoid acid contact with skin, eyes and clothing.
Do keep battery electrolyte level maintained.
Do avoid over-high charge rates.
Do avoid leaving the battery discharged.
Do avoid freezing.
Do use only distilled or demineralised water for topping up.

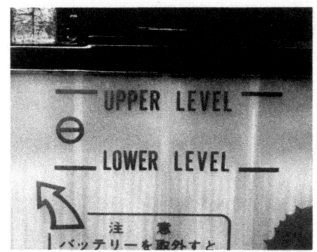

7.1 The battery electrolyte should be kept between the two level marks

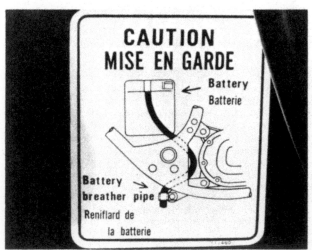

7.2 Note the instructions for correct routing of the battery vent pipe

9 Fuse: location, function and renewal

1 The fuse is contained within a plastic holder which is clipped to the plate covered by the left-hand sidepanel. A spare fuse of the same 15 amp rating is supplied with the machine when new and is contained within a second holder which is clipped to the wire leading from the fuse in circuit.

2 The fuse is fitted to protect the electrical system in the event of a short circuit or sudden surge; it is, in effect, an intentional 'weak link' which will blow, in preference to the circuit burning out.

3 Before replacing a fuse that has blown, check that no obvious short circuit has occurred, otherwise the replacement fuse will blow immediately it is inserted. It is always wise to check the electrical circuit thoroughly, to trace the fault and eliminate it.

4 When a fuse blows while the machine is running and no spare is available, a 'get you home' remedy is to remove the blown fuse and wrap it in silver paper before replacing it in the fuseholder. The silver paper will restore the electrical continuity by bridging the broken fuse wire. This expedient should **never** be used if there is evidence of a short circuit or other major electrical fault, otherwise more serious damage will be caused. Replace the 'doctored' fuse at the earliest possible opportunity, to restore full circuit protection. It follows that spare fuses that are used should be replaced as soon as possible to prevent the above situation from arising.

10 Starter motor: removal, examination, testing and fitting

1 An electric starter motor, operated from a small push-button on the right-hand side of the handlebars, provides the sole method of starting the engine. The starter motor is mounted within a compartment at the rear of the cylinder block, closed by an oblong, chromium plated cover. Current is supplied from the battery via a heavy duty solenoid switch and a cable capable of carrying the very high current demanded by the starter motor on the initial start-up.

2 The starter motor drives a free running clutch immediately behind the alternator rotor. The clutch ensures the starter motor drive is disconnected from the primary transmission immediately the engine starts. It operates on the roller and ramp principle.

3 To remove the starter motor from the engine unit, first disconnect the positive lead from the battery, to isolate the electrical system. Remove the cover plate which encloses the starter motor and detach the heavy duty cable from the terminal on the starter motor body. The starter motor is secured to the crankcase by two bolts which pass through a flange in its end cap. Remove these bolts and ease the starter motor across towards the right-hand side until it hits the chamber wall. The motor can then be lifted up at the right-hand end and away from the engine. If difficulty is encountered in moving the starter motor initially, a wooden lever may be used between the gear casing wall and starter motor front cover.

4 The parts of the starter motor most likely to require attention are the brushes. The end covers are secured to the motor body by two long screws which pass through the length of the motor. With these screws removed, the end cover can be separated from the motor body and the brush retaining plate exposed.

5 With the flat of a small screwdriver, lift the end of the spring clip bearing on the end of each brush and draw the brushes from their holders. Measure the length of each brush and if it is less than the service limit of 9.0 mm (0.4 in), renew the brush.

6 Before fitting the brushes, make sure that the commutator is clean. The commutator is the copper segments on which the brushes bear. Clean the commutator with a strip of glass paper. Never use emery cloth or 'wet-and-dry' as the small abrasive fragments may embed themselves in the soft copper of the commutator and cause excessive wear of the brushes. Finish off the commutator with metal polish to give a smooth surface and finally wipe the segments over with a methylated spirits soaked rag to ensure a grease free surface. Check that the mica insulators, which lie between the segments of the commutator, are undercut. If the amount of undercut is seen to be less than the service limit of 0.2 mm (0.008 in), then the armature should be withdrawn from the motor body and returned to an official Suzuki service agent or an experienced auto-electrician for re-cutting.

7 Fit the brushes in their holders and check that they slide quite freely. If the original brushes are being refitted, make sure that they are fitted in their original positions as they will have worn to the profile of the commutator.

8 If the motor has given indications of a more serious fault, withdraw the armature from the motor body, set a multimeter to its resistance function and carry out a check for equal resistance between the commutator segments and for good insulation between each commutator segment and the armature core. A fault in insulation or continuity will require renewal of the armature.

9 Before commencing reassembly of the motor, check the condition of the O-ring which forms a seal between the flanged end cap and the motor body and the condition of both the oil seal and O-ring located in the opposite end cap. If any of these items are seen to be damaged or deteriorating then they must be renewed. The seal can be removed from its location within the end cap by carefully easing it from position with the flat of a small screwdriver whilst taking care to avoid causing damage to the soft alloy mating surfaces of the cap. Ensure that, when fitting the new seal, the seal enters the end cap squarely. Lubricate the outer surface of the seal with a small amount of clean engine oil to ease fitting and smear a small amount of oil around the seal lip before reassembling the motor.

10 Check the condition of the shims, located one at either end of the armature, and commence reassembly of the starter motor whilst referring to the figure accompanying this text for the fitted positions of the components. It will be found that some difficulty will be experienced when refitting the flanged end cap over the brush holder assembly. Care must be taken to avoid interference of the cap with the brush assembly and to align correctly the protrusion cast in the motor body with the corresponding notch in the end cap. A similar method of alignment is used for the opposing cap.

11 Lubricate the O-ring fitted to the starter motor end cap boss with clean engine oil before inserting the motor into the engine casing. Push the motor fully home before fitting and tightening the two retaining bolts to the correct torque loading of 0.4 - 0.7 kgf m (3.0 - 5.0 lbf ft). Note that the threads of these bolts should be degreased and coated with a thread locking compound prior to the bolts being fitted. Remake the electrical connections at the starter motor terminal and battery terminal before checking that the motor functions correctly. Finally, refit the compartment cover plate.

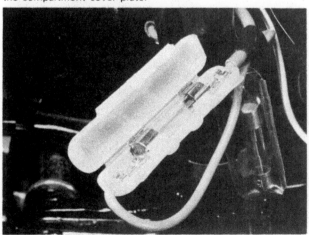

9.1 The fuse is contained within a plastic holder

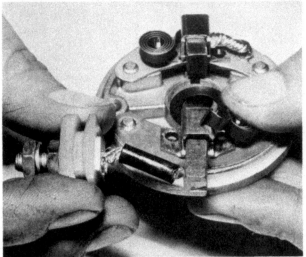
10.5 Draw each starter motor brush from its holder

10.6 Check the amount of undercut between each commutator segment

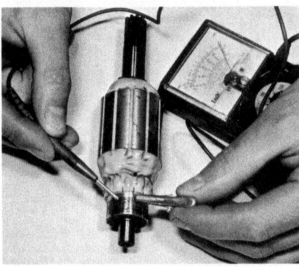
10.8a Check for insulation faults between the commutator segments...

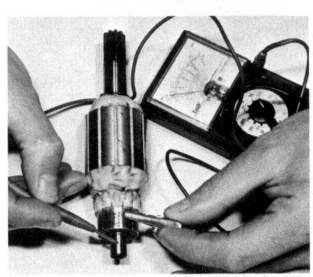
10.8b ...and for continuity between each commutator segment and the armature core

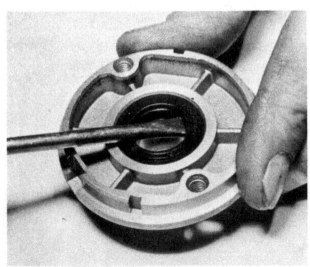
10.9 The end cap seal can be removed by careful use of a screwdriver

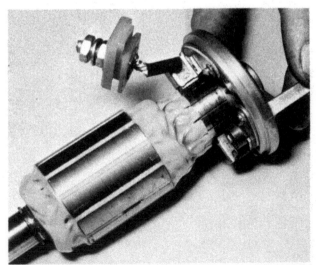
10.10a Commence reassembly of the starter motor by carefully locating the brushes over the commutator

10.10b Refit the end cap with its serviceable O-ring...

10.10c ...before tightening the assembly cap securing screws

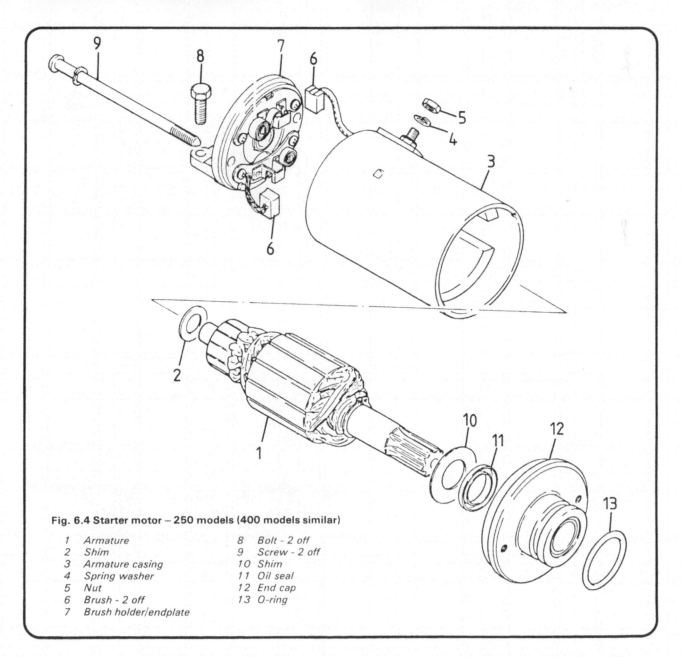

Fig. 6.4 Starter motor – 250 models (400 models similar)

1	Armature	8	Bolt - 2 off
2	Shim	9	Screw - 2 off
3	Armature casing	10	Shim
4	Spring washer	11	Oil seal
5	Nut	12	End cap
6	Brush - 2 off	13	O-ring
7	Brush holder/endplate		

11 Starter motor free running clutch: construction and renovation

1   Although a mechanical and not an electrical component, it is appropriate to include the free running clutch in this Chapter because it is an essential part of the electric starter system.

2   As mentioned in Chapter 1, the free running clutch is built into the alternator rotor assembly and will be found in the back of the rotor when the latter is removed from the left-hand end off the crankshaft. The only parts likely to require attention are the roller, spring and plunger assemblies along with the centre bush and teeth of the sprocket.

3   Check the condition of the three engagement rollers, springs and pushrods by removing each separate assembly from its location within the casing by carefully easing the roller from position with the flat of a small screwdriver. Carefully inspect each roller and plunger for signs of deterioration in the finish of their bearing surfaces and renew as necessary. Inspect each spring for signs of fatigue or failure and if possible, compare their length with that of a new item to determine whether or not they have taken a 'set' to a shorter length. If any one spring is found to be defective, it is advisable to renew all the springs as a set.

4   If it is found necessary to renew the rollers due to excessive wear on their bearing surfaces, then it is advisable to inspect closely the bearing faces of both the shim plate and starter clutch casing. This may be achieved initially by simply cleaning the roller location of any residue of oil and directing a strong light into the location; any excessive wear will show quite clearly, necessitating the need for the shim plate to be removed from the casing for further investigation and subsequent renewal.

5   If it has been found necessary during the course of examining and renovating the starter clutch assembly to detach the assembly from the alternator rotor, then ensure that each Allen bolt is thoroughly degreased, along with its corresponding thread in the rotor casting, before reassembly takes place. Coat the thread of each bolt with thread locking compound before inserting it through the starter clutch assembly, into the rotor and tightening it to the specified torque loading.

6   Inspect the condition of the sprocket boss where it comes into contact with the rollers. If the surface of the boss is scored or pitted or shows signs of excessive wear, then the sprocket should be replaced with a serviceable item. The sprocket should also be replaced if its teeth are chipped or badly worn. Suzuki do not list the bush contained within the sprocket as a separate item, nor do they give any indication as to the amount of wear allowed on the bush. If the bush seems excessively worn and the crankshaft surface with which it comes into contact is marked, return both items to an official Suzuki service agent who will be able to give advice on the matter.

7   To check whether the clutch is operating correctly, turn the driven sprocket clockwise. This should force the spring loaded rollers against the sprocket boss, locking the sprocket to the clutch hub, which is fixed to the alternator rotor.

12 Starter solenoid switch: function, location and testing

1   The starter motor switch is designed to work on the electromagnetic principle. When the starter motor button is depressed, current from the battery passes through windings in the switch solenoid and generates an electromagnetic force which causes a set of contact points to close. Immediately the points close, the starter motor is energised and a very heavy current is drawn from the battery.

2   This arrangement is used for at least two reasons. Firstly, the starter motor current is drawn only when the button is depressed and is cut off again when pressure on the button is released. This ensures minimum drainage on the battery. Secondly, if the battery is in a low state of charge, there will not

11.3a Carefully ease each starter clutch roller form its location...

11.3b ...to gain access to its pushrod and spring

11.6 Inspect the starter clutch sprocket boss and bush for wear or damage

be sufficient current to cause the solenoid contacts to close. In consequence it is not possible to place an excessive drain on the battery which, in some circumstances, can cause the plates to overheat and shed their coatings. If the starter will not operate, first suspect a discharged battery. This can be checked by trying the horn or switching on the lights. If this check shows the battery to be in good shape, suspect the starter switch which should come into action with a pronounced click.

3 The switch is located behind the left-hand sidepanel and can be identified by the heavy duty starter cable connected to it.

4 To test the switch system, disconnect the electrical lead from the switch terminal which serves to connect the switch to the starter motor and set a multimeter to its resistance function. With the ignition switch turned to the 'On' position, place the probes of the meter on the terminals of the switch and check for continuity between the switch terminals. Continuity should exist when the starter button is pressed; if this is the case, then the

switch is serviceable as well as the starter button control and the wiring between the two components.

5 If the above test proves to be unsatisfactory, carry out the following test on the switch to determine whether the switch itself is defective rather than the starter button control or the associated wiring. With the multimeter left set on its resistance function, connect the probes of the meter to the parts of the solenoid shown in the accompanying figure and check that the resistance reading obtained amounts to between 3 and 4 ohms. If this is the case, then the switch is serviceable and any fault must exist in either the starter button control or the wiring.

6 Physically inspect the wiring concerned for any obvious damage or corroded or dirty contacts before checking it for continuity with the multimeter, whilst referring to the wiring diagram at the end of this Chapter for details of the wire coding, etc. A test for the starter button control is given in Section 19 of this Chapter.

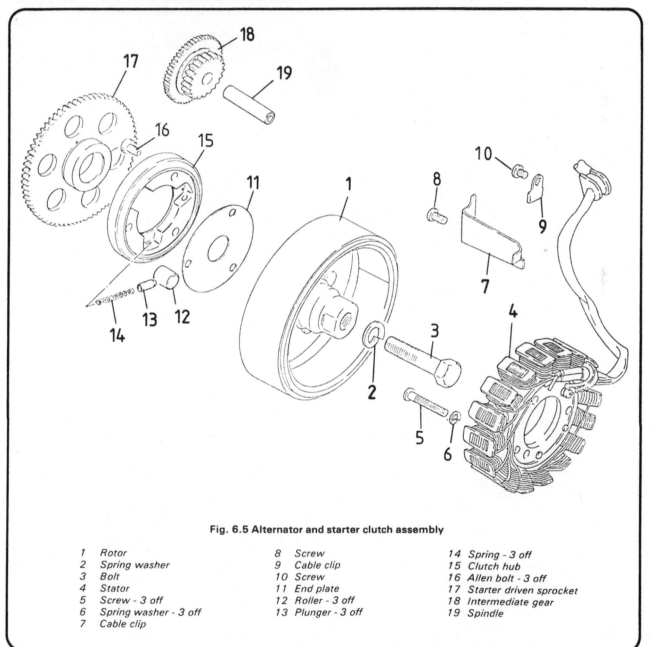

Fig. 6.5 Alternator and starter clutch assembly

1	Rotor	8	Screw	14	Spring - 3 off
2	Spring washer	9	Cable clip	15	Clutch hub
3	Bolt	10	Screw	16	Allen bolt - 3 off
4	Stator	11	End plate	17	Starter driven sprocket
5	Screw - 3 off	12	Roller - 3 off	18	Intermediate gear
6	Spring washer - 3 off	13	Plunger - 3 off	19	Spindle
7	Cable clip				

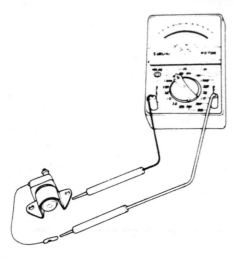

Fig. 6.6 Starter solenoid resistance test

13 Ignition switch: testing

1   The ignition switch is situated just below the centre of the handlebar mounted instrument console and may be tested for continuity at the block connector within the headlamp shell.
2   Remove the headlamp reflector unit and identify the block connector by the red, orange, grey and brown wires running into it from the switch. Disconnect the terminals by pulling apart the two halves of the block.
3   With the switch turned to the 'On' position, check for continuity between the red and orange and the grey and brown wire terminals. With the switch turned to the 'P' position, check for continuity between the red and brown wire terminals.
4   If continuity is found to exist in all of the above mentioned tests, then the switch is serviceable. If any one of the tests indicates non-continuity, then the switch is unserviceable and should be renewed. Before removing the switch for renewal, carry out a physical check of the wiring from the switch to the block connector to ensure that there is no indication of the wires having been cut or frayed by any of the associated cycle parts.

14 Oil pressure switch: location, testing and renewal

1   The oil pressure switch is located within the ATU housing on the right-hand side of the engine. Access to the switch may be gained by removing the circular outer cover from the right-hand crankcase cover. The switch is partially hidden by the pulse generator unit plate and is in roughly the 5 o'clock position in relation to that plate.
2   To test the switch, check that the engine oil level is correct, start the engine and allow it to idle. Set a multimeter to its resistance function and carry out a check for continuity between the switch terminal and earth. A reading which indicates no continuity should be obtained on the multimeter; this reading should change to one which shows continuity once the engine is stopped.
3   If the switch proves to be unserviceable, then it should be removed from the ATU housing and replaced with a serviceable item. This will almost certainly entail the removal of the ignition pick-up assembly from the machine in order to provide proper access to the switch. Full information on removal of the ignition pick-up assembly is contained within Section 9 of Chapter 1.
4   When fitting the replacement switch and refitting the disturbed components of the ignition pick-up assembly, refer to the appropriate paragraphs within Section 39 of Chapter 1.

15 Neutral indicator and gear position switches: location, testing and renewal

1   On 250 models the neutral indicator switch is located just above the gearbox sprocket on the left-hand side of the engine. Access to the switch may be gained by removing the rearmost section of the left-hand crankcase cover.
2   To test the switch, set a multimeter to its resistance function and carry out a check for continuity between the switch terminal and earth with the gearchange lever set in the neutral position. If continuity is found, then the switch is serviceable. If this is not the case, then the switch should be removed from its location by unscrewing its two crosshead retaining screws, disconnecting the switch electrical wire and pulling the switch clear of the crankcase whilst noting the position of the pin and spring located beneath it.
3   Before fitting a replacement switch, always check that the O-ring which forms a seal between the switch and crankcase is in good condition. Ideally, this O-ring should be renewed whenever the switch is removed.
4   GSX400 models are fitted with a gear position indicator switch, which incorporates a neutral indicator contact. The switch may be tested for continuity in each of the six different gear positions plus the neutral position in a manner similar to that detailed above for the neutral indicator switch.

16 Front brake stop lamp switch: testing

1   The front brake stop lamp switch is incorporated in the handlebar front brake lever assembly and is retained in position by two crosshead screws. No means of switch adjustment is provided.
2   To test the switch, trace the electrical leads from the switch to their terminal connections and disconnect the connections. It is most likely that these connections will be found inside the headlamp shell. Set a multimeter to the resistance function and with the probes of the meter connected one to each terminal, operate the switch by pulling the handlebar lever fully back. If continuity exists then the switch is serviceable. If no continuity is found then the switch must be renewed as it is not possible to effect a satisfactory repair.

17 Rear brake stop lamp switch: adjustment and testing

1   The stop lamp switch is located in a frame-mounted bracket which is situated on the frame downtube, just to the rear of the battery tray. It is operated by movement of the rear brake pedal, the two components being adjoined by an extension spring. The body of the switch is threaded to permit adjustment.
2   If the stop lamp is seen to be late in operating, raise the switch body by rotating its retaining nut in a clockwise direction whilst holding the switch steady. If the stop lamp is permanently on, then the switch body should be lowered in relation to its retaining bracket. As a guide to operation, the stop lamp should illuminate after the rear brake pedal has been depressed approximately 2 cm (0.75 in).
3   Testing the switch is a simple matter of checking for continuity between the switch wire terminals with the switch fully extended, that is in the 'On' position. Before testing, check the switch adjustment by referring to the above text. Trace the electrical leads from the switch to their terminal connections and disconnect the connections. Set a multimeter to the resistance function and with the probes of the meter connected one to each terminal, operate the switch. If no continuity is found then the switch must be renewed as it is not possible to effect a satisfactory repair.

14.1 The oil pressure switch is located within the ATU housing

15.3 The O-ring fitted beneath the neutral indicator switch should be renewed whenever the switch is removed

17.1 The rear brake stop lamp switch is mounted to the rear of the battery tray

18 Handlebar switches: general information

1   Generally speaking, the switches give little trouble, but if necessary they can be dismantled by separating the halves which form a split clamp around the handlebars.
2   Always disconnect the battery before removing any of the switches, to prevent the possibility of a short circuit. Most troubles are caused by dirty contacts, but in the event of the breakage of some internal part, it will be necessary to renew the complete switch.
3   Because the internal components of each switch are very small, and therefore difficult to dismantle and reassemble, it is suggested a special electrical contact cleaner be used to clean corroded contacts. This can be sprayed into each switch, without the need for dismantling.
4   It will be necessary to obtain a multimeter and set it to its resistance function in order to carry out the various continuity checks described in the following Sections.

19 Engine starter switch: testing

1   The engine starter switch takes the form of a push button and is incorporated in the handlebar switch unit located adjacent to the throttle twistgrip.
2   Trace the leads from the switch to the push connector located beneath the fuel tank and separate the two halves of the connector. It will be necessary to remove the fuel tank in order to achieve this. Full details of seat and tank removal are given in Section 15 of Chapter 4 and Section 2 of Chapter 2 respectively.
3   To test the switch, push the switch button fully in so that it is set to the 'On' position and check for continuity between the orange/white and yellow/green wire terminals of the connector. If continuity exists then the switch is serviceable.

20 Engine stop switch: testing

1   The engine stop switch is incorporated in the handlebar switch unit located adjacent to the throttle twistgrip. It takes the form of a simple two-position rocker switch, the two positions being marked 'Off' and 'Run'.
2   Trace the leads from the switch to the block connector located beneath the fuel tank and pull apart the two halves of the connector. It will be necessary to remove the fuel tank in order to achieve this. Full details of seat and tank removal are given in Section 15 of Chapter 4 and Section 2 of Chapter 2 respectively.
3   To test the switch, push the switch to the 'Run' position and check for continuity between the orange and orange/white wire terminals of the connector. If continuity exists then the switch is serviceable.

21 Lighting switch: testing

1   The lighting switch is incorporated in the handlebar switch unit located adjacent to the throttle twistgrip. Trace the leads running from the switch to their nearest connection point; separate the connection and carry out a check for continuity between the wire terminals. Continuity should be found as shown in the switch box connections for the lighting switch in the relevant wiring diagram.
2   If non-continuity is found between any wire where continuity is shown on the diagram the switch contacts should be cleaned and inspected for damage. See Section 18 for further details.

22 Headlamp dip switch: testing

1   The headlamp dip switch is incorporated in the switch assembly located next to the left-hand handlebar grip rubber. Trace the leads running from the switch to their nearest connection point and separate the connection.
2   To test the switch, move the switch button to the 'Hi' position and check for continuity between the yellow and yellow/white terminals. Move the switch to the 'Lo' position and check for continuity between the yellow/white and white wire terminals. If continuity exists in both of these tests then the switch is serviceable. If any one of these tests shows non-continuity, then the switch must be renewed, as any failure that occurs when changing from one beam to the other can plunge the lighting system into darkness, with disastrous consequences.

23 Direction indicator switch: testing

1   The direction indicator switch is incorporated in the switch assembly located adjacent to the left-hand handlebar grip rubber.
2   To test the switch, first trace the leads running from the switch to their nearest connection point and separate the connection. Move the switch button to the 'R' position and check for continuity between the light blue and light green wire terminals. Move the switch button to the 'L' position and check for continuity between the black and light blue wire terminals.
3   If continuity exists in both of these tests, then the switch is serviceable. If either one of these tests shows non-continuity, then the switch is unserviceable and should be cleaned or renewed.

24 Headlamp flashing switch: testing – UK models only

1   The headlamp flashing switch is incorporated in the switch assembly located adjacent to the left-hand handlebar rubber. To test the switch, first trace the leads running from the switch to

their nearest connection point and separate the connection. With the switch button moved to the 'P' position, check for continuity between the yellow and blue/white wire terminals (400 models) or between the yellow and orange/red terminals (250 models). If continuity is found to exist, then the switch is serviceable. If the test shows non-continuity, then the switch is unserviceable and should be cleaned or renewed.

25 Horn switch: testing

1   The horn switch takes the form of a push button which is incorporated in the lower part of the switch assembly located adjacent to the left-hand handlebar grip rubber.
2   Trace the leads running from the switch to their nearest connection point and separate the connection. With the horn button pushed fully in so that it is set to the 'On' position, carry out a check for continuity between the green and black/white wire terminals. If continuity exists then the switch is serviceable. If the test shows non-continuity, then the switch is unserviceable and should be renewed.

26 Instrument console assembly: testing the electrical components

1   It is not necessary to remove the instrument console assembly from the machine in order to test the electrical components contained within it. It is, however, necessary to remove the seat unit and fuel tank in order to gain access to the terminals which connect the console to the main electrical loom of the machine. Full details of seat and tank removal are given in Section 15 of Chapter 4 and Section 2 of Chapter 2 respectively.
2   With the fuel tank removed, trace the leads from the console and separate the terminal halves at their location beneath the tank. Using a multimeter set to its resistance function, refer to the diagram accompanying this text and check for continuity between the terminal ends shown for each component part within the console. If a circuit is found to be open, renew the component or wire concerned.

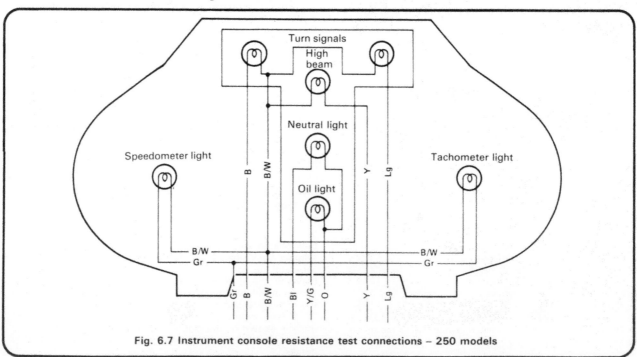

Fig. 6.7 Instrument console resistance test connections – 250 models

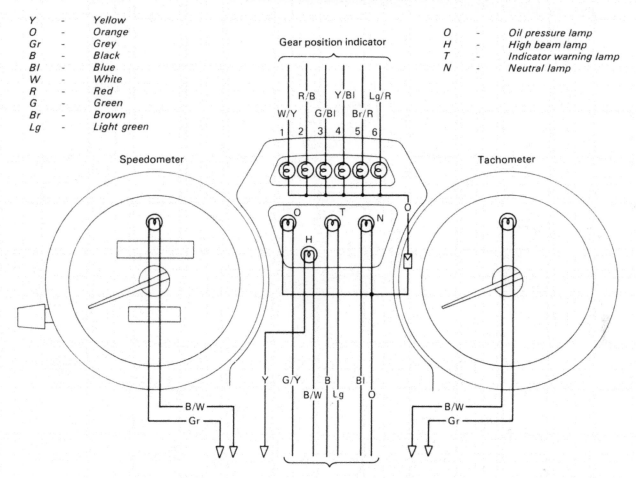

Y - Yellow
O - Orange
Gr - Grey
B - Black
Bl - Blue
W - White
R - Red
G - Green
Br - Brown
Lg - Light green

O - Oil pressure lamp
H - High beam lamp
T - Indicator warning lamp
N - Neutral lamp

Fig. 6.8 Instrument console resistance test connections – 400 models

27 Instrument console assembly: bulb renewal

1   Access may be gained to the various warning, gear in-dicator (where fitted) and instrument illumination bulbs by removing the instrument console top section and withdrawing the various bulb holders contained within. The top section of the console is retained by four crosshead securing screws. The rubber holder containing the various warning bulbs is contained within a plastic moulding which itself is attached to the underside of the top section of the console. With the rubber holder unplugged from its location in the plastic moulding, each warning bulb may be removed for inspection or renewal.

2   The instrument illumination bulbs are contained within the base of each instrument. In order to effect their removal, each instrument will have to be detached from the console casing by first unscrewing the knurled ring retaining the drive cable to the instrument base and then removing each of the two crosshead screws which pass through the console casing into the base of each instrument.

3   On machines where a gear indicator assembly is fitted to the top section of the console, remove the two screws retaining the cover to the back of the unit in order to enable withdrawal of the bulb holder located beneath it.

4   Take great care, when attempting to remove the various bulbs from their holders, to determine whether the type of bulbs fitted are of the earlier bayonet-fitting type, that is with a metal cap fitting, or of the later capless type. The difference between these two types of bulb is that whereas is necessary to push in and turn the bayonet type of bulb in order to free it from it

holder, it is only necessary to pull the capless type in order to effect its removal. Attempting to push in and rotate a capless type of bulb will not necessarily cause it any damage but attempting to pull a bayonet type of bulb directly out of its holder will not only prove very difficult but also expensive in terms of renewing damaged bulbs or holders.

28 Instrument console assembly: removal and refitting

1   If, for any reason, it is found necessary to remove the instrument console assembly in order to repair or renew the electrical components contained within it, then commence the removal procedure by gaining access to the connector terminal for the electrical wires which is located beneath the fuel tank. Full details for seat and tank removal are given in Section 15 of Chapter 4 and Section 2 of Chapter 2 respectively.

2   Separate the terminal halves and thread the wires forward until they are free to hang below the console. Disconnect the speedometer and tachometer cables from the base of the console by unscrewing their knurled retaining rings.

3   The instrument console may now be freed from its mount-ings on the upper fork yoke by removing the two mounting bolts and domed nuts with their associated washers. A note should be made of the fitted position of these washers for reference when refitting.

4   After having completed the necessary maintenance re-quired on the console, it may be refitted by using a direct reversal of the removal procedure.

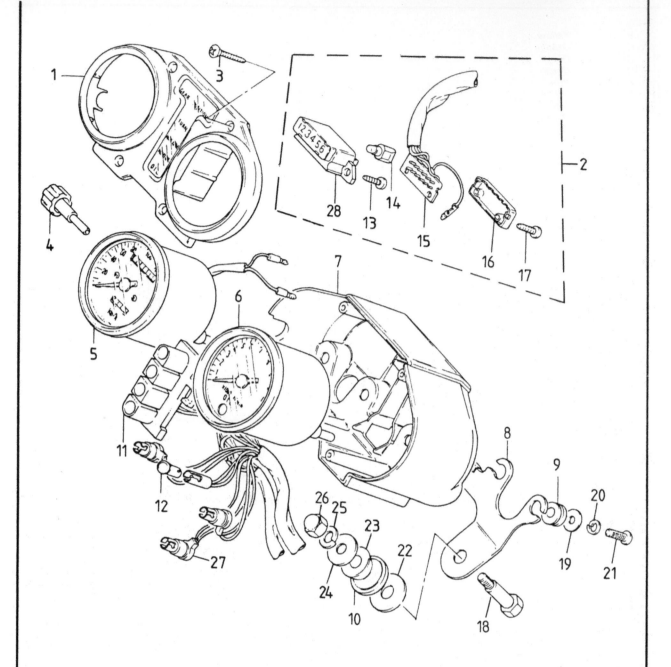

Fig. 6.9 Instrument console assembly – 250 models and GSX400 E model

1 Instrument console top section	11 Rubber bulb holder	20 Spring washer - 4 off
2 Gear position indicator assembly	12 Bulb - 6 off	21 Screw - 4 off
3 Screw - 4 off	13 Screw - 2 off	22 Washer - 2 off
4 Trip reset knob	14 Bulb - 6 off	23 Grommet - 2 off
5 Speedometer	15 Bulb holder	24 Washer - 2 off
6 Tachometer	16 Unit housing	25 Spring washer - 2 off
7 Instrument housing	17 Screw - 2 off	26 Nut - 2 off
8 Mounting bracket	18 Bolt - 2 off	27 Bulb holder - 6 off
9 Grommet - 4 off	19 Washer - 4 off	28 Gear position cover
10 Rubber washer - 2 off		

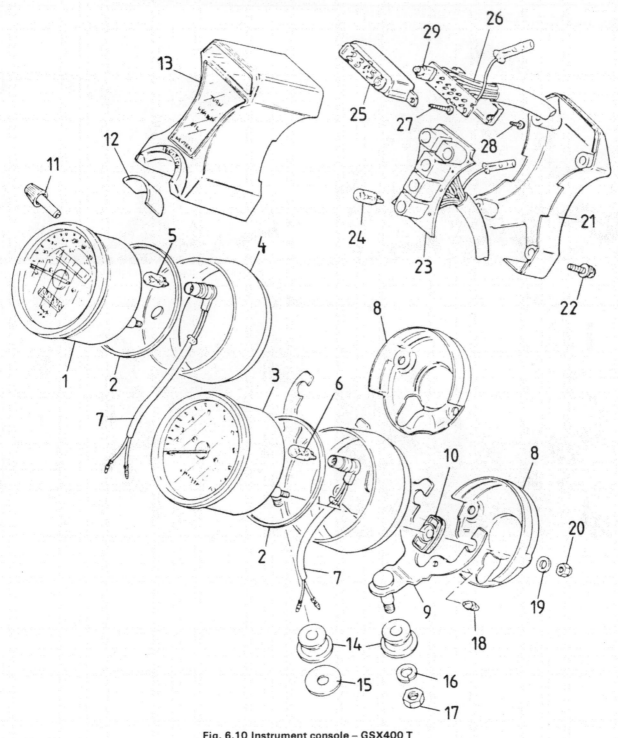

Fig. 6.10 Instrument console – GSX400 T

1	Speedometer	11	Trip reset knob	21	Pilot lamp housing
2	Seal - 2 off	12	Cushion	22	Screw and washer - 4 off
3	Tachometer	13	Pilot lamp console	23	Pilot lamp bulb holder
4	Instrument housing - 2 off	14	Grommet	24	Pilot lamp - 5 off
5	Bulb	15	Washer	25	Gear position cover
6	Bulb	16	Spring washer	26	Bulb holder
7	Bulb holder - 2 off	17	Nut	27	Screw - 2 off
8	Instrument holder - 2 off	18	Damper	28	Screw - 2 off
9	Mounting bracket	19	Washer	29	Bulb - 6 off
10	Grommet	20	Nut		

27.1a Remove the four securing screws...

27.1b ...to enable the top section of the instrument console to be fitted

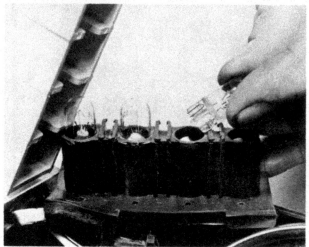

27.1c Each warning bulb may be removed for inspection or renewal

27.2 Detach each instrument head by removing its two retaining screws

29 Flasher unit: location and renewal

1   The flasher relay unit is located behind the left-hand side panel on the electrical components mounting plate, to the bottom right of the plate. The relay unit is supported on anti-vibration mountings made of rubber attached to the mounting plate.

2   If the flasher unit is functioning correctly, a series of audible clicks will be heard when the indicator lamps are in operation. If the unit malfunctions and all the bulbs are in working order, the usual symptom is one initial flash before the unit goes dead; it will be necessary to replace the unit complete if the fault cannot be attributed to any other cause.

3   If renewal is necessary, removal of the existing unit is simple. Unplug the block connector from the base of the unit and detach the unit from its plate mountings. When fitting a new unit, or handling the existing one, take great care; it is easily damaged if dropped.

30 Horn: location, adjustment and replacement

1   The horn is located on the right-hand side of the machine and is attached to a frame mounting plate by a single nut and plain washer.

2   On some models the horn is adjustable by means of a small grub screw at the back of the body so that the volume can be varied if necessary. To adjust the volume, loosen the locknut and turn the screw about half a turn either way until the desired tone is required. After adjustment tighten the locknut.

3   If necessary, the horn unit may be removed from the machine by simply removing the retaining nut and washer and disconnecting the two electrical wires from the horn terminals, after first having noted their fitted positions. Fitting a horn unit is a direct reversal of the removal procedure.

31 Headlamp: bulb renewal and alignment

1   In order to gain access to the headlamp bulbs, it is necessary first to remove the rim, complete with the reflector and headlamp glass. The rim is retained by three crosshead screws (GSX models) or two crosshead screws (GS models) equally spaced around the headlamp shell. Remove the screws completely and draw the rim from the headlamp shell.

2   UK models have a main headlamp bulb which is a push fit into the central bulb holder of the reflector. The bulb holder can be fitted in one position only to ensure the bulb is always

correctly focussed. It is retained by a bayonet fitting. A bulb of the twin filament type is fitted, the rating of which may be found in the Specifications Section of this Chapter. The pilot lamp bulb is bayonet fitting and fits within a bulb holder which has the same form of attachment to the headlamp reflector.

3   US models have a sealed beam type of headlamp unit fitted which contains no provision for a pilot lamp. If one filament blows, then the complete unit must be renewed. To release the lamp unit, remove the horizontal adjusting screw, detach the headlamp rim from the headlamp shell and then the lamp unit from its collar. Make a note of the setting of the adjusting screw, otherwise it will be necessary to re-adjust the beam height after installing the new light unit by reversing the dismantling procedure.

4   Beam height on all models is effected by tilting the headlamp shell after the mounting bolts have been loosened slightly. On sealed beam units the horizontal alignment of the

beam can be adjusted by altering the position of the screw which passes through the headlamp rim. The screw is fitted at the 9 o'clock position when viewed from the front of the machine. Turning the screw in a clockwise direction will move the beam direction over to the left-hand side.

5   In the UK, regulations stipulate that the headlamp must be arranged so that the light will not dazzle a person standing at a distance greater than 25 feet from the lamp, whose eye level is not less than 3 feet 6 inches above that plane. It is easy to approximate this setting by placing the machine 25 feet away from a wall, on a level road, and setting the dip beam height so that it is concentrated at the same height as the distance of the centre of the headlamp from the ground. The rider must be seated normally during this operation and also the pillion passenger, if one is carried regularly.

6   Most other areas have similar regulations controlling headlamp beam alignment, and these should be checked before any adjustment is made.

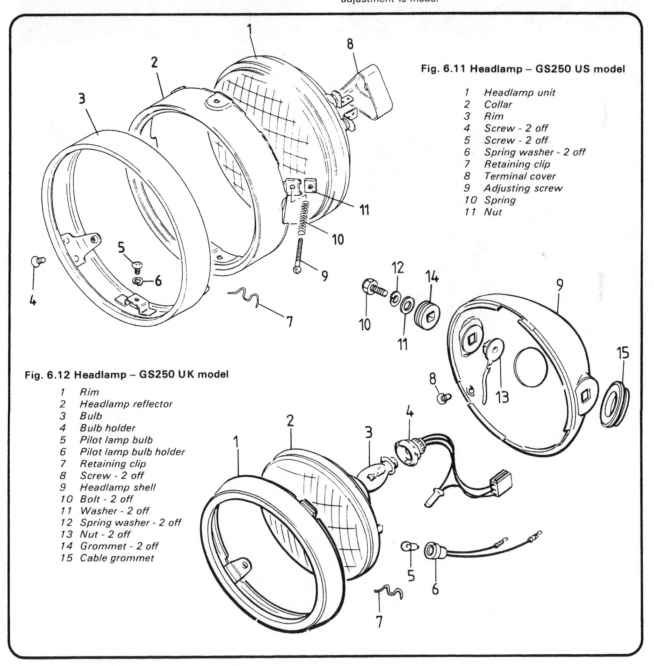

Fig. 6.11 Headlamp – GS250 US model

1   Headlamp unit
2   Collar
3   Rim
4   Screw - 2 off
5   Screw - 2 off
6   Spring washer - 2 off
7   Retaining clip
8   Terminal cover
9   Adjusting screw
10 Spring
11 Nut

Fig. 6.12 Headlamp – GS250 UK model

1   Rim
2   Headlamp reflector
3   Bulb
4   Bulb holder
5   Pilot lamp bulb
6   Pilot lamp bulb holder
7   Retaining clip
8   Screw - 2 off
9   Headlamp shell
10 Bolt - 2 off
11 Washer - 2 off
12 Spring washer - 2 off
13 Nut - 2 off
14 Grommet - 2 off
15 Cable grommet

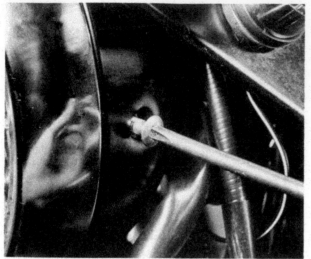

31.1 The headlamp rim is retained by three crosshead screws

31.2a Detach the bulb holder from the reflector...

31.2b ...to allow removal of the main headlamp bulb

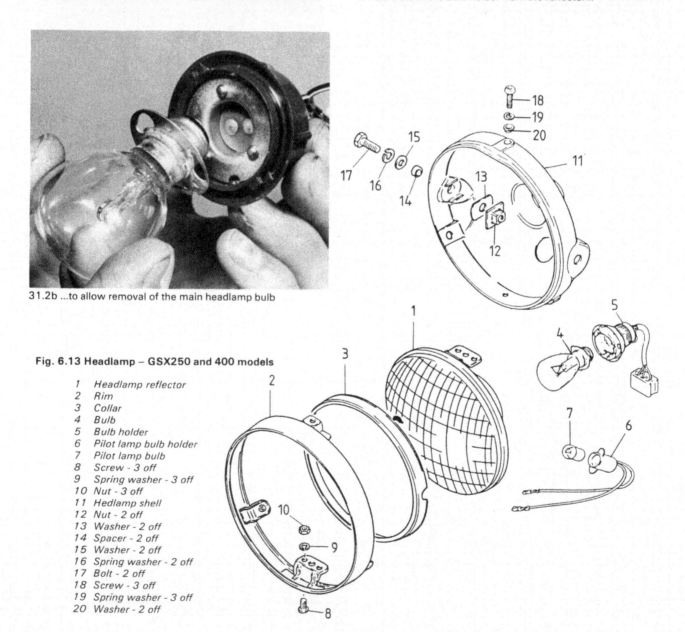

Fig. 6.13 Headlamp – GSX250 and 400 models

1   Headlamp reflector
2   Rim
3   Collar
4   Bulb
5   Bulb holder
6   Pilot lamp bulb holder
7   Pilot lamp bulb
8   Screw - 3 off
9   Spring washer - 3 off
10 Nut - 3 off
11 Hedlamp shell
12 Nut - 2 off
13 Washer - 2 off
14 Spacer - 2 off
15 Washer - 2 off
16 Spring washer - 2 off
17 Bolt - 2 off
18 Screw - 3 off
19 Spring washer - 3 off
20 Washer - 2 off

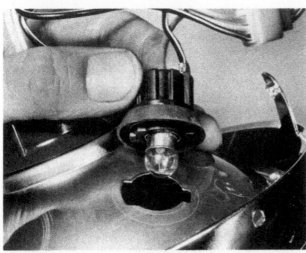

31.2c UK models have a pilot bulb fitting

31.2d The pilot bulb is a bayonet fitting in its holder

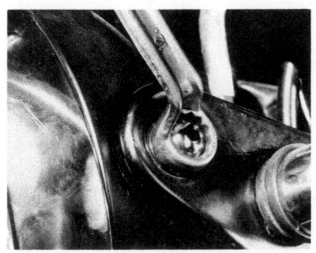

31.4 Loosen each mounting bolt before tilting the headlamp shell

32 Stop and tail lamp: bulb renewal

1 There are two basic types of stop/tail lamp units fitted to the various models covered within this Manual; the main difference being that one type of unit has only the one twin filament bulb fitted, whereas the other type of unit has two twin filament bulbs fitted. The procedure for changing the bulbs in both types of unit is essentially the same.

2 The bulb ratings are listed in the Specifications Section of this Chapter. One filament of the bulb(s) will serve to illuminate the rear of the machine, including the number plate, whilst the other filament serves as a warning that the rear brake has been applied. On some models the stop lamp also operates in conjunction with the front brake; a stop lamp switch may be incorporated in the front brake lever to meet the statutory requirements of the country or state to which the machine is exported.

3 To gain access to the bulb(s), unscrew the crosshead screws which serve to retain the plastic lens cover to the lamp unit and remove the lens whilst taking care not to tear the sealing gasket located between the lens and its baseplate. This gasket is essential to keep moisture and dirt from entering the various electrical contacts within the unit and to stop corrosion or dulling of the reflector plate. These screws will have their heads facing either to the rear or to the front of the machine and be two or four in number depending on the type of lamp unit fitted.

4 Each bulb has a bayonet fitting with offset pins so that the stop lamp filament cannot be inadvertently connected with the tail lamp and vice versa. The bulb may be released by pressing inwards and turning anti-clockwise; once fully turned the bulb will come out of its holder.

5 Refitting of the bulb and lens is a reversal of the removal procedure. Check the inside of the bulb holder for any signs of corrosion or moisture ensuring that the contacts are free to move when depressed. When refitting the lens securing screws, take care not to overtighten them as it is possible that the lens may crack.

33 Flashing indicator lamp: bulb renewal

1 Flashing indicator lamps are fitted to the front and rear of the machine. They are mounted on short stalks through which the wires pass. Access to each bulb is gained by removing the two screws holding the plastic lens cover.

2 The bulbs fitted are of the bayonet type and may be released by pushing in, turning anti-clockwise and pulling out of the holder.

3 Refitting of the bulbs and lens is a reversal of the removal procedure. Check the inside of the bulb holder for any signs of corrosion or moisture, ensuring that the contact is free to move when depressed. Take great care not to overtighten the lens securing screws as it is possible to crack the lens by doing so.

34 Wiring: layout and examination

1 The wiring harness is colour-coded and will correspond with the accompanying diagrams. Where socket connectors are used they are designed so that reconnection can be made only in the one correct position.

2 Visual inspection will show whether any breaks or frayed outer coverings are giving rise to short circuits. Another source of trouble may be the snap connectors and sockets, where the connector has not been pushed home fully in the outer housing.

3 Intermittent short circuits can often be traced to a chafed wire that passes through or is close to a metal component, such as a frame member. Avoid tight bends in the wire or situations where the wire can become trapped between casings.

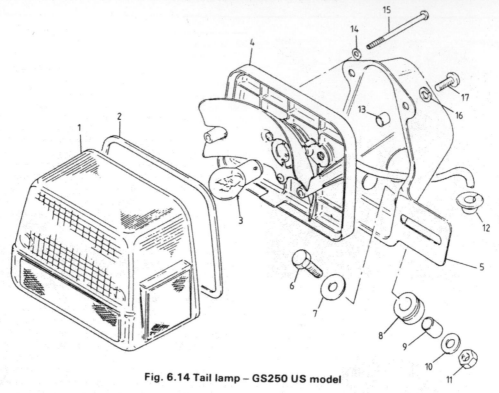

Fig. 6.14 Tail lamp – GS250 US model

1	Lens	7	Washer - 3 off	13	Spacer - 2 off
2	Sealing ring	8	Grommet - 3 off	14	Washer - 2 off
3	Bulb	9	Spacer - 3 off	15	Screw - 2 off
4	Reflector	10	Washer - 3 off	16	Spring washer - 2 off
5	Mounting bracket	11	Nut - 3 off	17	Screw - 2 off
6	Bolt - 3 off	12	Cable grommet		

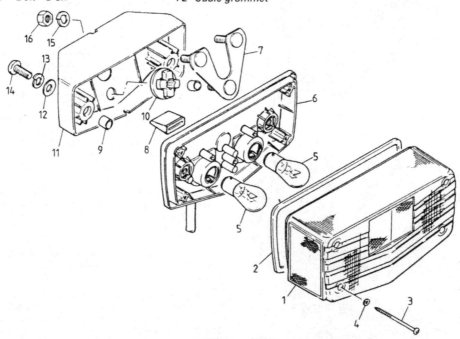

Fig. 6.15 Tail lamp – GSX250 and GSX400 E models

1	Lens	7	Mounting bracket	12	Washer - 2 off
2	Sealing ring	8	Rubber cushion	13	Spring washer - 2 off
3	Screw - 4 off	9	Spacer - 2 off	14	Screw - 2 off
4	Washer - 4 off	10	Rubber cushion	15	Spring washer - 3 off
5	Bulb - 2 off	11	Reflector housing	16	Nut - 3 off
6	Baseplate				

32.3 Remove the plastic lens cover to gain access to the tail lamp bulb

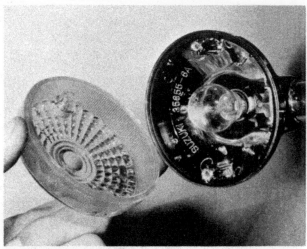

33.1 Access to each indicator bulb is gained by removing the plastic lens cover

35 Fault diagnosis: electrical system

Symptom	Cause	Remedy
Complete electrical failure	Blown fuse	Check wiring and electrical component for short circuit before fitting new fuse.
	Isolated battery	Check battery connections, also whether connections show signs of corrosion.
Dim lights, horn and starter inoperative	Discharged battery	Remove battery and charge with battery charger. Check generator output and voltage regulator settings.
Constantly blowing bulbs	Vibration or poor earth connection	Check security of bulb holders. Check earth return connections.
Starter motor sluggish	Worn brushes	Remove starter motor. Renew brushes.
Parking lights dims rapidly	Battery will not hold charge	Renew battery at earliest opportunity.
Flashing indicators do not operate	Blown bulb Damaged flasher unit	Renew bulb. Renew flasher unit.

The GSX250 EZ UK model

The GS450 LF US model

Chapter 7 The UK
GSX250/400 EZ and US GS450 models

Contents

Specifications

UK GSX250 EZ model
Note: *Specifications are as given for the GSX250 model at the beginning of each Chapter unless shown below*

Specifications relating to Chapter 1

Valves
Stem to guide clearance – service limit:
Inlet .... 0.35 mm (0.014 in)
Exhaust .... 0.35 mm (0.014 in)

Valve springs
Free length service limit .... 28.0 mm (1.102 in)

Camshafts
Bearing surface ID .... 22.012 – 22.025 mm (0.8666 – 0.8671 in)

Clutch
Spring free length service limit .... 40.1 mm (1.578 in)
Friction plate tongue width (service limit) .... 11.2 mm (0.44 in)
Plain plate thickness .... 1.55 – 1.65 mm (0.061 – 0.065 in)

Torque wrench settings

Component	lbf ft	kgf m
Cylinder head bolt	5.8 – 8.7	0.8 – 1.2
Cylinder head cover bolt	5.8 – 8.7	0.8 – 1.2
Cam chain tensioner shaft assembly	21.7 – 25.3	3.0 – 3.5
Camshaft sprocket retaining bolts	12.3 – 13.7	1.7 – 1.9
Rocker arm shaft stopper bolts	5.8 – 8.7	0.8 – 1.2
Primary drive gear pinion nut	65.1 – 79.6	9.0 – 11.0
Starter clutch Allen bolts	16.6 – 20.3	2.3 – 2.8
Clutch centre retaining nut	28.9 – 43.4	4.0 – 6.0
Alternator rotor retaining bolt	65.1 – 79.6	9.0 – 10.0
Gearbox sprocket retaining nut	57.9 – 79.6	8.0 – 10.0
Sump cover bolts	5.8 – 8.7	0.8 – 1.2
Exhaust pipe clamp bolts	7.2 – 11.6	1.0 – 1.6
Power chamber bolt	14.5 – 25.3	2.0 – 3.0
Exhaust pipe/silencer clamp bolt	6.5 – 10.1	0.9 – 1.4

Specifications relating to Chapter 2

Fuel tank

Overall capacity	16 lit (3.5 Imp gal)
Reserve capacity	3.5 lit (0.8 Imp gal)

Carburettors

Make	Mikuni
Type	B30SS
Bore size	30 mm
Fuel level	4.0 ± 0.5 mm (0.16 ± 0.02 in)
Float height	21.4 ± 1.0 mm (0.84 ± 0.04 in)
Main jet	117.5
Main air jet	1.0
Jet needle	5DFT76
Needle clip position	3rd from top
Needle jet	P-1
Throttle valve	85
Pilot jet	17.5
Bypass	0.9,0.9,0.8
Pilot outlet	0.8
Float valve seat	2.0
Starter jet	35
Pilot screw	Preset (11/2 turns out)
Pilot air jet	132.5
Throttle cable free play	0.5 – 1.0 mm
Idle speed	1250 ± 50 rpm

Lubrication

Type	Wet sump, pumped
Engine oil capacity:	
At oil change	2.0 lit (3.6 Imp pint)
Oil and filter change	2.6 lit (4.6 Imp pint)
After engine overhaul	2.6 lit (4.6 Imp pint)

Oil pump

Pump drive reduction ratio	1.905 : 1 (75/24 x 25/41)
Oil pressure at 60°C (140°F):	
Minimum	1.5 kg/cm^2 (21.4 psi) @ 3000 rpm
Maximum	5.0 kg/cm^2 (71.2 psi) @ 3000 rpm
Pump internal clearances	Not available

Specifications relating to Chapter 3

Ignition coil

Primary winding resistance	3 – 5 ohms
Secondary winding resistance	20 – 30 K ohms

Pulse generator coil

Resistance	300 – 460 ohms

Specifications relating to Chapter 4

Front fork
Spring free length service limit .... 485.2 mm (19.1 in)
Fork oil level .... 190 mm (7.48 in)
Fork oil capacity:
    Right leg .... 165 cc (5.8 Imp fl oz)
    Left leg .... 187 cc (6.6 Imp fl oz)
Oil type .... SAE 15W fork oil

Torque wrench settings

Component	lbf ft	kgf m
Steering stem top bolt	25.3 – 36.2	3.5 – 5.0
Swinging arm pivot shaft nut	36.2 – 57.8	5.0 – 8.0

Specifications relating to Chapter 5

Torque wrench settings

Component	lbf ft	kgf m
Rear wheel sprocket retaining nuts	28.9 – 43.4	4.0 – 6.0
Anti-dive modulator plunger bolt	2.2 – 3.6	0.3 – 0.5
Anti-dive modulator bleed valve bolt	4.3 – 6.5	0.6 – 0.9

Specifications relating to Chapter 6

Battery
Type YB12B-B2

Bulbs
Headlamp 12V 60/55W
Fuel gauge lamp 12V 1.7W

UK GSX400 EZ model
Note: *Specifications are as given for GSX400 EX model at the beginning of each Chapter unless shown below*

Specifications relating to Chapter 1

Pistons
Gudgeon pin OD .... 17.992 – 17.995 mm (0.7084 – 0.7085 in)
Service limit .... 17.980 mm (0.7079 in)

Piston rings
End gap (free):
    Top (N marking) .... About 9.5 mm (0.374 in)
    Service limit .... 7.6 mm (0.299 in)
    Top (R marking) .... About 9.0 mm (0.354 in)
    Service limit .... 7.2 mm (0.284 in)
    2nd (N or R marking) .... About 10.0 mm (0.374 in)
    Service limit .... 8.0 mm (0.315 in)

Valves
Stem end length service limit .... 3.4 mm (0.13 in)
Stem to guide clearance service limit (inlet and exhaust) .... 0.35 mm (0.014 in)

Valve springs
Free length service limit:
    Inner .... 32.2 mm (1.27 in)
    Outer .... 35.5 mm (1.40 in)

Camshafts
Bearing surface ID .... 22.012 – 22.025 mm (0.8666 – 0.8671 in)

Rockers
Rocker arm spindle OD .... 11.970 – 11.980 mm (0.4713 – 0.4717 in)

Crankshaft
Thrust clearance .... 0.04 – 0.30 mm (0.0016 – 0.0118 in)
Service limit .... Not available
Thrust bearing thickness .... 2.950 – 2.975 mm (0.1161 – 0.1171 in)
Service limit .... 2.85 mm (0.112 in)

Clutch

Friction plate thickness	2.65 – 2.95 mm (0.104 – 0.116 in)
Service limit	2.50 mm (0.098 in)
Friction plate maximum warpage	Not available
Friction plate tongue width service limit	15.5 mm (0.61 in)

Specifications relating to Chapter 2

Fuel tank

Overall capacity	16 lit (3.5 Imp gal)
Reserve capacity	3.5 lit (0.8 Imp gal)

Carburettors

Make	Mikuni
Type	BS34SS
Bore	34 mm (1.34 in)
ID number	44230
Fuel level	4.0 ± 1.0 mm (0.16 ± 0.04 in)
Float height	23.0 ± 1.0 mm (0.91 ± 0.04 in)
Main jet	120
Main air jet	0.7
Jet needle	5D69
Needle clip position	4th groove from top
Needle jet	P-2
Pilot jet	42.5
Bypass	0.9, 0.8, 0.8
Pilot outlet	0.8
Float valve seat	2.0
Starter jet	35
Pilot screw (turns out)	Preset (2.0)
Pilot air jet	125
Throttle cable free play	0.5 – 1.0 mm
Engine idle speed	1100 ± 100 rpm

Oil pump

Oil pump reduction ratio	2.071 : 1 (76/28 x 29/38)
Oil pressure at 60°C (140°F):	
Minimum	2.0 kg/cm² (28.5 psi) at 3000 rpm
Maximum	4.0 kg/cm² (56.9 psi) at 3000 rpm
Pump internal clearances	Not available

Specifications relating to Chapter 3

Ignition coil

Primary winding resistance	3 – 5 ohms
Secondary winding resistance	20 – 30 K ohms

Pulse generator coil

Resistance	60 – 80 ohms

Specifications relating to Chapter 4

Front forks

Spring free length service limit	485.2 mm (19.10 in)
Fork oil capacity:	
Right leg	165 cc (5.8 Imp fl oz)
Left leg	187 cc (6.6 Imp fl oz)
Oil type	SAE 15W fork oil

Rear suspension

Travel	100 mm (3.94 in)

Specifications relating to Chapter 5

Brakes

Disc thickness	5.0 ± 0.2 mm (0.20 ± 0.0079 in)
Service limit	Not available
Caliper piston diameter	42.770 – 42.820 mm (1.6839 – 1.6858 in)

Tyre pressures (cold)

Normal riding, solo:
Front	25 psi (1.75 kg/cm^2)
Rear	28 psi (2.00 kg/cm^2)

Normal riding, dual:
Front	25 psi (1.75 kg/cm^2)
Rear	32 psi (2.25 kg/cm^2)

Continuous high speed, solo:
Front	28 psi (2.00 kg/cm^2)
Rear	32 psi (2.25 kg/cm^2)

Continuous high speed, dual:
Front	28 psi (2.00 kg/cm^2)
Rear	36 psi (2.50 kg/cm^2)

Specifications relating to Chapter 6

Battery
Type	TB12B-B2

Bulbs
Headlamp	12V 60/55W
High beam indicator lamp	12V 1.7W
Gear position indicator lamp	12V 1.12W
Fuel gauge lamp	12V 1.7W

US GS450 models

Note: *Specifications are as given for the 400 model at the beginning of each Chapter unless shown below*

Specifications relating to Chapter 1

Engine
Bore	71.0 mm (2.795 in)
Capacity	448cc (27.3 cu in)
Compression ratio	9.0 : 1

Cylinder bores
Standard bore diameter	71.000 – 71.015 mm (2.7951 – 2.7959 in)
Service limit	71.080 mm (2.7984 in)
Maximum ovality	0.10 mm (0.004 in)

Pistons
Piston outside diameter	70.945 – 70.960 mm (2.7931 – 2.7959 in)
Service limit	71.880 mm (2.7905 in)

Ring to groove clearance:
Top	0.020 – 0.055 mm (0.0008 – 0.0022 in)
Service limit	0.180 mm (0.0071 in)
2nd	0.020 – 0.060 mm (0.0008 – 0.0024 in)
Service limit	0.150 mm (0.0059 in)

Ring free end gap:
Top (N)*	9.0 mm (0.35 in) approx
Service limit	7.2 mm (0.28 in)
Top (R) *	9.0 mm (0.35 in) approx
Service limit	7.2 mm (0.28 in)
2nd (N)*	9.5 mm (0.37 in) approx
Service limit	7.6 mm (0.30 in)
2nd (R)*	9.0 mm (0.35 in) approx
Service limit	7.2 mm (0.28 in)

*Note: *Piston rings may be supplied by one of two manufacturers and are stamped (N) or (R) on the top surface to denote this. Check the letter before making the free end gap measurement*

Valves

Stem OD:
Inlet	6.960 – 6.975 mm (0.2740 – 0.2746 in)
Exhaust	6.954 – 6.960 mm (0.2743 – 0.2740 in)
Stem end length	Not available
Service limit	4.0 mm (0.16 in)
Valve guide ID (inlet and exhaust)	7.000 – 7.015 mm (0.2756 – 0.2762 in)

Stem to guide clearance:
    Inlet .... 0.025 – 0.055 mm (0.0009 – 0.0022 in)
    Service limit (ED, LD, LF,TXD) .... 0.35 mm (0.0138 in)
    Service limit (others) .... 0.09 mm (0.0035 in)
    Exhaust .... 0.040 – 0.070 mm (0.0016 – 0.0028 in)
    Service limit (ED, LD, LF, TXD) .... 0.35 mm (0.0138 in)
    Service limit (others) .... 0.10 mm (0.0039 in)
Valve head diameter:
    Inlet .... 35.9 – 36.1 mm (1.41 – 1.42 in)
    Exhaust .... 29.9 – 30.1 mm (1.17 – 1.18 in)
Valve seat width .... 1.0 – 1.2 mm (0.04 – 0.05 in)
Valve lift:
    Inlet .... 8.5 mm (0.33 in)
    Exhaust .... 8.0 mm (0.31 in)

Valve springs
Free length service limit:
    Inner .... 35.52 mm (1.398 in)
    Outer .... 40.51 mm (1.595 in)

Valve clearances (cold engine)
Inlet and exhaust .... 0.03 – 0.08 mm (0.001 – 0.003 in)

Camshafts
Cam height:
    Inlet .... 36.782 – 36.812 mm (1.4481 – 1.4493 in)
    Service limit .... 36.490 mm (1.4366 in)
    Exhaust .... 36.283 – 36.313 mm (1.4285 – 1.4296 in)
    Service limit .... 35.990 mm (1.4169 in)
Camshaft journal oil clearance:
    Inlet and exhaust .... 0.032 – 0.066 mm (0.0013 – 0.0026 in)
    Service limit .... 0.150 mm (0.0059 in)
Camshaft journal OD – inlet and exhaust .... 21.959 – 21.980 mm (0.8645 – 0.8654 in)
Camshaft bearing surface ID – inlet and exhaust .... 22.012 – 22.025 mm (0.8666 – 0.8671 in)

Clutch
Friction plate thickness – ED,LD,LF,TXD .... 2.65 – 2.95 mm (0.104 – 0.116 in)
Service limit – ED,LD,LF,TXD .... 2.40 mm (0.09 in)

Gearbox
Final reduction ratio:
    GS450 L models .... 2.625 : 1 (41/16T)
    GS450 TX,TXD .... 2.687 : 1 (43/16T)
    All other models .... 2.812 : 1 (45/16T)

Specifications relating to Chapter 2

Fuel tank
Overall capacity:
    ET,EX,ST,SX .... 15 lit (4.0 US gal)
    LT,LX,LZ .... 11 lit (2.9 US gal)
    TX,TZ,TXZ .... 12 lit (3.2 US gal)
    EZ .... 14.5 lit (2.8 US gal)
    ED .... 16.0 lit (3.5 US gal)
    LD,TXD,LF .... 13.0 lit (2.9 US gal)
Reserve capacity:
    ET,EX,EZ,ST,SX,LD,TXD,LF .... 3.0 lit (0.79 US gal)
    LT,LX,LZ .... 2.0 lit (0.53 US gal)
    TX,TZ,TXZ .... 4.0 lit (1.06 US gal)
    ED .... 3.5 lit (0.92 US gal)

Carburettors
Make .... Mikuni
Type BS34SS
Venturi size .... 34 mm (1.34 in)
ID number:
    ED 44400
    LD,TXD,LF 44460
    Other models 44110

Idle speed:
    ED,LD,TXD .... 1200 ± 50 rpm
    Other models .... 1200 ± 100 rpm
Fuel level:
    ED,LD,TXD,LF .... 5.0 ± 0.5 mm (0.20 ± 0.02 in)
    Other models .... 6.5 ± 0.5 mm (0.26 ± 0.02 in)
Float height:
    ED,LD,TXD,LF .... 23.0 ± 1.0 mm (0.89 ± 0.04 in)
    Other models .... 22.4 ± 1.0 mm (0.88 ± 0.04 in)
Main jet:
    ED 120
    Other models 115
Main air jet:
    ED,LD,TXD,LF 0.7
    Other models 0.6
Jet needle:
    ED 5DX87
    LD,TXD,LF 5CDT60
    Other models 4C2
Needle jet:
    ED,LD,TXD,LF 0-6
    Other models Y-6
Throttle valve:
    LF 95
    Other models .... Not available
Pilot jet:
    ED,LD,TXD,LF 45
    Other models 17.5
Bypass:
    ED,LD,TXD,LF 1.0,0.8,0.9
    Other models 0.8
Pilot outlet:
    ED,LD,TXD,LF 0.7
    Other models 1.1
Float valve seat 2.0
Starter jet:
    ED,LD,TXD,LF 32.5
    Other models 30.0
Pilot screw setting:
    ED,LD,TXD .... Preset (2.0 turns out)
    Other models .... Preset
Pilot air jet:
    ED 135
    LD,TXD 145
    LF .... Not available
    Other models 1.6

Specifications relating to Chapter 3

Ignition timing
    Below 1650 ± 100 rpm:
        TX .... 20° BTDC
        Other models .... 10° BTDC
    Above 3500 ± 100 rpm .... 40° BTDC

Spark plugs
    Type .... NGK B8ES or ND W24ES-U
    Gap .... 0.6 – 0.8 mm (0.024 – 0.031 in)

Ignition coil
    Primary winding resistance:
        ET,EX .... 3.5 – 4.5 ohms
        TX .... 4.0 – 5.0 ohms
        ED .... 3.5 – 5.0 ohms
        Other models .... 3.0 – 5.0 ohms
    Secondary winding resistance:
        ET,EX .... 23 – 25 K ohms
        TX .... 20 – 23 K ohms
        ED,LD,TXD,LF .... 20 – 30 K ohms
        Other models .... 21 – 25 K ohms

Pulser coil

Resistance:

ED .... 300 – 400 ohms
LD,TXD,LF .... 300 – 460 ohms
Other models .... 60 – 80 ohms

Specifications relating to Chapter 4

Frame

Type .... Welded tubular steel, cradle

Front forks

Type .... Oil damped telescopic
Travel .... 140 mm (5.5 in)
Spring free length service limit:

EZ,ED,TZ,TXZ .... 494 mm (19.4 in)
LT,LX .... 506 mm (19.9 in)
LZ .... 503 mm (19.8 in)
LD .... 504 mm (19.8 in)
LF .... 505 mm (19.9 in)
TXD .... 493 mm (19.4 in)
Other models .... 496 mm (19.5 in)
Oil capacity per leg:

EZ,ED .... 178 cc (6.0 US fl oz)
TZ .... 181 cc (6.1 US fl oz)
TXZ .... 143 cc (4.8 US fl oz)
LZ .... 189 cc (6.4 US fl oz)
LD,LF .... 277 cc (9.5 US fl oz)
TXD .... 155 cc (5.2 US fl oz)
Other models .... 145 cc (4.9 US fl oz)
Oil level:

EZ,ED .... 131 mm (5.2 in)
LT,LX,TXZ .... 207 mm (8.1 in)
TZ .... 125 mm (4.9 in)
LZ .... 117 mm (4.6 in)
LD,LF .... 118.5 mm (4.67 in)
TXD .... 208 mm (8.2 in)
Other models .... 197 mm (7.8 in)
Oil grade .... SAE 15W fork oil
Air pressure:

EZ,ED,TZ,LZ,LD,LF .... 0.5 kg/cm² (7.11 psi)
Other models .... Not applicable

Rear suspension

Type .... Swinging arm fork, controlled by oil damped suspension units
Travel:

EZ,LT,LX,ED,LF .... 100 mm (3.9 in)
Other models .... 95 mm (3.7 in)
Pivot shaft runout (max) .... 0.30 mm (0.012 in)

Specfications relating to Chapter 5

Wheels

Type .... Cast aluminium alloy
Rim runout service limit:

Axial .... 2.0 mm (0.08 in)
Radial .... 2.0 mm (0.08 in)
Wheel spindle runout (max) .... 0.25 mm (0.010 in)

Front brake – except TXD model

Type .... Single hydraulically-operated disc brake
Disc thickness .... 5 ± 0.1 mm (0.20 ± 0.004 in)
Service limit .... 4.5 mm (0.18 in)
Disc runout (max) .... 0.3 mm (0.012 in)
Caliper bore diameter .... 42.850 – 42.926 mm (1.6870 – 1.6900 in)
Caliper piston diameter .... 42.770 – 42.820 mm (1.6839 – 1.6858 in)

Master cylinder bore diameter .... 14.000 – 14.043 mm (0.5512 – 0.5529 in)
Master cylinder piston diameter .... 13.957 – 13.984 mm (0.5495 – 0.5506 in)
Hydraulic fluid type .... DOT 3

Front brake – TXD model
Type .... Twin leading shoe (tls) drum
Drum diameter service limit .... 200.7 mm (7.90 in)
Lining thickness service limit .... 1.5 mm (0.06 in)

Rear brake
Type .... Single leading shoe (sls) drum
Drum diameter service limit:
    TXZ .... 180.7 mm (7.11 in)
    Other models .... 160.7 mm (6.33 in)
Lining thickness service limit .... 1.5 mm (0.06 in)

Tyres
Size, front:
    ET,EX,EZ,LT,LX,ST,SX 3.00S18-4PR
    TX,TZ,TXZ 90/90-19 52S
    LZ 3.60S19-4PR
Size, rear:
    ET,EX,EZ,LT,LX,ST,SX 3.50S18-4PR
    TX,TZ,TXZ 110/90-17 60S
    LZ 4.60S16-4PR

Tyre pressures, psi (kg/cm²)

	Front	Rear
ET,EX,EZ,LT,LX,ST,SX:		
Normal riding:		
Solo	25 (1.75)	28 (2.00)
Dual	25 (1.75)	32 (2.25)
Continuous high speed:		
Solo	28 (2.00)	32 (2.25)
Dual	32 (2.25)	36 (2.50)
TX,TZ,TXZ,TXD:		
Normal riding:		
Solo	22 (1.50)	24 (1.75)
Dual	24 (1.75)	32 (2.25)
Continuous high speed:		
Solo	24 (1.75)	28 (2.00)
Dual	28 (2.00)	36 (2.50)
LZ,ED:		
Normal riding:		
Solo	25 (1.75)	28 (2.00)
Dual	25 (1.75)	32 (2.25)
Continuous high speed:		
Solo	28 (2.00)	32 (2.25)
Dual	28 (2.00)	36 (2.50)
LD,LF:		
Normal riding:		
Solo	28 (2.00)	32 (2.25)
Dual	28 (2.00)	36 (2.50)

Specifications relating to Chapter 6

Battery
Capacity .... 12 volt 12 Ah (Ampere-hour)
Type:
    ED,LD YB12B-B2
    Other models YB10L-A2
Earth (ground) .... Negative (–)
Electrolyte specific gravity .... 1.28 @ 20°C (68°F)

Alternator
No load voltage .... More than 75V (ac) @ 5000 rpm
Regulated voltage .... 14 – 15V @ 5000 rpm

Starter motor
Brush length service limit .... 9 mm (0.4 in)
Commutator undercut service limit .... 0.2 mm (0.008 in)

Starter relay
Resistance .... 3 – 4 ohms (approx)

Main fuse
Rating .... 15 amp

Bulbs (all at 12 volts)
Headlamp:
 LT,LX,LZ,LD,LF,TXD 50/35W
 ED 60/55W
 Other models 45/40W
Tail/brake lamp 8/23W
Turn signals 23W
Speedometer illumination 3.4W
Tachometer illumination 3.4W
Turn signal warning 3.4W
High beam warning:
 LD,TXD,LF 1.7W
 Other models 3.4W
Neutral indicator 3.4W
Oil pressure warning 3.4W
Side stand warning – LD,TXD,LF 3.4W
Gear position indicator – ED,LD,TXD,LF 1.12W
Fuel gauge illumination:
 ED,LD 1.7W
 LF 3.4W

Torque wrench settings – Chapters 1-6

Engine	lbf ft	kgf m
Cylinder head bolt	5.0 – 8.0	0.7 – 1.1
Cylinder head nuts	25.5 – 30.0	3.5 – 4.0
Cylinder head cover bolts	6.5 – 7.0	0.9 – 1.0
Cam chain tensioner bolts	4.5 – 6.0	0.6 – 0.8
Cam chain tensioner shaft locknut	6.5 – 10.0	0.9 – 1.4
Cam chain tensioner shaft assembly	22.5 – 25.5	3.1 – 3.5
Camshaft cap retaining bolts	6.0 – 8.5	0.8 – 1.2
Camshaft sprocket retaining bolts	6.0 – 8.5	0.8 – 1.2
Balancer shaft centre bolt	25.5 – 32.5	3.5 – 5.5
Connecting rod cap nuts	21.5 – 24.5	3.0 – 3.4
Crankshaft pinion nut	65.1 – 79.5	9.0 – 11.0
Starter clutch Allen bolts	11.0 – 14.5	1.5 – 2.0
Starter motor retaining bolts	3.0 – 5.0	0.4 – 0.7
Clutch centre nut	30.0 – 43.5	4.0 – 6.0
Clutch spring retaining bolts	3.0 – 4.5	0.4 – 0.6
Gearchange stopper arm pivot bolt	11.0 – 16.5	1.5 – 2.3
Alternator rotor bolt:		
EZ,TZ,TXZ,LZ	65.0 – 72.5	9.0 – 10.0
Other models	43.5 – 50.5	6.0 – 7.0
Automatic timing unit bolt	9.5 – 16.5	1.3 – 2.3
Crankcase holding bolts:		
Nos 1 to 8	14.5 – 17.5	2.0 – 2.4
Nos 9 to 12	6.5 – 9.5	0.9 – 1.3
6 mm	7.0	1.0
8 mm	14.5	2.0
Gearbox sprocket retaining nut:		
EZ,TZ,TXZ,LZ	58.0 – 72.0	8.0 – 10.0
Other models	36.0 – 50.5	5.0 – 7.0
Sump (oil pan) bolts	7.0	1.0
Oil pressure switch	9.5 – 12.5	1.3 – 1.7
Neutral stopper housing	13.0 – 20.0	1.8 – 2.8
Engine mounting bolts:		
8 mm	14.5 – 21.5	2.0 – 3.0
10 mm	21.5 – 26.8	3.0 – 3.7
Exhaust pipe clamp bolts	6.5 – 10.0	0.9 – 1.4
Chassis		
Front wheel spindle	26.0 – 37.5	3.6 – 5.2
Front wheel spindle clamp	11.0 – 18.0	1.5 – 2.5
Front fork damper rod bolt	11.0 – 18.0	1.5 – 2.5
Brake disc mounting bolts	11.0 – 18.0	1.5 – 2.5
Caliper mounting bolts	18.0 – 29.0	2.5 – 4.0
Caliper pivot bolt	11.0 – 14.5	1.5 – 2.0
Caliper bleed nipple	4.5 – 6.5	0.6 – 0.9

Component	lbf ft	kgf m
Fork upper yoke pinch bolt	14.5 – 21.5	2.0 – 3.0
Fork lower yoke pinch bolt	18.0 – 29.0	2.5 – 4.0
Fork cap bolt – where fitted	11.0 – 21.5	1.5 – 3.0
Steering stem nut	29.0 – 36.0	4.0 – 5.0
Steering stem pinch bolt	11.0 – 18.0	1.5 – 2.5
Steering stem top bolt	26.0 – 37.5	3.6 – 5.2
Handlebar clamp bolt	8.5 – 14.5	1.2 – 2.0
Hydraulic hose union bolt	14.5 – 18.0	2.0 – 2.5
Master cylinder clamp bolt	3.5 – 6.0	0.5 – 0.8
Brake pedal mounting bolt	7.0 – 11.0	1.0 – 1.5
Swinging arm pivot nut	36.0 – 42.0	5.0 – 5.8
Front footrest mounting bolts:		
10 mm	19.5 – 31.0	2.7 – 4.3
8 mm	11.0 – 18.0	1.5 – 2.5
Silencer bracket bolt	19.5 – 31.0	2.7 – 4.3
Rear footrest mounting bolt	19.5 – 31.0	2.7 – 4.3
Silencer to bracket bolt	14.5 – 21.5	2.0 – 3.0
Rear brake torque arm bolt	14.5 – 21.5	2.0 – 3.0
Rear suspension unit nut	14.5 – 21.5	2.0 – 3.0
Rear wheel spindle nut	36.0 – 58.0	5.0 – 8.0
Rear sprocket nut	18.0 – 29.0	2.5 – 4.0
Rear brake arm bolt	3.5 – 6.0	0.5 – 0.8
Spoke nipple	3.0 – 3.5	0.4 – 0.5

1 Introduction

This update Chapter covers the US GS450 models and the UK GSX 250 EZ and GSX400 EZ models, and should be used in conjunction with the main Chapters of this manual. In most respects the above models are broadly similar to the earlier versions, but where differences exist they are discussed below.

The GS450 models can be regarded as an amalgamation of the GS and GSX range featured in the earlier Chapters, and the older GS400/425 series. A summary of the GS450 range is shown below:

GS450 ET Basic model produced for 1980
GS450 EX Basic model produced for 1981
GS450 EZ Basic model produced for 1982
GS450 ED Basic model produced for 1983
GS450 ST Sports, or cafe version of the 1980 ET
GS450 SX Sports, or cafe version of the 1981 EX
GS450 LT Custom version of the 1980 ET
GS450 LX Custom version of the 1981 EX
GS450 LZ Custom version of the 1982 EZ
GS450 LD Custom version of the 1983 ED
GS450 LF 1985 Custom model
GS450 TX Semi-Custom version of the 1981 EX
GS450 TZ Semi-Custom version of the 1982 EZ
GS450 TXZ As TZ, with detail changes
GS450 TXD 1983 version of the TXZ

In the case of the UK models, the GSX250 and 400 EX models remained available until 1984. Meanwhile, the Katana-inspired EZ versions were introduced in 1982 and were sold alongside the earlier version at a higher price. No machines were officially listed for 1985, though some may have been registered well after the end of 1984.

2 Cylinder head: removal – 450 models

1 The breather cover, fitted to the cylinder head cover, need not be removed unless on subsequent inspection the engine is found to have suffered extreme sludging or carbon build-up. If this is the case, the oil separator mesh inserts fitted beneath the cover should be removed and cleaned.
2 Remove the four caps which enclose the projecting ends of the camshaft and cylinder head castings. Each cap is secured by two screws. Loosen and remove the bolts securing the cylinder head cover. The bolts should be slackened evenly, in a diagonal sequence, to help prevent distortion. If necessary, use a rawhide mallet to break the seal between the cylinder head cover and the gasket, and then lift the cover away.
3 Before attention is given to the camshafts and chain, the automatic chain tensioners must be detached from the rear of the cylinder block. The tensioner plunger within the body of the tensioner is spring-loaded. To prevent the plunger flying out when the body is removed, loosen the locknut on the left-hand side of the body and tighten the stop screw which runs inside the locknut. This will clamp the tensioner plunger in position. Once locked, slacken and remove the two bolts which retain the tensioner body to the cylinder block and lift the tensioner assembly away.
4 Withdraw the cam chain guide from its recess in the front of the chain tunnel, and place it to one side. Slacken the camshaft cap retaining bolts evenly and progressively, turning each one by about $\frac{1}{2}$ a turn at a time. Once loose, remove the bolts and caps, placing them to one side. Note that each cap and the camshafts are clearly identifiable, so it is not necessary to mark these components. The camshafts can now be lifted away from the cylinder head, disengaged from the cam chain, and removed. Note that unless the engine is to be stripped completely it is important to avoid letting the cam chain drop down into the crankcase. This can be accomplished by passing a bar or similar item through the chain loop, allowing it to rest across the cylinder head. The tachometer driven gear should be removed at this stage and must not be refitted until **after** the camshafts have been installed. Slacken and remove the single bolt and retainer plate which secures it to the cylinder head and pull the assembly clear.
5 The cylinder head is retained by eight 10 mm domed nuts, plus a 6 mm bolt located between the exhaust ports. Start by removing the 6 mm bolt. Each of the 10 mm nuts has a number cast into the cylinder head next to it. This indicates the tightening sequence, and the nuts should be removed in the **reverse** of this sequence. Slacken each one by about $\frac{1}{2}$ of a turn at a time until all are loose and can be run off the holding studs.
6 If necessary, use a rawhide mallet to break the seal between the cylinder head and gasket. Strike only those parts of the casting which are well supported by lugs or webs. Under no circumstances should levers be used to raise the head; this will result in broken fins. Where required, the cam chain should be prevented from falling down by rearranging the temporary securing wire or dowel. After lifting the cylinder head from position remove the two locating dowels and the gasket.

2.4 Withdraw the cam chain guide from its recess in the cylinder head

2.6 Removing the cylinder head

3 Cylinder head: installation and valve timing – 450 models

1   Check that the gasket faces of the cylinder head and cylinder block are clean and dry. Place a new cylinder head gasket over the holding studs. Fit the two locating dowels into their respective drillings in the cylinder head, tapping them home. Offer up the cylinder head, feeding the cam chain through the central tunnel as it is lowered into position.

2   Fit new copper sealing washers over the eight holding studs. In an emergency the old washers may be re-used if they are undamaged, but should be annealed first. This is accomplished by heating them to a dull red colour in a blowlamp flame, then quenching in water. This process will soften the work-hardened copper, allowing it to seal over any small irregularities. Fit the domed nuts finger-tight only at this stage.

3   Each of the cylinder head nuts is identified by a number cast into the cylinder head. This indicates the order in which the nuts must be tightened to prevent the cylinder head from becoming warped. Set the torque wrench to 3.5 – 4.0 kgf m (25.5 – 29.0 lbf ft) and tighten each nut progressively and in the indicated sequence until this figure is reached. Fit the 6 mm bolt between the exhaust ports and tighten it to 0.7 – 1.1. kgf m (5.0 – 8.0 lbf ft).

4   Remove the three screws which retain the inspection cover to the right-hand side of the crankcase, lifting the cover away to expose the ignition pickup assembly. Have ready a 19 mm, socket or ring spanner with which the crankshaft can be rotated via the large hexagon provided for this purpose. Note that on no account should any attempt be made to turn the crankshaft by means of the 12 mm retaining bolt.

5   Lift the bar supporting the cam chain until any slack has been removed. This will obviate any risk of the chain's bunching and jamming the crankshaft sprocket. Rotate the crankshaft clockwise until the T mark is aligned with the fixed pointer. This can be viewed through the inspection window in the backplate. The crankshaft must remain in this position while the camshafts are fitted. If at any stage there is some reason to suspect that it has moved, stop and re-check the setting.

6   Each camshaft has identification marks cast into it, the inlet shaft being marked 'IN', 'L' and 'R' and the exhaust shaft being marked 'EX', 'L' and 'R'. In addition, the exhaust camshaft can be identified by its integral tachometer drive gear. When fitting the shafts check that the 'L' and 'R' marks are on the left and right respectively. Coat all camshaft bearing surfaces with molybdenum disulphide grease prior to installation.

7   Fit the exhaust camshaft first. As it is positioned, pull the cam chain taut between it and the crankshaft sprocket. When fitted, the '1' mark on the camshaft sprocket **must** point forward and parallel to the gasket face. The '2' mark will point directly upwards. Note the cam chain roller pin at which the '2' arrow is pointing. If possible, this should be marked using a spirit-base felt marker or similar. Count the pins along the chain towards the inlet camshaft area, and mark the 18th one. Engage the inlet camshaft sprocket in the chain, ensuring that the '3' arrow points towards this pin. The inlet camshaft can now be located in its recess in the cylinder head. Note that the rectangular notches in the right-hand end of the two camshafts should face each other and lie parallel with the gasket face. Re-check each timing mark, referring to Fig. 7.1 to confirm that everything has been aligned properly.

8   Fit the camshaft bearing caps, noting that the letter in the cast-in triangle on each one must correspond with that cast into the cylinder head nearby. Note also that the point of the triangle should face forward. Drop the camshaft cap bolts into positon, checking that each has a '9' on its head. This indicates that the bolts are the correct high tensile type; do not use any other bolt to retain the caps or breakage may occur. When tightening the bolts, make sure that the caps are drawn down evenly and progressively to avoid any risk of distortion. The correct torque setting for the bolts is 0.8 – 1.2 kgf m (6.0 – 8.5 lbf ft).

9   The tensioner unit should now be set up ready for installation. Holding the end of the plunger to prevent it from flying out of the body, slacken the locknut and stop screw. The plunger will be under pressure from the mechanism and will move outwards if it is allowed to. To retract the plunger, turn the large knurled knob anticlockwise against spring pressure. The plunger should now be pushed inwards and held in this position. Tighten the stop screw to retain the plunger in the retracted position. Do not release the stop screw until the tensioner has been installed.

10 Fit the cam chain guide at the front of the tunnel, making sure that it locates correctly. Check that the slack run in the cam chain is near the back of the chain tunnel so that the tensioner assembly can be installed. If necessary, rotate the crankshaft slowly until slack appears. Fit the tensioner body using a new gasket and tighten its retaining bolts.

11 Slacken the stop bolt by $\frac{1}{2}$ turn. An audible click should indicate that the plunger has moved forward to take up slack in the chain. Hold the stop bolt in this position with a screwdriver, and secure the locknut. Check the operation of the tensioner mechanism by turning the knurled wheel anticlockwise. Holding this position, turn the crankshaft anticlockwise so that the cam

chain is pulled taut at the rear of the tunnel, thus pushing the plunger into the housing. Release the knurled wheel and then turn the crankshaft clockwise whilst watching the knurled wheel. What should happen is that the wheel should rotate slowly indicating that the plunger is moving inwards to take up slack in the chain. If the wheel sticks or moves jerkily it will be necessary to remove and check the operation of the tensioner mechanism before proceeding further. Note that once the mechanism has been set up and checked as described above, the knurled wheel should be left undisturbed until the tensioner is next dismantled.

12 Check, and where necessary adjust, the valve clearances as described in Section 4 of this Chapter. Fit the tachometer drive gearbox into its recess in the cylinder head, retaining it with its locating plate and bolt. It should be noted that installation has been left until now to obviate any risk of damage to the gear teeth during camshaft installation. It is far easier to ensure that the teeth mesh properly by fitting the gearbox with the camshaft already in position.

13 Before fitting the cylinder head cover, pour a small quantity of engine oil into the camshaft lobe recesses to ensure lubrication of the cams and valve gear when the engine is first started. Check that the gasket faces are clean and dry, then fit and secure the cylinder head cover followed by the camshaft end caps. Use new semi-circular end seals and a new gasket, ensuring that both locate correctly.

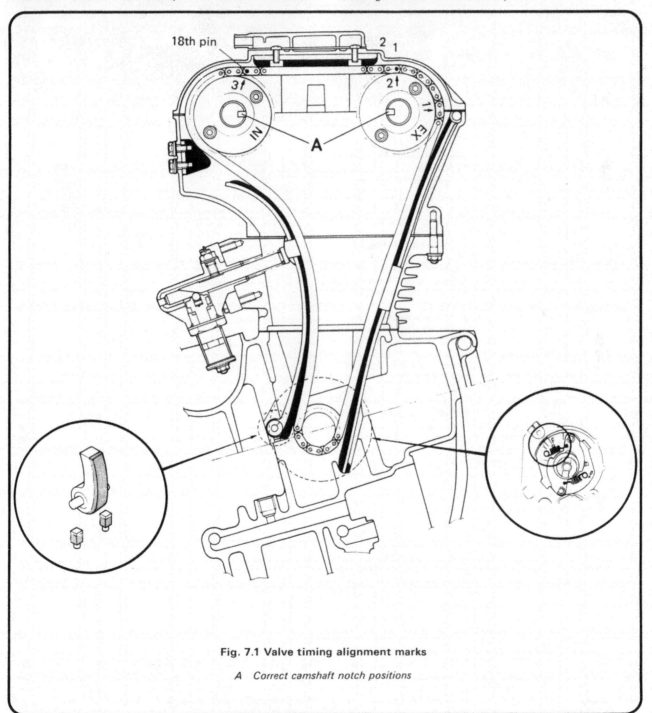

Fig. 7.1 Valve timing alignment marks

A   Correct camshaft notch positions

3.2 Place new copper washers over the eight holding studs

3.8 Bearing caps are marked and should be fitted accordingly

4   Valve clearance adjustment: 450 models

1   Unlike the models featured in the earlier Chapters of this manual, valve clearance adjustment on the GS450 models involves the removal and fitting of shims, or adjuster pads, between the cam and cam follower or tappet.

2   To gain access to the camshafts and cam followers the petrol tank must be removed and the camshaft cover detached. After removal of the cover bolts, the seal between the cover and the gasket may be broken by the judicious use of a rawhide mallet. Strike only those parts of the cover which are well supported by lugs.

3   Unscrew the spark plugs and remove the inspection cover from the right-hand side of the engine. The clearance between each cam and cam follower must be checked and if necessary adjusted by removal of the existing adjuster pad and replacement of a pad of suitable thickness. Make the clearance check and adjustment of each valve in sequence and then continue with the next valve. Commence by rotating the engine so that the exhaust camshaft right-hand lobe is pointing vertically up. The two exhaust valve clearances can now be checked. Having done this rotate the engine a further 180° until the inlet camshaft left-hand lobe is pointing vertically up. The two inlet valve clearances can now be checked. Rotate the engine in a forward direction by means of the engine turning hexagon on the right-hand end of the crankshaft using a 19 mm socket or ring spanner.

4   Using a feeler gauge determine and record the clearance at the first valve. If the clearance is incorrect, not being within the range of 0.03 – 0.08 mm (0.001 – 0.003 in), the adjuster pad must be removed and replaced by one of suitable thickness. A special tool is available (Suzuki part No. 09916 – 64510) which may be pushed between the camshaft adjacent to the cam lobe and the raised edge of the cam follower, to allow removal of the shim. If the special tool is not available, a simple substitute may be fabricated from a portion of steel plate. The final form of the tool which has a handle about 6 inches long, can be seen in the accompanying photograph.

5   The Suzuki tool may be pushed into position, depressing the cam follower and securing it in a depressed position, in one operation. Where a home-made tool is used, the cam follower may be depressed using a suitable lever placed between the adjuster pad and the cam lobe. The tool may then be inserted to secure the cam follower whilst the adjuster pad is removed. Before installing either type of tool, rotate the cam follower so that the slot in the raised edge is not obscured by the camshaft.

Insert a small screwdriver through the slot to displace the adjuster shim.

6   Adjustment pads are available in 20 sizes ranging from 2.15 mm to 3.10 mm in increments of 0.05 mm. Each pad is identified by a three digit number etched on the reverse face. The number (eg. 235) indicates that the pad thickness is 2.35 mm thick. To select the correct pad subtract 0.03 mm from the measured clearance and add the resultant figure to that of the existing pad. Select the largest available pad whose thickness is slightly smaller than the final figure. Refer to the accompanying table for the selection of available pads.

7   Although the adjuster pads are available as a complete set their price is prohibitive. It is suggested that pads are purchased individually, after an accurate assessment of requirement has been made. It is possible that some Suzuki service agents will be prepared to exchange needed pads for others of the correct size, providing that the original pads are not worn.

8   Before installing a replacement pad, lubricate both sides thoroughly with engine oil. Always fit the pad with the identification number downwards, so that it does not become obliterated by the action of the cam. After fitting new adjuster pads, rotate the engine forwards a number of times and then recheck the clearances to verify that no errors have occurred.

9   Before refitting the cam cover, together with a new gasket, lubricate the camshafts with copious quantities of clean engine oil.

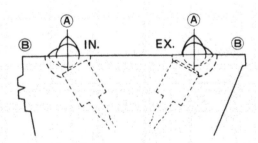

Fig. 7.2 Correct camshaft positions for valve clearance check

Move camshafts to positions A or B to check the clearance and remove the shim

4.4a Checking the valve clearance

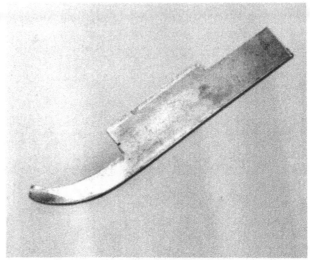
4.4b Fabricated tool to remove adjustment shim

4.5a Depress the cam follower to permit ...

4.5b ... removal of shim

4.6 Adjustment shims are marked for size

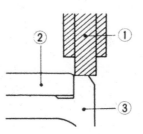
Fig. 7.3 Manufacturers tappet shim removal tool in use

1   Tool (pt no 09916-64510)   3   Cam follower
2   Shim

PART NUMBER – PREFIX 12892
PRESENT SHIM SIZE – mm

P/N SUFFIX-	45000	45001	45002	45003	45004	45005	45006	45007	45008	45009	45010	45011	45012	45013	45014	45015	45016	45017	45018	45019
Tappet Clearance (mm)	2.15	2.20	2.25	2.30	2.35	2.40	2.45	2.50	2.55	2.60	2.65	2.70	2.75	2.80	2.85	2.90	2.95	3.00	3.05	3.10
0.00~0.02		2.15	2.20	2.25	2.30	2.35	2.40	2.45	2.50	2.55	2.60	2.65	2.70	2.75	2.80	2.85	2.90	2.95	3.00	3.05
0.03~0.08	SPECIFIED CLEARANCE: NO ADJUSTMENT REQUIRED																			
0.09~0.13	2.20	2.25	2.30	2.35	2.40	2.45	2.50	2.55	2.60	2.65	2.70	2.75	2.80	2.85	2.90	2.95	3.00	3.05	3.10	
0.14~0.18	2.25	2.30	2.35	2.40	2.45	2.50	2.55	2.60	2.65	2.70	2.75	2.80	2.85	2.90	2.95	3.00	3.05	3.10		
0.19~0.23	2.30	2.35	2.40	2.45	2.50	2.55	2.60	2.65	2.70	2.75	2.80	2.85	2.90	2.95	3.00	3.05	3.10			
0.24~0.28	2.35	2.40	2.45	2.50	2.55	2.60	2.65	2.70	2.75	2.80	2.85	2.90	2.95	3.00	3.05	3.10				
0.29~0.33	2.40	2.45	2.50	2.55	2.60	2.65	2.70	2.75	2.80	2.85	2.90	2.95	3.00	3.05	3.10					
0.34~0.38	2.45	2.50	2.55	2.60	2.65	2.70	2.75	2.80	2.85	2.90	2.95	3.00	3.05	3.10						
0.39~0.43	2.50	2.55	2.60	2.65	2.70	2.75	2.80	2.85	2.90	2.95	3.00	3.05	3.10							
0.44~0.48	2.55	2.60	2.65	2.70	2.75	2.80	2.85	2.90	2.95	3.00	3.05	3.10								
0.49~0.53	2.60	2.65	2.70	2.75	2.80	2.85	2.90	2.95	3.00	3.05	3.10									
0.54~0.58	2.65	2.70	2.75	2.80	2.85	2.90	2.95	3.00	3.05	3.10										
0.59~0.63	2.70	2.75	2.80	2.85	2.90	2.95	3.00	3.05	3.10											
0.64~0.68	2.75	2.80	2.85	2.90	2.95	3.00	3.05	3.10												
0.69~0.73	2.80	2.85	2.90	2.95	3.00	3.05	3.10													
0.74~0.78	2.85	2.90	2.95	3.00	3.05	3.10														
0.79~0.83	2.90	2.95	3.00	3.05	3.10															
0.84~0.88	2.95	3.00	3.05	3.10																
0.89~0.93	3.00	3.05	3.10																	
0.94~0.98	3.05	3.10																		
0.99~1.03	3.10																			

I.   Measure tappet clearance. "ENGINE is COLD"
II. Measure present shim size.
III. Match clearance in vertical column with present shim size in horizontal column.

EXAMPLE
Tappet clearance is   – 0.55 mm
Present shim size   – 2.40 mm
Shim size to be used   – 2.90 mm

Fig. 7.4 Tappet shim selection chart

5 Refitting the clutch and alternator covers: all models

1   To eliminate any possibility of oil leakage, Suzuki recommend that the following points are noted when refitting the engine side covers: Always use new gaskets as a matter of course. Apply a good quality sealant across the crankcase joint area. The sealant should be applied in strips about 20 – 30 mm (0.8 – 1.2 in) in length to ensure a good seal. Use Suzuki Bond No. 4 or an RTV silicone-rubber sealant.

6 Cylinder head modification: GSX400 EZ model

1   There was a minor change to the layout of the cylinder head cover retaining bolts effective from Engine No. GSX400 – 128131 onwards. This was designed to improve sealing between the two components. Whilst this has little effect on the procedure for removing or refitting the cover, note that if at some stage the new type head is fitted to an older model, it will also be necessary to fit the new type cover. Details of the changes are shown in the accompanying line drawing.

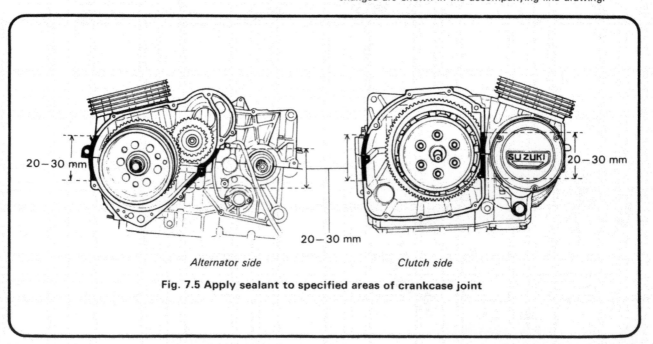

Alternator side   Clutch side

Fig. 7.5 Apply sealant to specified areas of crankcase joint

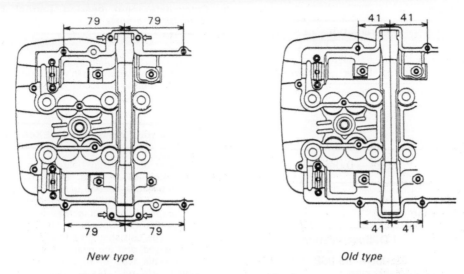

New type   *Old type*

Fig. 7.6 Modified cylinder head and cover bolt positions – GSX400 EZ model

7 Valve rocker arm modification: GSX400 EZ model

1 The later model GSX400 engine employed a revised exhaust valve rocker arm in which the oil groove on the upper face was omitted. The plain rocker arm fitted on the later machines is identical to the inlet rocker. The modification was effective from Engine No. 128037 onwards.

2 When fitting new exhaust rockers to the earlier engine, the later plain type should be fitted, these being supplied as standard by Suzuki. Note that the two rockers are interchangeable without modification.

8 Big-end bearings: examination and renovation – GSX250/400 EZ models

1 The big-end bearing shells were modified in the case of the models covered by this update, this being effective for the GSX400 from Engine No. 102254 onwards and for all GSX250 EZ models. The earlier models made use of an aluminium-based shell material, whilst on later machines this was changed to a copper-based type. The new type shells can be identified by the light silver-coloured plating on their outer surfaces. Four alternate standard size shells are available to match crankshafts of various tolerances, and in addition, two

undersizes can be obtained. Big-end bearing wear can be assessed as described in Chapter 1, Section 17, noting the following points.

2 Check the crankpin journals visually, noting that scoring or other damage will necessitate renewal of the crankshaft. Measure the clearance between the bearing surface and the crankpin using Plastigage, this being supplied by Suzuki dealers as part number 09900-22301. Tighten the bearing caps down in two stages, first to 1.2 – 1.8 kgf m (8.7 – 13.0 lbf ft), and finally to 3.0 – 3.4 kgf m (21.7 – 24.6 lbf ft). Remove the cap and measure the compressed Plastigage strip at its widest point against the scale on the envelope. Note that the service limit is 0.080 mm (0.003 in).

3 If the clearance exceeds this figure, check the connecting rod ID code which is printed across the machined face of the rod and cap and which will be either 1 or 2. Check the corresponding crankpin OD code which is marked on the crankshaft in the position shown in the accompanying line drawing. Select new shells according to the accompanying table. Always replace bearings as a set, never singly.

4 Note that when assembling the connecting rods the bearing shells should be coated with a thin film of molybdenum disulphide grease. The oil hole in the connecting rod should face towards the rear of the engine. To allow the new shells to bed in correctly, keep the engine speed below 4000 rpm for the first 1000 km (620 miles) and 5000 rpm up to 2000 km (1240 miles).

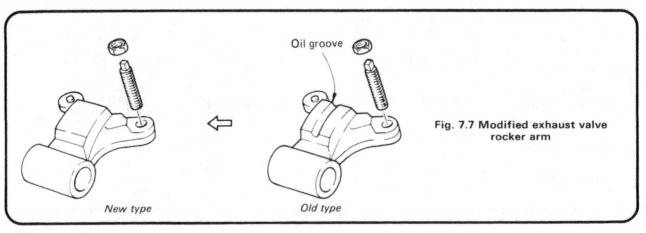

Oil groove

New type   *Old type*

Fig. 7.7 Modified exhaust valve rocker arm

Bearing selection table

		Crank pin		
	Code	1	2	3
Conrod	1	Green	Black	Brown
	2	Black	Brown	Yellow

Oil clearance

Standard	0.024 – 0.048 mm

Conrod I.D. specification

Code	I.D. Specification
1	35.000 – 35.008 mm
2	35.008 – 35.016 mm

Crank pin O.D. specification

Code	O.D. Specification
1	31.992 – 32.000 mm
2	31.984 – 31.992 mm
3	31.976 – 31.984 mm

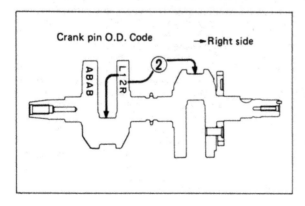

Bearing thickness

Color (Part No.)	Thickness
Green (12164-33210-010)	1.484 – 1.488 mm
Black (12164-33210-020)	1.488 – 1.492 mm
Brown (12164-33210-030)	1.492 – 1.496 mm
Yellow (12164-33210-040)	1.496 – 1.500 mm

Fig. 7.8 Big-end bearing shell selection tables – GSX250 and 400 EZ models

9 Crankshaft main bearing journal code location: GSX250/400 EZ models

1   The code letters indicating the diameter of each main bearing journal will be found stamped in a line along the left-hand crankshaft web as shown in the accompanying illustration. On the earlier models the code for each journal was located adjacent to it (see Fig. 1.6).

10 Alternator rotor removal: all GSX400 models

1   The design of the rotor holding bolt was changed from Engine No. 127536 onwards. The old arrangement used a conventional hexagon-headed bolt with a spring washer. This was changed to a flange bolt used without a spring washer. The rotor bolt torque setting was increased for both types from 9.0 – 10.0 kgf m (65 – 72 lbf ft) to 11.0 – 13.0 kgf m (80 – 94 lbf ft). The new type bolt should be fitted to all models if it or the rotor are renewed, and the new torque setting should be used regardless of the type of bolt fitted. Use a thread locking compound such as Suzuki type 1305 (part number 99000-32100) or an equivalent type.

11 Clutch modifications: GSX400 EZ

1   The clutch pushrod, pushrod oil seal and release mechanism were modified from Engine No. 128041 onwards. The original 251 mm one-piece pushrod was replaced by a two-piece arrangement, the longer (151 mm) part being fitted on the clutch side and the shorter (100 mm) section towards the release mechanism. Note that the longer section is fitted from the clutch side of the shaft and thus must be installed before the clutch pressure plate is fitted.
2   The oil seal was changed to a type having a spring band around the lip to ensure an oil-tight seal on the pushrod. The later pattern seal was supplied as the standard replacement part for all models, the earlier type being replaced by it. The clutch release mechanism adjuster screw was changed from 6.5 mm to 6.2 mm in diameter.
3   With the exception of the oil seal, there is no inter-changeability of parts between the early and later pattern clutches, and dealers stock, or can order, parts for either type. If it becomes necessary to renew a complete clutch release assembly on an early machine it may be worthwhile fitting the later type complete. If only some parts are required, the original type must be ordered. Note that when ordering parts it is particularly important to specify the engine number and whether a one-piece or two-piece pushrod is fitted.

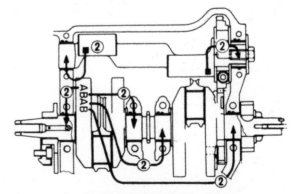

Fig. 7.9 Location of crankshaft main bearing journal code – GSX250 and 400 EZ models

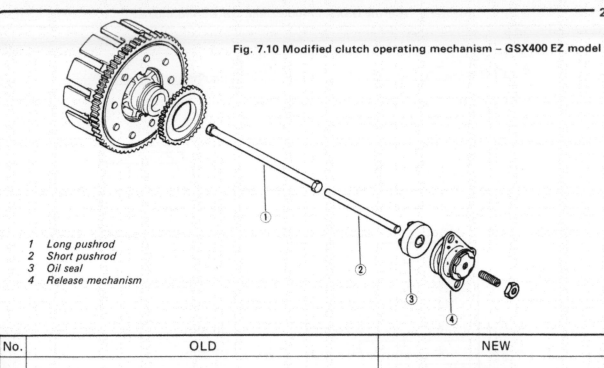

Fig. 7.10 Modified clutch operating mechanism – GSX400 EZ model

1 Long pushrod
2 Short pushrod
3 Oil seal
4 Release mechanism

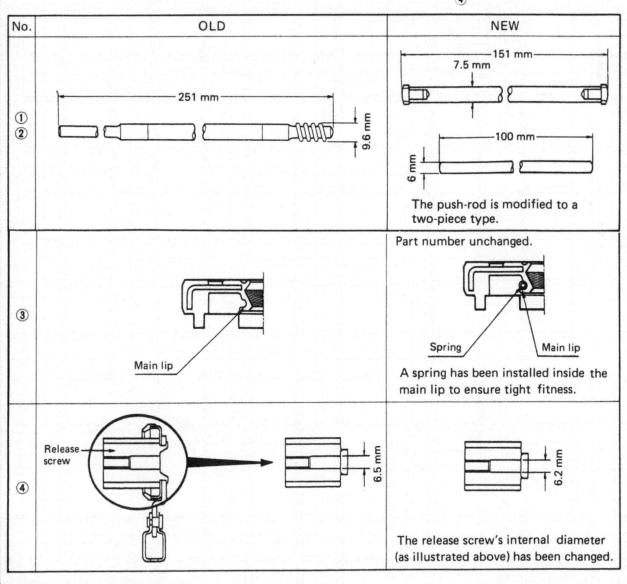

No.	OLD	NEW
①②	251 mm / 9.6 mm	151 mm, 7.5 mm / 100 mm, 6 mm — The push-rod is modified to a two-piece type.
③	Main lip	Part number unchanged. Spring, Main lip — A spring has been installed inside the main lip to ensure tight fitness.
④	Release screw / 6.5 mm	6.2 mm — The release screw's internal diameter (as illustrated above) has been changed.

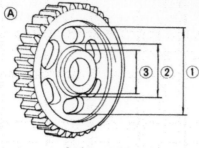

2nd gear pinion

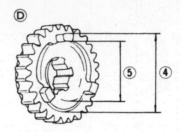

Top gear pinion

	OLD	NEW
①	54 mm	56 mm
②	34.5 mm	36 mm
③	33.5 mm	35 mm
④	50 mm	51 mm
⑤	36 mm	37 mm

Old type   New type

2nd gear pinion bush modification

Fig. 7.11 Gearbox modification – GSX250 and 400 EZ models

12 Gearbox modifications: GSX250/400 EZ models

1   In the case of the GSX400 EZ model, the output shaft 2nd and top gear pinions are modified, effective from Engine No.127534 onwards. The modifications relate to the engagement dogs of the two gears and are designed to improve gear selection. The dimensions of the two pinions were altered, and this is shown in the accompanying line drawing.

2   Another change to the 2nd gear pinion was the adoption of a headed bush in place of a plain bush and separate thrust washer, this change also being applicable to the GSX250 EZ model.

13 Carburettor modifications: GSX400 EZ model

1   The design of the synchronisation screw (item 49 in Fig. 2.2) was changed on the EZ model, the new type screw having a larger head. This will normally have no real consequence unless the adjustment tool, part number 09913-14910 is being used. This will will not fit the later screw, for which a new tool, part number 09913-14911 is available. The later tool will fit either screw head.

14 Ignition system modifications: GSX250/400 EZ, GS450 ED, LD, LF, TXD models

1   The ignition pickup assembly was modified on the above models. The mechanical automatic timing unit (ATU) used on all previous models was omitted and a fixed rotor, or reluctor, used in its place. Little information is available on this particular change, but it can be assumed that the ignition advance is carried out electronically. This modification leaves the system entirely non-mechanical in operation and thus it is no longer subject to wear. In consequence, the periodic need to check the ignition timing is obviated. The newer-type fixed rotor is retained by a centre bolt in much the same way as the older ATU. A large hexagon is provided which can be used to hold the rotor while the retaining bolt is removed or tightened.

15 Front forks: general

1   The models covered in this update Chapter use forks of four basic types. Starting with the US models, the GS450 ET, EX, LT, LX, ST and SX models use straightforward oil-damped coil spring units as described in Chapter 4, Section 4. The GS450 EZ, ED, LZ, TZ and TXZ models have forks which are

broadly similar to the earlier models, but which have Schraeder-type air valves built into the fork top plugs. This allows the coil springs to be assisted by air pressure, and thus provides a degree of spring rate adjustment.

2 The GS450 LD, TXD and LF models employ yet another fork type, again featuring air assistance. Unlike the earlier models, this type of fork includes renewable bushes. This means that wear between the stanchion and lower leg can be corrected by fitting new bushes, rather than by renewing the forks complete. Bushed forks also tend to provide a more precise ride than the earlier types.

3 The UK GSX250 and 400 EZ models use forks which are different again to those described above. These are bushed, but without air assistance. The left-hand lower leg incorporates an anti-dive unit. This device regulates the damping effect according to the pressure in the front brake hydraulic circuit. This means that whilst under braking the damping effect can be stiffened to resist the tendency for the machine to pitch forward.

16 Front forks: overhaul – GS450 EZ, ED, TZ, TXZ and LZ models

1 The procedure for fork overhaul is generally similar to that described in Chapter 4, Section 4, noting the additional points described in this Section. The fork legs are removed in the normal way and should then be clamped between soft vice jaws to allow the removal of the top plug.

2 Remove the air valve cap and use a small screwdriver or a piece of wire to depress the valve plunger. Keep the valve open for a second or two to allow all of the air to escape. Removal of the top plug is a fiddly job which benefits from two pairs of hands. It is necessary to depress the plug against fork spring pressure using a large screwdriver or similar. Holding the plug in this position, use a small electrical screwdriver to dislodge and remove the wire internal circlip. The clip is rather inaccessible and its removal calls for a degree of patience. Once the clip has been freed, release pressure on the plug which will be pushed out by the fork spring.

3 The fork oil seal is of a slightly different design to that used in the standard fork, and is inclined to be a tight fit in the top of the lower leg. It is retained by a circlip, removal of which requires the use of a pair of circlip pliers. Note that removal of the seal will almost invariably destroy it, so do not attempt removal unless a new pair of seals is available. Warming the top of the lower leg in boiling water will often ease removal and fitting of the seal. Take care not to damage the seal during installation and check that it enters its recess squarely and fully. A large socket can be used as a drift when tapping the seal home. The seal lip should be lubricated thoroughly before it is eased over the stanchion, and care must be taken to avoid forcing the stanchion through the seal and damaging the sealing faces; a tiny scratch will allow leakage of air pressure.

4 Before further reassembly takes place, the fork should be filled with damping oil to the prescribed level. This operation is described in Section 8 of this Chapter.

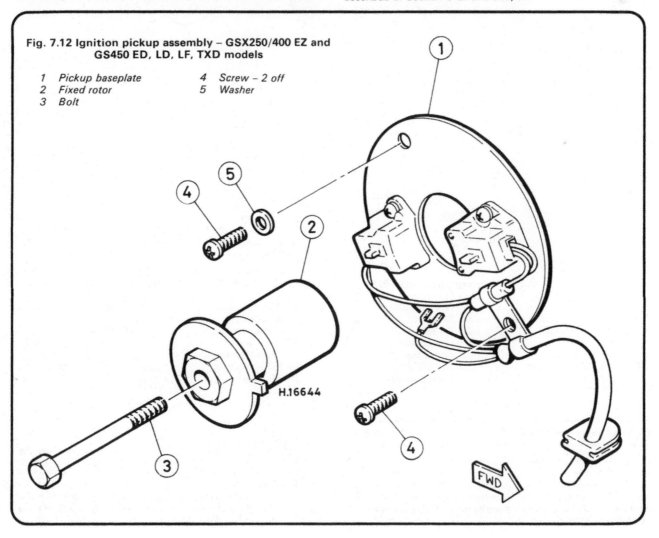

Fig. 7.12 Ignition pickup assembly – GSX250/400 EZ and GS450 ED, LD, LF, TXD models

1 Pickup baseplate
2 Fixed rotor
3 Bolt
4 Screw – 2 off
5 Washer

H.16644

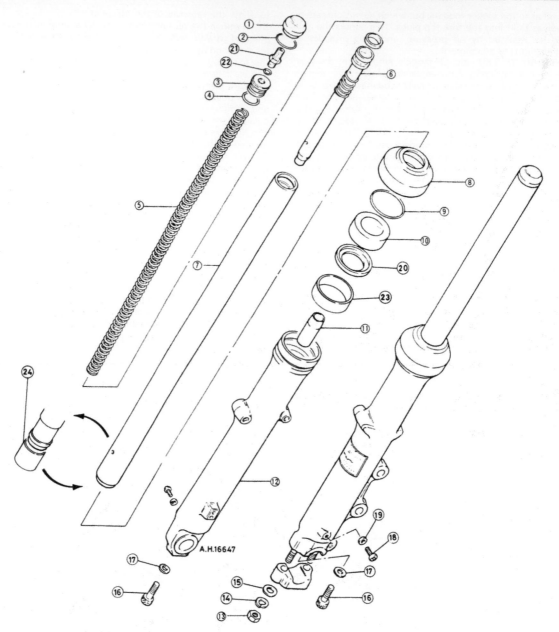

Fig. 7.13 Front forks – GS450 EZ, ED, TZ, TXZ, TXD, LZ, LD and LF models

1	Cap	8	Dust excluder	15	Washer – 2 off	22	O-ring
2	Wire circlip	9	Circlip	16	Allen bolt	23	Top bush
3	Top plug	10	Oil seal	17	Sealing washer	24	Bottom bush
4	O-ring	11	Damper rod seat	18	Drain screw		Items 23 and 24 fitted to LD,
5	Spring	12	Lower leg	19	Sealing washer		LF and TXD models only
6	Damper rod	13	Nut – 2 off	20	Backing ring		Items 21 and 22 not fitted to
7	Stanchion	14	Spring washer – 2 off	21	Air valve		TXZ and TXD models

---

17 Front forks: overhaul – GS450 LD, TXD and LF models

1   These can be dealt with in much the same way as described above, with the exception of the renewable top and bottom bushes. Unfortunately, Suzuki provide little information, so checking for wear must be restricted to a visual check of the condition of the bush surfaces and feeling for play between the stanchion and lower leg.

2   When separating the stanchion from the lower leg its removal will be impeded by the bottom bush striking the top bush. This can be used to advantage by providing a simple method of removing the top bush. Start by removing the damper rod Allen bolt and circlip which retains the oil seal and top bush. Pull the stanchion sharply outwards until the two bushes contact each other. Repeat this process until the top bush is tapped out of the lower leg.

3 Once the fork stanchion and lower leg have been separated the top bush can be slid off the stanchion and its condition checked. Wear is unlikely to be even and will show up as worn patches on the inner surface of the bush. If there is obvious play or wear present, the bushes should be renewed as a set. The lower bush can be checked in position, and should not be removed unless renewal is deemed necessary. The old bush is removed by opening the longitudinal split with a screwdriver so that it may be slid off the end of the stanchion, but note that this will destroy the bush. When fitting a new lower bush do not distort it by more than is absolutely necessary to ease it over the stanchion.

18 Front forks: overhaul – GSX250 and 400 EZ

1 The front forks of the UK EZ models can be dealt with as described in Section 17 above with the exception of the anti-dive mechanism fitted to the left-hand lower leg. When removing the left-hand leg note that it will be necessary to disconnect the hydraulic hose from the anti-dive modulator union. Provision should be made to catch the hydraulic fluid which will drain from the system. To this end it is probably best to pump the brake lever until the entire system is drained, taking care not to allow fluid to contact paintwork or plastic parts. Place the open end of the hose inside a plastic bag to prevent the ingress of dirt.

2 The modulator unit can be removed from the fork lower leg after the four Allen bolts have been released. Separate the modulator valve assembly from the main plunger unit by removing the two retaining screws. Check each component for signs of leakage of either hydraulic fluid or fork oil, and check that the plunger moves smoothly in its bore. If any defect is noted it will be necessary to renew the plunger unit and valve assembly as one part; they are not supplied separately. When refitting the anti-dive unit, use new O-rings between the unit and the fork lower leg. Use a thread locking compound on the retaining screws and tighten them securely. Remember that the hydraulic system must now be filled and bled to remove air. This procedure is described in Section 9 of Chapter 5, but note that the anti-dive unit is fitted with a second bleed nipple to allow air to be removed from the unit and connecting hose.

19 Changing the fork oil: GSX250/400 EZ and GS450 EZ, TZ, TXZ, LD, TXD, LF models

1 If the forks are to perform correctly it is essential that the fork oil level is correct and identical in both fork legs, noting that the oil level is far more significant than the precise quantity. To ensure that this accuracy is achieved the job must be done with the fork legs removed and clamped in turn between soft vice jaws. Remove the air valve dust cap and depress the valve core to release air pressure. The top plug can be removed by depressing it using a large screwdriver. Hold the plug down against spring pressure and displace and remove the wire circlip with the aid of an electrical screwdriver. This operation is best undertaken with the help of an assistant. Once the clip has been freed the plug can be released and removed. On models without air assisted forks, the top plug is screwed into position. It is a good idea to slacken this a few turns before removing the fork leg from the yokes.

2 Withdraw the fork spring and place it to one side. Invert the fork leg over a drain bowl and leave the oil to drain. The process can be speeded if the fork is 'pumped' to force the old oil out. The fork drain plug need not be disturbed.

3 The fork oil level is measured from the top of the stanchion with the fork leg vertical. It is important that the fork spring is removed and that the fork is at full compression, ie with the stanchion pushed fully home in the lower leg. Suzuki advise the use of a fork oil level gauge, Part Number 09943–74111. This device consists of a syringe, connected by a length of flexible

tubing to a rigid tube. The rigid tube carries an adjustable stop which can be rested on the top of the stanchion. If the stop has been set to leave the prescribed length of tube below it, any excess oil can now be drawn off using the syringe. Few owners will want the expense of an infrequently used special tool, so it is suggested that a cheap syringe is obtained from a chemist or veterinary supplier. Connect a length of aquarium or car screen washer hose to the syringe, making a mark on the tube the appropriate distance from the bottom. Hold the mark next to the top edge of the stanchion and draw off the excess oil.

4 As a starting point, add the amount of oil shown in the Specifications, noting that the fork should be pumped a few times to make sure that any air is dispelled. The resulting oil level will probably be a little above that recommended, so use the apparatus described above to reduce it to the correct figure. Note that on the UK models described in this Chapter the oil quantity varies between the individual fork legs. This is due to the anti-dive unit fitted to the left-hand leg only.

5 Fit the fork spring to each leg, noting that the reduced diameter coils face downwards. Refit the top plugs, securing them with new wire circlips. Fit the fork legs into the yokes, ensuring that the top of the stanchion is flush with the upper edge of the top yoke. Tighten the upper and lower pinch bolts to the figures shown in the Specifications.

20 Checking and adjusting fork air pressure – GS450 EZ, ED, LZ, TZ, TXZ, LD, TXD and LF models

1 It will be necessary to check the fork air pressure periodically, or every time the fork plugs are removed for servicing or oil changes. Start by placing the machine on its centre stand, then arrange a jack or wooden blocks beneath the front of the crankcase to raise the front wheel just clear of the ground. Release the handlebar clamps and pull the assembly clear of the tops of the stanchion. The handlebar can be placed on some cloths to protect the machine's paintwork.

2 Suzuki prescribe the use of a special pump and gauge assembly which is not likely to be available to the owner. A suitable alternative is a hand-held pressure gauge of the plunger or clock type. The gauge must be accurate, easy to read and should take the minimum of air when taking the reading. This is important because the volume of air in the fork is very small, and even taking a pressure reading will itself allow pressure to be lost. It is advisable to practice taking the pressure readings a number of times so that the drop during each check can be allowed for. Also required is an air pump. This must be of the manual type, either a bicycle-type pump or one of the syringe-type pumps such as that produced by S & W specifically for suspension use. **On no account** use an air line. This will increase pressure so rapidly that it will be almost impossible to control, and can easily destroy the fork seals by applying too much pressure. Under no circumstances must the pressure exceed 36 psi (2.5 kg/cm^2).

3 The recommended working pressure is 7.11 psi (0.5 kg/cm^2) and each fork leg must be within 1.42 psi (0.1 kg/cm^2) of the other. The latter condition is not easy to achieve, but is arguably more important than the pressure figure. Any significant imbalance will lead to very unpredictable handling and could prove dangerous. Pump each leg to more than the required pressure, say about 10 – 12 psi. Using the pressure gauge, take successive readings until the pressure drops to the required level. Repeat on the other fork leg until the two are equal.

4 It is quite in order to experiment with pressure settings, bearing in mind that the maximum figure must never be exceeded. Owners whose riding is mostly gentle commuting may wish to run at a lower, or zero, pressure in the interests of comfort, whilst those with sporting inclination may find a slightly higher than standard pressure to be desirable. When adjustment is complete, refit the air valve dust caps and fit the handlebar assembly. Tighten the clamp bolts to 8.5 – 14.5 lbf ft (1.2 – 2.0 kgf m).

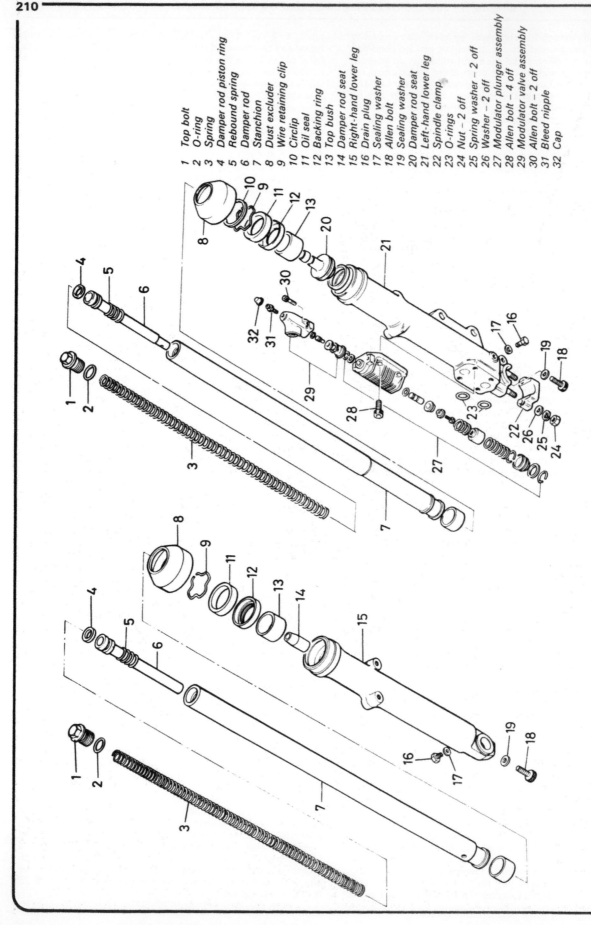

1 Top bolt
2 O-ring
3 Spring
4 Damper rod piston ring
5 Rebound spring
6 Damper rod
7 Stanchion
8 Dust excluder
9 Wire retaining clip
10 Circlip
11 Oil seal
12 Backing ring
13 Top bush
14 Damper rod seat
15 Right-hand lower leg
16 Drain plug
17 Sealing washer
18 Allen bolt
19 Sealing washer
20 Damper rod seat
21 Left-hand lower leg
22 Spindle clamp
23 O-rings
24 Nut – 2 off
25 Spring washer – 2 off
26 Washer – 2 off
27 Modulator plunger assembly
28 Allen bolt – 4 off
29 Modulator valve assembly
30 Allen bolt – 2 off
31 Bleed nipple
32 Cap

Fig. 7.14 Front forks – GSX250 and 400 EZ models

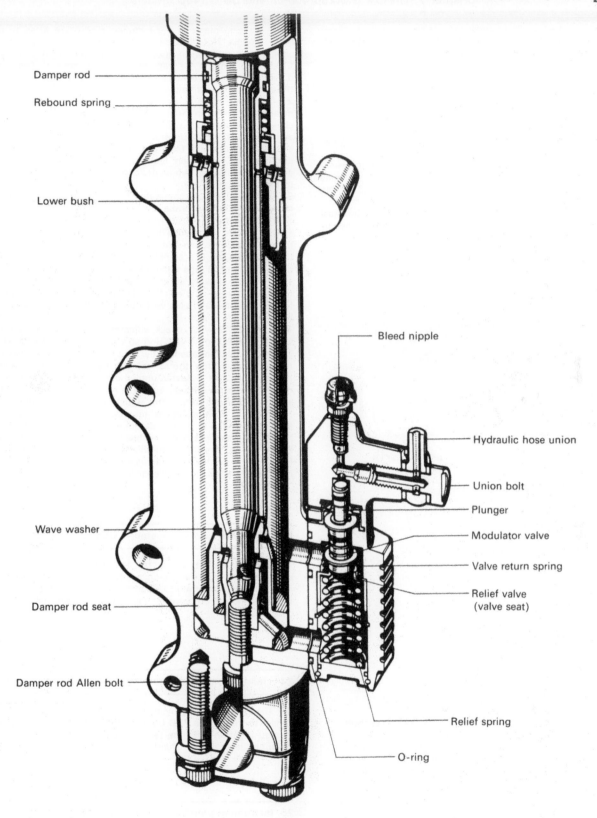

Damper rod

Rebound spring

Lower bush

Wave washer

Damper rod seat

Damper rod Allen bolt

Bleed nipple

Hydraulic hose union

Union bolt

Plunger

Modulator valve

Valve return spring

Relief valve
(valve seat)

Relief spring

O-ring

Fig. 7.15 Sectioned view of left-hand fork leg showing anti-dive assembly – GSX250 and 400 EZ models

21 Handlebar fairing: removal and refitting – GS450 S models

1   The GS450 S model is equipped with a small handlebar-mounted fairing. It is retained by a single bolt to a bracket fitted to the centre of the bottom yoke, and by two similar mountings on either side of the top yoke. To remove the fairing for access to the headlamp and steering head withdraw the central bolt first, followed by the two upper mounting bolts. The fairing can now be lifted away.

2   The component parts are available separately in the event of damage. The screen is retained by seven screws and nuts passing through rubber bushes to prevent cracking. When fitting these take care not to overtighten them or the pressure on the screen may cause it to fracture. The rubber bushes must be renewed if damaged and on no account should the screen be bolted rigidly to the fairing.

3   When cleaning the fairing and screen, take care not to scratch it. Use water and a soft sponge or brush to remove road dirt, polishing with car wax or similar when dry.

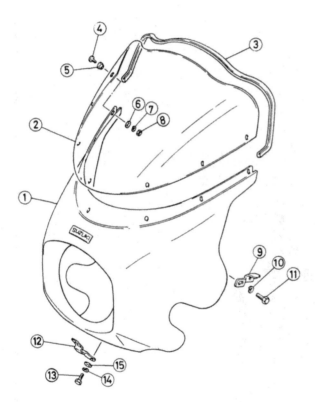

Fig. 7.16 Fairing – GS450 S model

1 Fairing	9 Top mounting bracket
2 Screen	– 2 off
3 Trim	10 Spring washer – 2 off
4 Screw – 7 off	11 Bolt – 2 off
5 Bush – 7 off	12 Lower mounting bracket
6 Washer – 7 off	13 Bolt
7 Spring washer – 7 off	14 Spring washer
8 Nut – 7 off	15 Washer

22 Front drum brake: GS450 TXD

1   The US GS450 TXD model differs from the rest of the range in its use of a 200 mm (7.9 in) twin leading shoe front drum brake in place of the hydraulic disc brake fitted to the other models. Not surprisingly, the front drum brake is fairly similar to the rear drum brake and the procedures described in Chapter 5, Section 13 can be applied in general. The main differences are due to the twin leading shoe (tls) operation and these are discussed below.

2   The rear brake is of the single leading shoe type. This means that whilst the leading edge of one shoe is at the brake cam end, the remaining shoe's leading edge is at the pivot end, and thus overall braking effort is reduced because the trailing shoe does not benefit from the 'self-servo' effect acting on the leading shoe. The twin leading shoe brake overcomes this drawback by having two pivots and two cams at opposite sides of the brake backplate.

3   The two cams are connected by an external linkage, so that movement from the brake lever can be applied to both cams and shoes simultaneously. It will be noted that the length of the linkage can be adjusted. This allows the brake to be set up so that both shoes touch the drum surface at exactly the same time. This setting is a significant factor in the efficiency of the brake; if one shoe lags behind the other it is doing no effective work and the braking effort is significantly impaired. It follows that it is essential to check the adjustment whenever new shoes have been fitted and also when braking performance seems to have fallen off.

4   The best way to check adjustment is to disconnect the linkage at the shorter link end so that the effect of each shoe can be assessed independently. Support the front wheel clear of the ground so that the wheel can be spun, then gradually tighten the cable adjuster until the first shoe just begins to contact the drum surface. Apply the second shoe by turning the short link by hand until it too is just in contact with the drum. Holding this position, check that the hole in the linkage will align exactly with the corresponding hole in the end of the short link. If necessary, back off the locknut and adjust the effective length of the linkage until this is achieved. Reconnect the linkage, then set the cable adjuster so that the brake begins to operate just after the handlebar lever is moved.

23 Instrument panel assembly: GS450 TXD model

1   The instrument panel on the TXD model is slightly redesigned. Whilst the changes are mostly cosmetic and will not seriously affect the procedures for removal, refitting or bulb renewal, the accompanying line drawing should be used for guidance.

24 Instrument panel assembly: GS450 LD and LF models

1   The LD and LF employ an instrument panel which is broadly similar to that of the TXD but with the addition of a fuel gauge housed together with the tachometer head. The fuel gauge is described later in this Chapter.

25 Instrument panel assembly: GS450 ED model

1   The GS450 ED employs yet another variation of instrument panel, similar to the earlier GSX400 T but with the addition of a fuel gauge housed between the two main instrument heads and below the gear position indicator panel. For details refer to the accompanying line drawing.

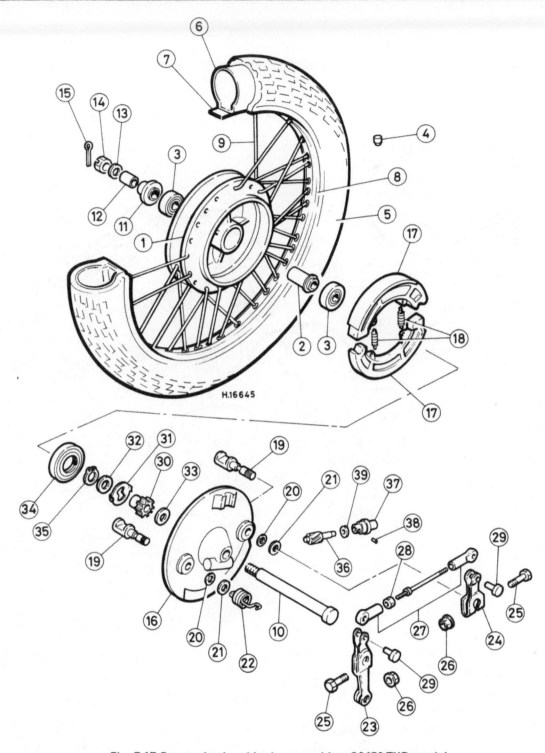

Fig. 7.17 Front wheel and brake assembly – GS450 TXD model

1	Hub	10	Wheel spindle	20	O-ring – 2 off	30	Speedometer drive gear
2	Spacer	11	Right-hand spacer	21	Washer – 2 off	31	Speedometer drive plate
3	Bearing	12	Spacer	22	Spring	32	Washer
4	Balance weight – as	13	Washer	23	Front operating lever	33	Washer
	required	14	Nut	24	Rear operating lever	34	Oil seal
5	Tyre	15	Split pin	25	Pinch bolt – 2 off	35	Circlip
6	Inner tube	16	Brake panel	26	Nut – 2 off	36	Speedometer driven gear
7	Rim tape	17	Brake shoe – 2 off	27	Linkage	37	Driven gear housing
8	Rim	18	Return spring – 2 off	28	Locknut	38	Grub screw
9	Spoke	19	Operating cam – 2 off	29	Pivot pin – 2 off	39	Seal

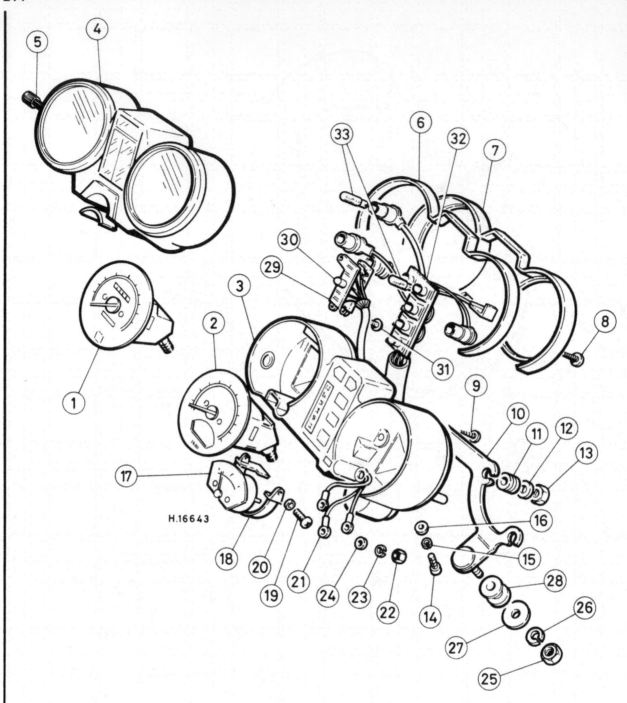

H.16643

Fig. 7.18 Instrument panel assembly – GS450 TXD, LD and LF models

1	Speedometer	10	Mounting bracket	19	Screw – 2 off	28	Damping rubber – 2 off
2	Tachometer	11	Grommet – 4 off	20	Spring washer – 2 off	29	Gear position indicator
3	Instrument housing	12	Washer – 2 off	21	Fuel gauge wiring		panel
4	Top cover	13	Nut – 2 off	22	Nut – 3 off	30	Bulb – 6 off
5	Trip reset knob	14	Screw – 4 off	23	Spring washer – 3 off	31	Screw – 2 off
6	Cover ring	15	Spring washer – 4 off	24	Washer – 3 off	32	Warning lamp bulbholder
7	Lower cover	16	Washer – 4 off	25	Nut – 2 off	33	Bulb – 9 off
8	Screw – 4 off	17	Fuel gauge	26	Spring washer – 2 off		
9	Screw – 4 off	18	Fuel gauge housing	27	Washer – 2 off		

Items 17 to 24 fitted to GS450 L models only
Items 22-24 replaced by screw on GS450 LF models

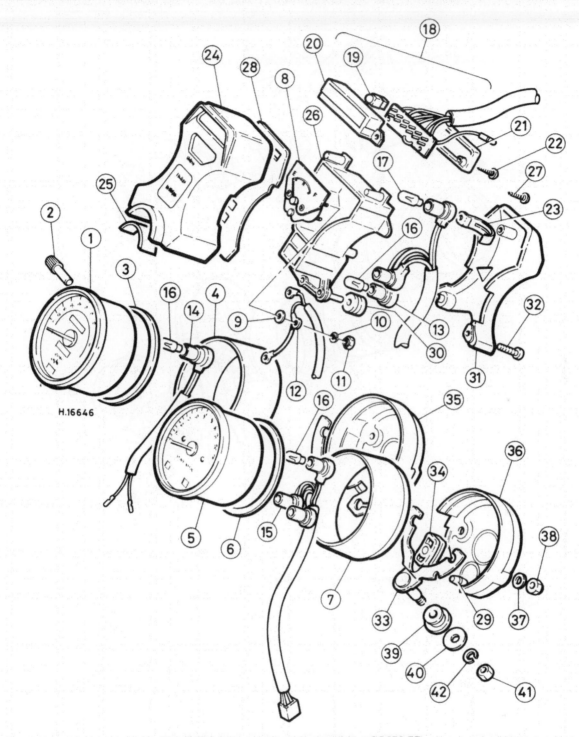

Fig. 7.19 Instrument panel assembly – GS450 ED

1	Speedometer	12	Fuel gauge wiring	21	Lower cover	32	Screw – 3 off
2	Trip reset knob	13	Warning lamp bulbholders	22	Screw – 2 off	33	Mounting bracket
3	Damping ring	14	Speedometer lamp bulbholder	23	Clamp	34	Grommet – 4 off
4	Speedometer housing			24	Warning lamp top cover	35	Speedometer lower cover
5	Tachometer	15	Tachometer lamp bulbholder	25	Cushion	36	Tachometer lower cover
6	Damping ring			26	Warning lamp housing	37	Washer – 4 off
7	Tachometer housing	16	Bulb – 6 off	27	Screw – 5 off	38	Nut – 4 off
8	Fuel gauge	17	Bulb	28	Panel	39	Grommet – 2 off
9	Washer – 3 off	18	Gear position indicator	29	Rubber damper – 4 off	40	Washer – 2 off
10	Spring washer – 3 off	19	Bulb – 6 off	30	Grommet – 2 off	41	Nut – 2 off
11	Nut – 3 off	20	Cover	31	Warning lamp lower cover	42	Spring washer – 2 off

H.16646

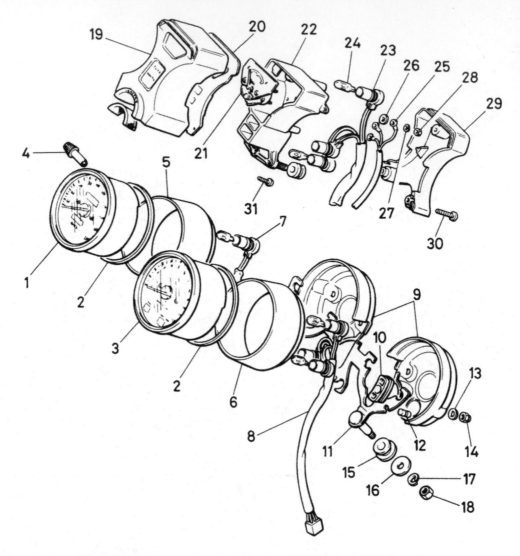

Fig. 7.20 Instrument panel assembly – GSX250 and 400 EZ models

1 Speedometer	8 Tachometer lamp	15 Grommet – 2 off	24 Bulb – 7 off
2 Damping ring – 2 off	bulbholder	16 Washer – 2 off	25 Fuel gauge wiring
3 Tachometer	9 Lower covers	17 Spring washer – 2 off	26 Washer – 3 off
4 Trip reset knob	10 Grommet – 4 off	18 Nut – 2 off	27 Spring washer – 3 off
5 Speedometer housing	11 Mounting bracket	19 Warning lamp top cover	28 Nut – 3 off
6 Tachometer housing	12 Damping rubber – 4 off	20 Panel	29 Warning lamp housing
7 Speedometer lamp	13 Washer – 4 off	21 Fuel gauge	lower cover
bulbholder	14 Nut – 4 off	22 Warning lamp housing	30 Screw – 3 off
		23 Bulbholders	31 Screw – 3 off

26 Instrument panel assembly: GSX250/400 EZ

1   The instrument panel fitted to the UK GSX250 and 400 EZ models is similar to that fitted to the US GS450 ED, but with the gear position indicator omitted.

27 Fuel gauge: examination and renovation – GSX250/400 EZ and GS450 LD, LF, and ED models

1   The above models are equipped with a fuel gauge system comprising a tank-mounted sender unit which controls the

gauge meter movement in the instrument panel. In the case of the US 'L' models the gauge is located in the tachometer face, whilst on the remaining machines it is housed between the two main instrument heads.

2   In the event of a fault in the circuit, check first that the battery is fully charged; the system will not operate normally with a discharged battery. If this fails to resolve the problem, trace the wiring from the meter and disconnect the Black/white and Yellow/black leads below the tank. Connect the Yellow/black lead to earth (ground) and switch on the ignition. If the meter movement is working, it should indicate 'F'. Note that the meter movement is damped and will take a few seconds to respond. It is possible to check meter accuracy, but this test requires the use of resistors and so is best left to a

Fig. 7.21 Fuel gauge sender unit test

dealer to perform; in practice it is more usual to check by substitution. If the meter seems to work normally, attention should be turned to the sender unit.

3 Drain and remove the fuel tank and place it inverted on some soft cloth on the workbench. Remove the screws which retain the drip tray and remove it, together with its pipe. Release the sender unit retaining bolts and lift it away from the tank, taking care not to twist or bend the float arm as it is manoeuvred out of the tank.

4 Connect a multimeter to the two sender leads and set it to the resistance (ohms) range. Note the readings with the arm fully raised (Full) and fully lowered (Empty). The standard figures are as follows:

Full	4 – 7 ohms
Empty	95 – 115 ohms

If the readings are significantly outside this range, the sender unit should be renewed. Before refitting the sender, check that the float arm moves freely and easily, and that the sealing gasket is intact. Refit the sender and drip tray, then refill the tank to check for leaks. If all is well, refit the fuel tank and reconnect the fuel gauge wiring.

28 Headlamp bulb renewal: Quartz halogen type – all models

1 Most of the later models employ an uprated headlamp utilising a 60/55W H4 quartz halogen bulb. This replaces the previous tungsten bulb (UK) or tungsten sealed beam unit (US). Bulb renewal is mostly obvious, but the following points should be noted concerning the handling of quartz halogen bulbs. When in use the quartz envelope of the bulb gets very hot indeed, so do not attempt to change a bulb until it has cooled down for a few minutes. When removing the bulb, always handle it by the metal cap, **never** by the quartz envelope. This will be marked if touched by the fingers and the resulting hot spot will shorten the life of the bulb significantly. If touched accidentally, remove fingerprints from the envelope using soft rag moistened in alcohol or methylated spirit. Allow the surface to dry before refitting the bulb.

29 Electrical system detail changes – all models

1 Apart from the specific modifications discussed in the previous Sections there have been numerous detail alterations to the electrical system components. These are mostly of a cosmetic nature and form part of overall restyling exercises, and affect items such as switches, lamps etc. In the case of switches, additional information will be found in the wiring diagrams which follow this Chapter. These show details of connections and wiring colours.

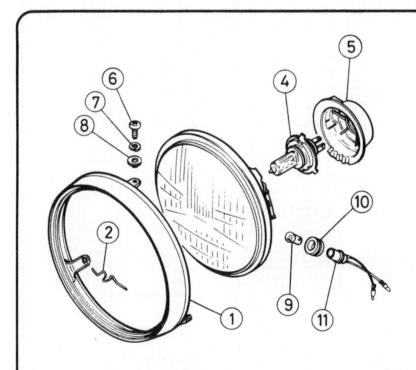

1 Rim
2 Spring clip – 3 off
3 Reflector
4 Headlamp bulb
5 Cover
6 Screw – 3 off
7 Spring washer – 3 off
8 Collar – 3 off
9 Parking lamp bulb
10 Grommet
11 Bulbholder

Fig. 7.22 Headlamp assembly – later models with Quartz halogen bulb

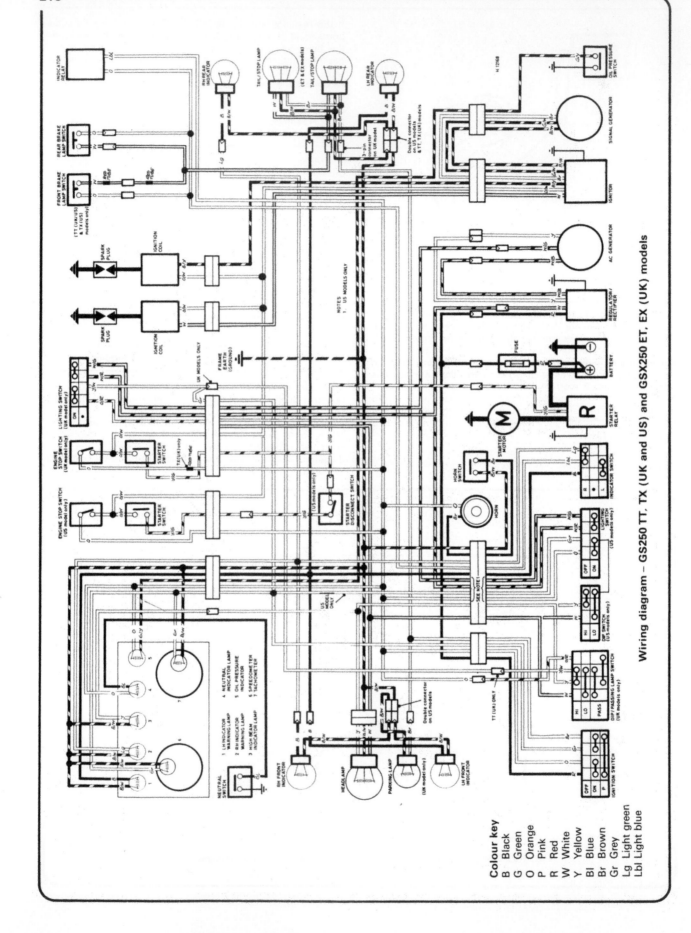

Wiring diagram – GS250 TT, TX (UK and US) and GSX250 ET, EX (UK) models

Colour key

B   Black
G   Green
O   Orange
P   Pink
R   Red
W   White
Y   Yellow
Bl Blue
Br Brown
Gr Grey
Lg Light green
Lbl Light blue

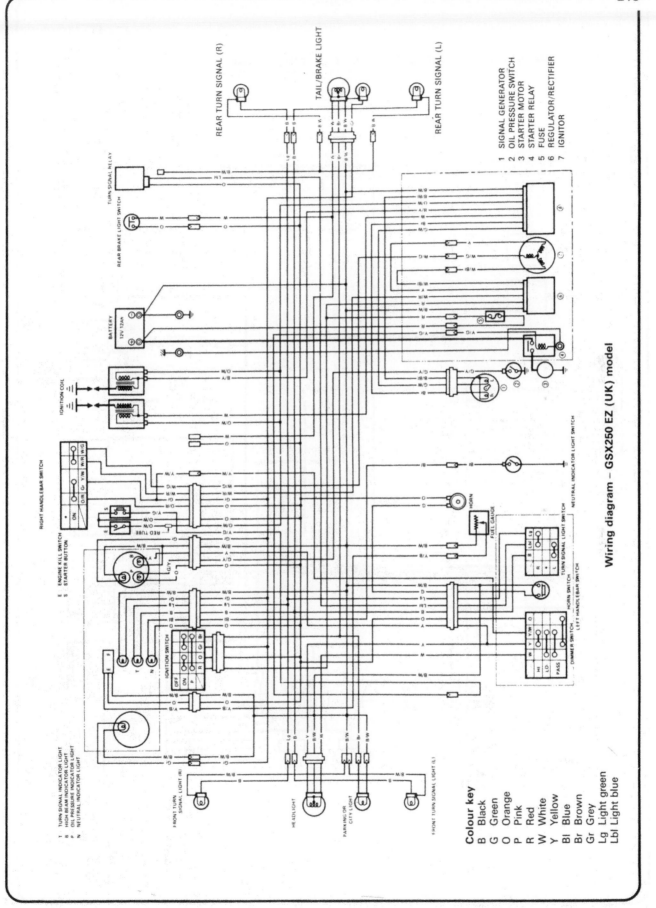

Wiring diagram – GSX250 EZ (UK) model

Colour key

B	Black
G	Green
O	Orange
P	Pink
R	Red
W	White
Y	Yellow
Bl	Blue
Br	Brown
Gr	Grey
Lg	Light green
Lbl	Light blue

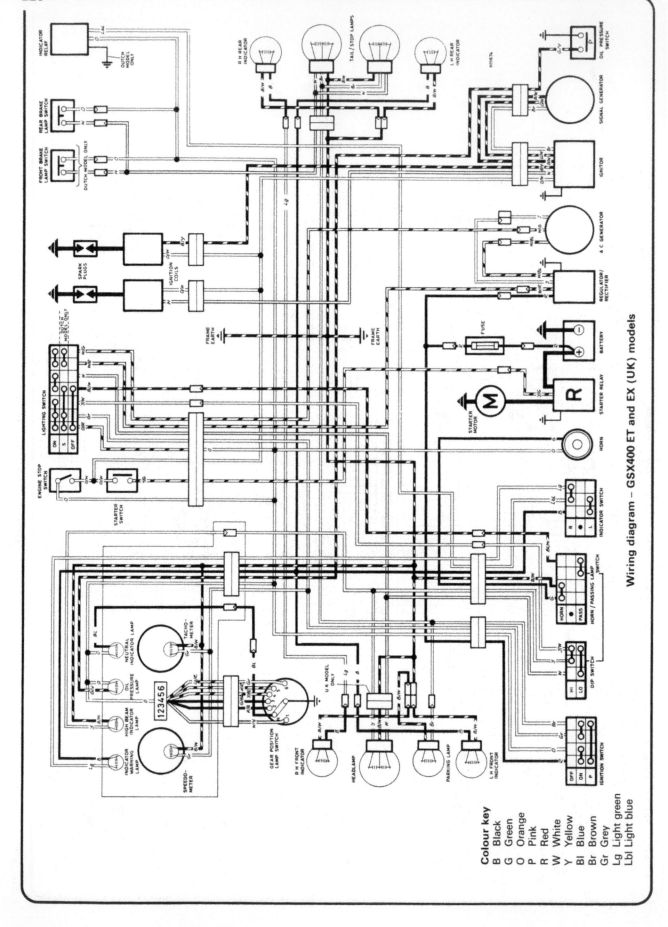

Wiring diagram – GSX400 ET and EX (UK) models

Colour key
B   Black
G   Green
O   Orange
P   Pink
R   Red
W   White
Y   Yellow
Bl Blue
Br Brown
Gr Grey
Lg Light green
Lbl Light blue

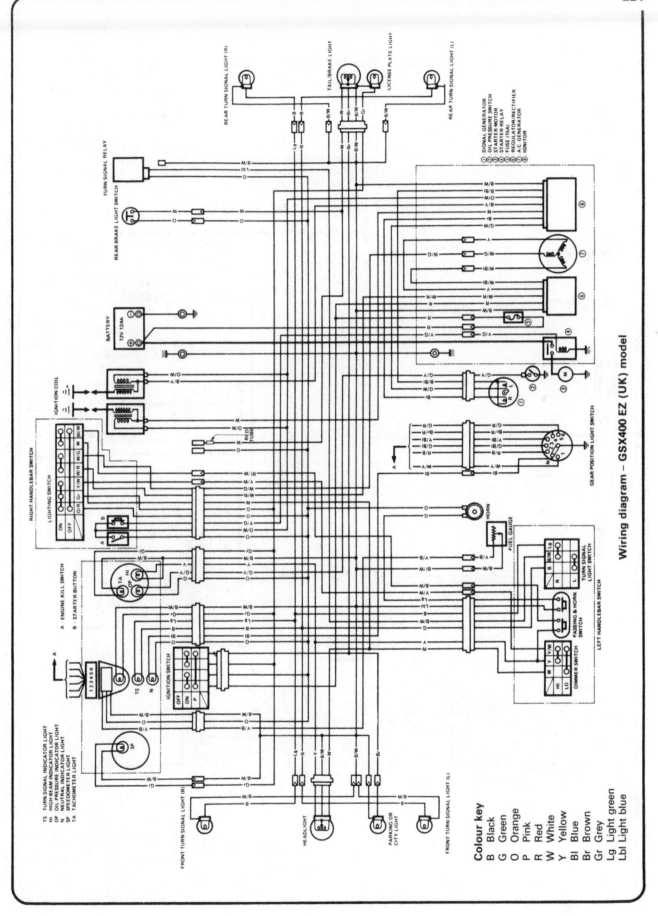

Wiring diagram – GSX400 EZ (UK) model

Colour key

B	Black
G	Green
O	Orange
P	Pink
R	Red
W	White
Y	Yellow
Bl	Blue
Br	Brown
Gr	Grey
Lg	Light green
Lbl	Light blue

TS TURN SIGNAL INDICATOR LIGHT
HI HIGH BEAM INDICATOR LIGHT
OP OIL PRESSURE INDICATOR LIGHT
N   NEUTRAL INDICATOR LIGHT
SP SPEEDOMETER LIGHT
TA TACHOMETER LIGHT

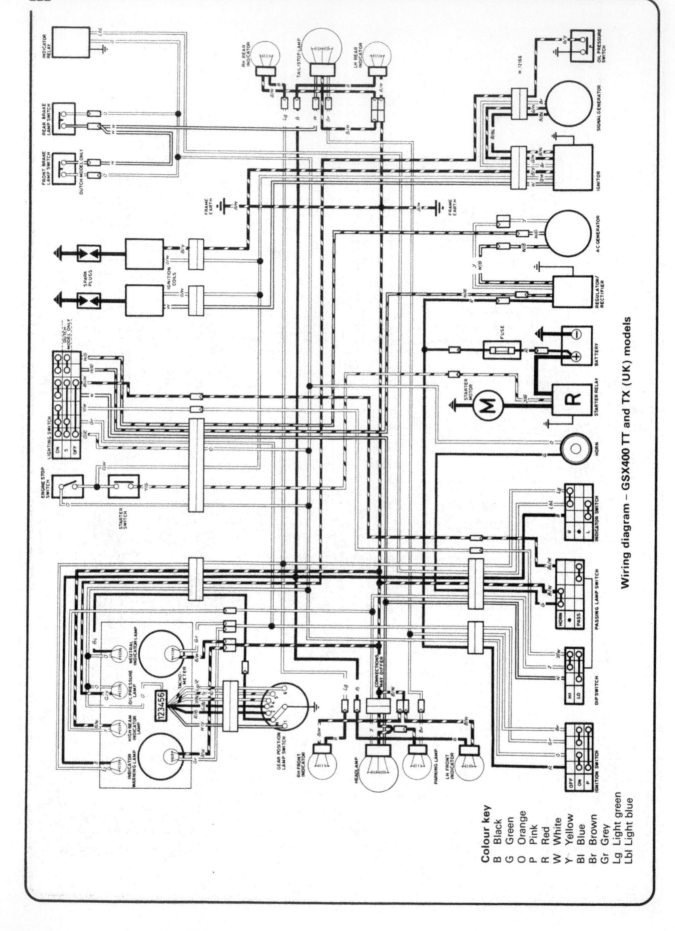

Wiring diagram – GSX400 TT and TX (UK) models

Colour key
B Black
G Green
O Orange
P Pink
R Red
W White
Y Yellow
Bl Blue
Br Brown
Gr Grey
Lg Light green
Lbl Light blue

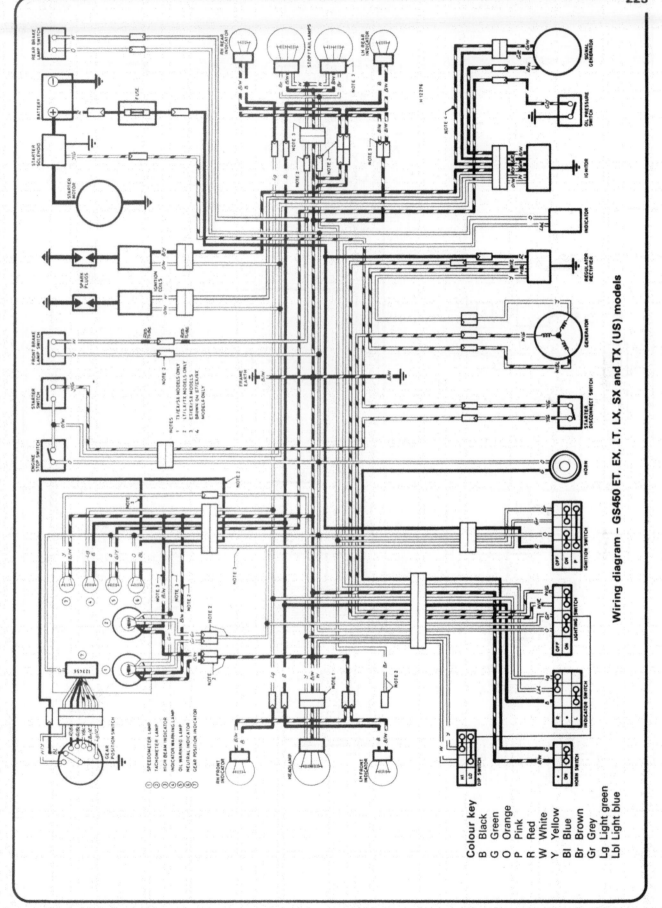

Wiring diagram – GS450 ET, EX, LT, LX, SX and TX (US) models

Colour key
B Black
G Green
O Orange
P Pink
R Red
W White
Y Yellow
Bl Blue
Br Brown
Gr Grey
Lg Light green
Lbl Light blue

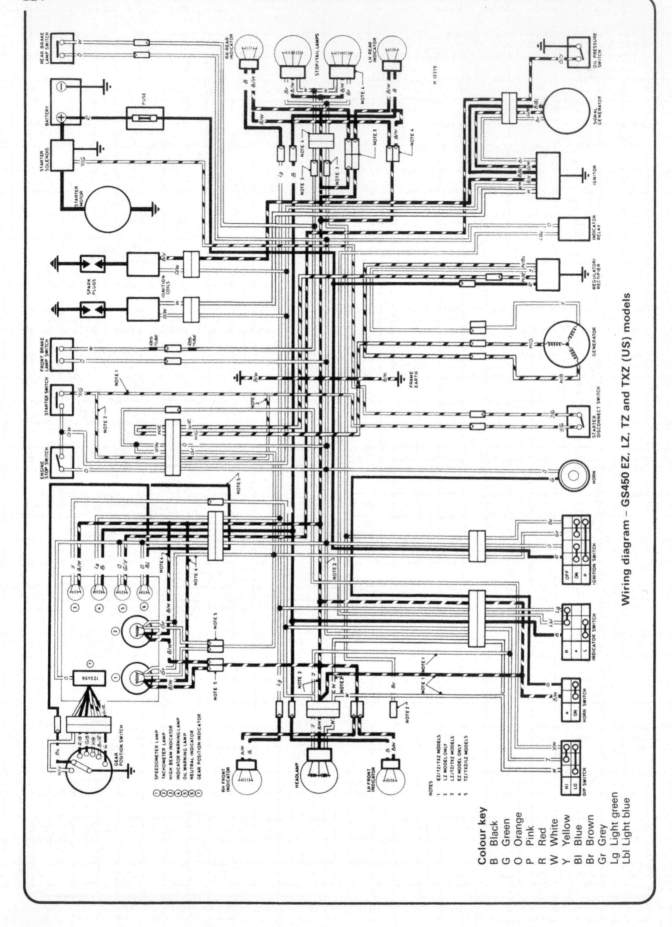

Wiring diagram – GS450 EZ, LZ, TZ and TXZ (US) models

Colour key

B Black
G Green
O Orange
P Pink
R Red
W White
Y Yellow
Bl Blue
Br Brown
Gr Grey
Lg Light green
Lbl Light blue

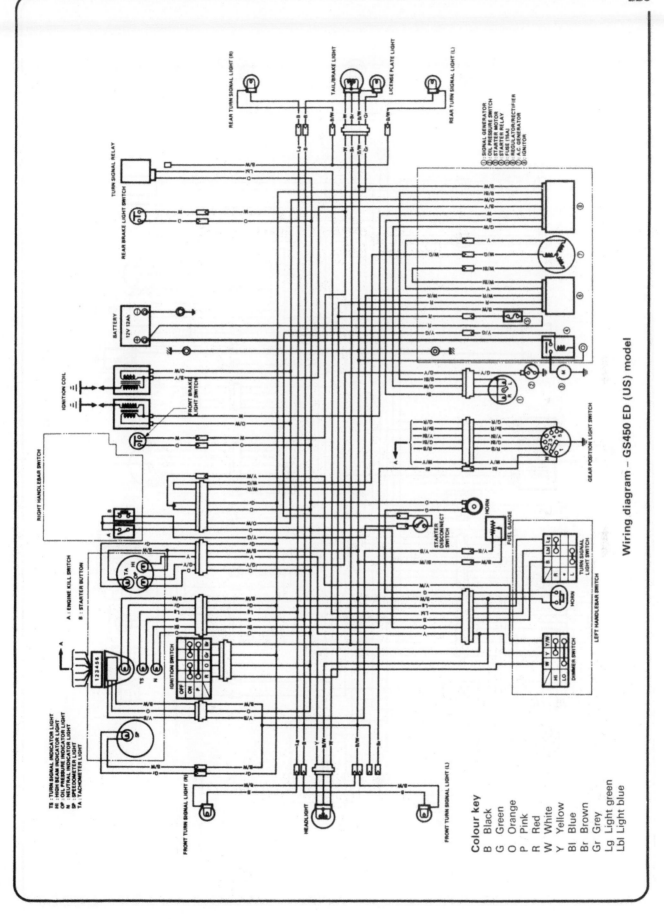

Wiring diagram – GS450 ED (US) model

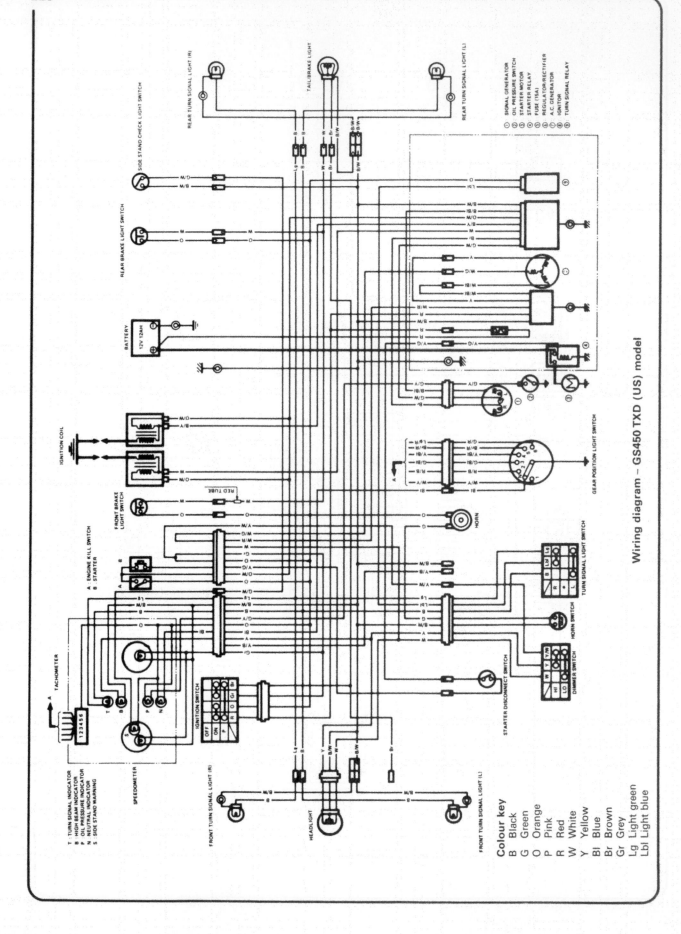

Wiring diagram – GS450 TXD (US) model

Colour key

B	Black
G	Green
O	Orange
P	Pink
R	Red
W	White
Y	Yellow
Bl	Blue
Br	Brown
Gr	Grey
Lg	Light green
Lbl	Light blue

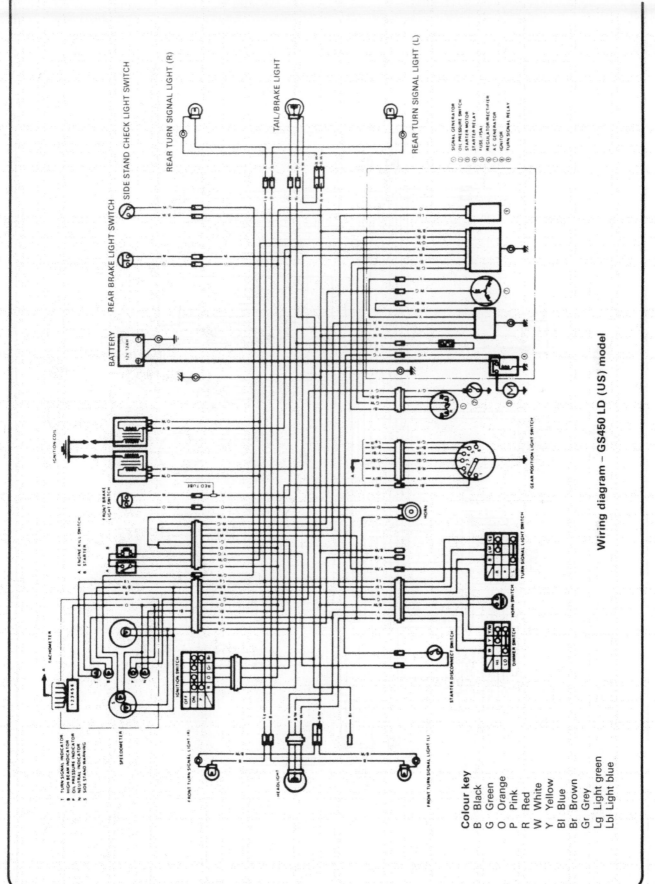

Wiring diagram – GS450 LD (US) model

SIDE STAND CHECK LIGHT SWITCH

REAR TURN SIGNAL LIGHT (R)

TAIL/BRAKE LIGHT

REAR TURN SIGNAL LIGHT (L)

REAR BRAKE LIGHT SWITCH

BATTERY
12V 12Ah

IGNITION COIL

RED TUBE

FRONT BRAKE LIGHT SWITCH

1 SIGNAL GENERATOR
2 OIL PRESSURE SWITCH
3 STARTER MOTOR
4 STARTER RELAY
5 FUSE (15A)
6 REGULATOR-RECTIFIER
7 A.C GENERATOR
8 IGNITOR
9 TURN SIGNAL RELAY

GEAR POSITION LIGHT SWITCH

HORN

TURN SIGNAL LIGHT SWITCH

HORN SWITCH

STARTER DISCONNECT SWITCH

DIMMER SWITCH

A ENGINE KILL SWITCH
B STARTER

TACHOMETER

SPEEDOMETER

IGNITION SWITCH
OFF
ON

FRONT TURN SIGNAL LIGHT (R)

HEADLIGHT

FRONT TURN SIGNAL LIGHT (L)

1 TURN SIGNAL INDICATOR
2 HIGH BEAM INDICATOR
3 OIL PRESSURE INDICATOR
N NEUTRAL INDICATOR
S SIDE STAND WARNING

Colour key

B	Black
G	Green
O	Orange
P	Pink
R	Red
W	White
Y	Yellow
Bl	Blue
Br	Brown
Gr	Grey
Lg	Light green
Lbl	Light blue

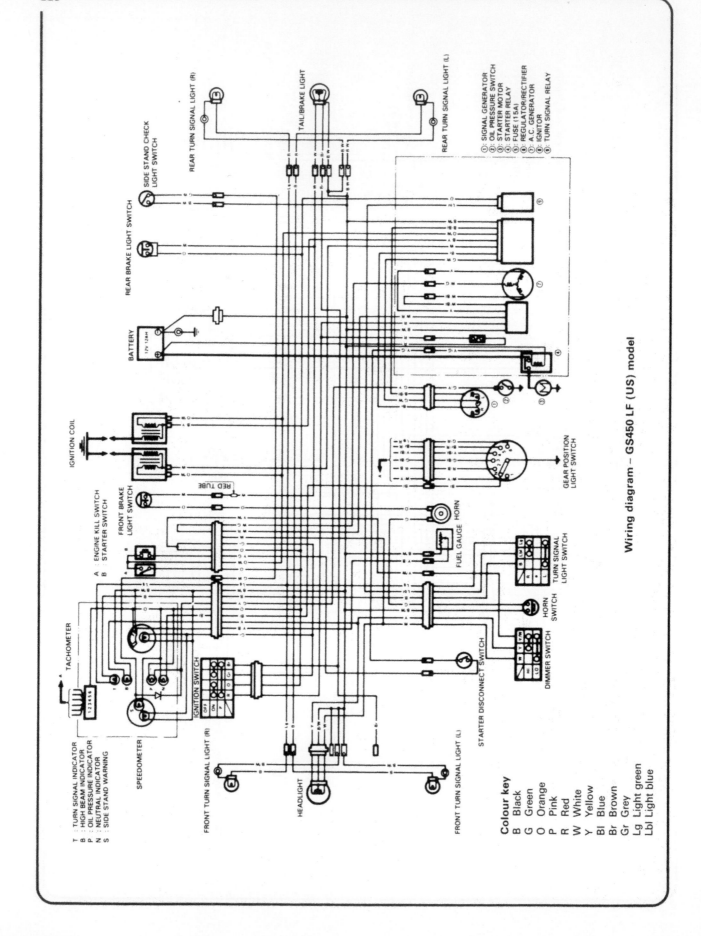

Wiring diagram – GS450 LF (US) model

REAR TURN SIGNAL LIGHT (R)

REAR TURN SIGNAL LIGHT (L)

TAIL/BRAKE LIGHT

SIDE STAND CHECK LIGHT SWITCH

REAR BRAKE LIGHT SWITCH

BATTERY
12V 12AH

IGNITION COIL

A : ENGINE KILL SWITCH
B : STARTER SWITCH

FRONT BRAKE LIGHT SWITCH

RED TUBE

GEAR POSITION LIGHT SWITCH

HORN

FUEL GAUGE

TURN SIGNAL LIGHT SWITCH

HORN SWITCH

DIMMER SWITCH

STARTER DISCONNECT SWITCH

TACHOMETER

SPEEDOMETER

IGNITION SWITCH

FRONT TURN SIGNAL LIGHT (R)

HEADLIGHT

FRONT TURN SIGNAL LIGHT (L)

① : SIGNAL GENERATOR
② : OIL PRESSURE SWITCH
③ : STARTER MOTOR
④ : STARTER RELAY
⑤ : FUSE (15A)
⑥ : REGULATOR/RECTIFIER
⑦ : A.C. GENERATOR
⑧ : IGNITOR
⑨ : TURN SIGNAL RELAY

T : TURN SIGNAL INDICATOR
B : HIGH BEAM INDICATOR
P : OIL PRESSURE INDICATOR
N : NEUTRAL INDICATOR
S : SIDE STAND WARNING

Colour key
B Black
G Green
O Orange
P Pink
R Red
W White
Y Yellow
Bl Blue
Br Brown
Gr Grey
Lg Light green
Lbl Light blue

English/American terminology

Because this book has been written in England, British English component names, phrases and spellings have been used throughout. American English usage is quite often different and whereas normally no confusion should occur, a list of equivalent terminology is given below.

English	American	English	American
Air filter	Air cleaner	Number plate	License plate
Alignment (headlamp)	Aim	Output or layshaft	Countershaft
Allen screw/key	Socket screw/wrench	Panniers	Side cases
Anticlockwise	Counterclockwise	Paraffin	Kerosene
Bottom/top gear	Low/high gear	Petrol	Gasoline
Bottom/top yoke	Bottom/top triple clamp	Petrol/fuel tank	Gas tank
Bush	Bushing	Pinking	Pinging
Carburettor	Carburetor	Rear suspension unit	Rear shock absorber
Catch	Latch	Rocker cover	Valve cover
Circlip	Snap ring	Selector	Shifter
Clutch drum	Clutch housing	Self-locking pliers	Vise-grips
Dip switch	Dimmer switch	Side or parking lamp	Parking or auxiliary light
Disulphide	Disulfide	Side or prop stand	Kick stand
Dynamo	DC generator	Silencer	Muffler
Earth	Ground	Spanner	Wrench
End float	End play	Split pin	Cotter pin
Engineer's blue	Machinist's dye	Stanchion	Tube
Exhaust pipe	Header	Sulphuric	Sulfuric
Fault diagnosis	Trouble shooting	Sump	Oil pan
Float chamber	Float bowl	Swinging arm	Swingarm
Footrest	Footpeg	Tab washer	Lock washer
Fuel/petrol tap	Petcock	Top box	Trunk
Gaiter	Boot	Torch	Flashlight
Gearbox	Transmission	Two/four stroke	Two/four cycle
Gearchange	Shift	Tyre	Tire
Gudgeon pin	Wrist/piston pin	Valve collar	Valve retainer
Indicator	Turn signal	Valve collets	Valve cotters
Inlet	Intake	Vice	Vise
Input shaft or mainshaft	Mainshaft	Wheel spindle	Axle
Kickstart	Kickstarter	White spirit	Stoddard solvent
Lower leg	Slider	Windscreen	Windshield
Mudguard	Fender		

Conversion factors

Length (distance)

	X			X		
Inches (in)	X	25.4	= Millimetres (mm)	X	0.0394	= Inches (in)
Feet (ft)	X	0.305	= Metres (m)	X	3.281	= Feet (ft)
Miles	X	1.609	= Kilometres (km)	X	0.621	= Miles

Volume (capacity)

Cubic inches (cu in; in^3)	X	16.387	= Cubic centimetres (cc; cm^3)	X	0.061	= Cubic inches (cu in; in^3)
Imperial pints (Imp pt)	X	0.568	= Litres (l)	X	1.76	= Imperial pints (Imp pt)
Imperial quarts (Imp qt)	X	1.137	= Litres (l)	X	0.88	= Imperial quarts (Imp qt)
Imperial quarts (Imp qt)	X	1.201	= US quarts (US qt)	X	0.833	= Imperial quarts (Imp qt)
US quarts (US qt)	X	0.946	= Litres (l)	X	1.057	= US quarts (US qt)
Imperial gallons (Imp gal)	X	4.546	= Litres (l)	X	0.22	= Imperial gallons (Imp gal)
Imperial gallons (Imp gal)	X	1.201	= US gallons (US gal)	X	0.833	= Imperial gallons (Imp gal)
US gallons (US gal)	X	3.785	= Litres (l)	X	0.264	= US gallons (US gal)

Mass (weight)

Ounces (oz)	X	28.35	= Grams (g)	X	0.035	= Ounces (oz)
Pounds (lb)	X	0.454	= Kilograms (kg)	X	2.205	= Pounds (lb)

Force

Ounces-force (ozf; oz)	X	0.278	= Newtons (N)	X	3.6	= Ounces-force (ozf; oz)
Pounds-force (lbf; lb)	X	4.448	= Newtons (N)	X	0.225	= Pounds-force (lbf; lb)
Newtons (N)	X	0.1	= Kilograms-force (kgf; kg)	X	9.81	= Newtons (N)

Pressure

Pounds-force per square inch (psi; lbf/in^2; lb/in^2)	X	0.070	= Kilograms-force per square centimetre (kgf/cm^2; kg/cm^2)	X	14.223	= Pounds-force per square inch (psi; lbf/in^2; lb/in^2)
Pounds-force per square inch (psi; lbf/in^2; lb/in^2)	X	0.068	= Atmospheres (atm)	X	14.696	= Pounds-force per square inch (psi; lbf/in^2; lb/in^2)
Pounds-force per square inch (psi; lbf/in^2; lb/in^2)	X	0.069	= Bars	X	14.5	= Pounds-force per square inch (psi; lbf/in^2; lb/in^2)
Pounds-force per square inch (psi; lbf/in^2; lb/in^2)	X	6.895	= Kilopascals (kPa)	X	0.145	= Pounds-force per square inch (psi; lbf/in^2; lb/in^2)
Kilopascals (kPa)	X	0.01	= Kilograms-force per square centimetre (kgf/cm^2; kg/cm^2)	X	98.1	= Kilopascals (kPa)
Millibar (mbar)	X	100	= Pascals (Pa)	X	0.01	= Millibar (mbar)
Millibar (mbar)	X	0.0145	= Pounds-force per square inch (psi; lbf/in^2; lb/in^2)	X	68.947	= Millibar (mbar)
Millibar (mbar)	X	0.75	= Millimetres of mercury (mmHg)	X	1.333	= Millibar (mbar)
Millibar (mbar)	X	0.401	= Inches of water (inH$_2$O)	X	2.491	= Millibar (mbar)
Millimetres of mercury (mmHg)	X	0.535	= Inches of water (inH$_2$O)	X	1.868	= Millimetres of mercury (mmHg)
Inches of water (inH$_2$O)	X	0.036	= Pounds-force per square inch (psi; lbf/in^2; lb/in^2)	X	27.68	= Inches of water (inH$_2$O)

Torque (moment of force)

Pounds-force inches (lbf in; lb in)	X	1.152	= Kilograms-force centimetre (kgf cm; kg cm)	X	0.868	= Pounds-force inches (lbf in; lb in)
Pounds-force inches (lbf in; lb in)	X	0.113	= Newton metres (Nm)	X	8.85	= Pounds-force inches (lbf in; lb in)
Pounds-force inches (lbf in; lb in)	X	0.083	= Pounds-force feet (lbf ft; lb ft)	X	12	= Pounds-force inches (lbf in; lb in)
Pounds-force feet (lbf ft; lb ft)	X	0.138	= Kilograms-force metres (kgf m; kg m)	X	7.233	= Pounds-force feet (lbf ft; lb ft)
Pounds-force feet (lbf ft; lb ft)	X	1.356	= Newton metres (Nm)	X	0.738	= Pounds-force feet (lbf ft; lb ft)
Newton metres (Nm)	X	0.102	= Kilograms-force metres (kgf m; kg m)	X	9.804	= Newton metres (Nm)

Power

Horsepower (hp)	X	745.7	= Watts (W)	X	0.0013	= Horsepower (hp)

Velocity (speed)

Miles per hour (miles/hr; mph)	X	1.609	= Kilometres per hour (km/hr; kph)	X	0.621	= Miles per hour (miles/hr; mph)

Fuel consumption

Miles per gallon, Imperial (mpg)	X	0.354	= Kilometres per litre (km/l)	X	2.825	= Miles per gallon, Imperial (mpg)
Miles per gallon, US (mpg)	X	0.425	= Kilometres per litre (km/l)	X	2.352	= Miles per gallon, US (mpg)

Temperature

Degrees Fahrenheit = (°C x 1.8) + 32   Degrees Celsius (Degrees Centigrade; °C) = (°F – 32) x 0.56

It is common practice to convert from miles per gallon (mpg) to litres/100 kilometres (l/100km), where mpg (Imperial) x l/100 km = 282 and mpg (US) x l/100 km = 235

Metric conversion tables

Inches	Decimals	Millimetres	Millimetres to Inches		Inches to Millimetres	
			mm	Inches	Inches	mm
1/64	0.015625	0.3969	0.01	0.00039	0.001	0.0254
1/32	0.03125	0.7937	0.02	0.00079	0.002	0.0508
3/64	0.046875	1.1906	0.03	0.00118	0.003	0.0762
1/16	0.0625	1.5875	0.04	0.00157	0.004	0.1016
5/64	0.078125	1.9844	0.05	0.00197	0.005	0.1270
3/32	0.09375	2.3812	0.06	0.00236	0.006	0.1524
7/64	0.109375	2.7781	0.07	0.00276	0.007	0.1778
1/8	0.125	3.1750	0.08	0.00315	0.008	0.2032
9/64	0.140625	3.5719	0.09	0.00354	0.009	0.2286
5/32	0.15625	3.9687	0.1	0.00394	0.01	0.254
11/64	0.171875	4.3656	0.2	0.00787	0.02	0.508
3/16	0.1875	4.7625	0.3	0.01181	0.03	0.762
13/64	0.203125	5.1594	0.4	0.01575	0.04	1.016
7/32	0.21875	5.5562	0.5	0.01969	0.05	1.270
15/64	0.234375	5.9531	0.6	0.02362	0.06	1.524
1/4	0.25	6.3500	0.7	0.02756	0.07	1.778
17/64	0.265625	6.7469	0.8	0.03150	0.08	2.032
9/32	0.28125	7.1437	0.9	0.03543	0.09	2.286
19/64	0.296875	7.5406	1	0.03937	0.1	2.54
5/16	0.3125	7.9375	2	0.07874	0.2	5.08
21/64	0.328125	8.3344	3	0.11811	0.3	7.62
11/32	0.34375	8.7312	4	0.15748	0.4	10.16
23/64	0.359375	9.1281	5	0.19685	0.5	12.70
3/8	0.375	9.5250	6	0.23622	0.6	15.24
25/64	0.390625	9.9219	7	0.27559	0.7	17.78
13/32	0.40625	10.3187	8	0.31496	0.8	20.32
27/64	0.421875	10.7156	9	0.35433	0.9	22.86
7/16	0.4375	11.1125	10	0.39370	1	25.4
29/64	0.453125	11.5094	11	0.43307	2	50.8
15/32	0.46875	11.9062	12	0.47244	3	76.2
31/64	0.48375	12.3031	13	0.51181	4	101.6
1/2	0.5	12.7000	14	0.55118	5	127.0
33/64	0.515625	13.0969	15	0.59055	6	152.4
17/32	0.53125	13.4937	16	0.62992	7	177.8
35/64	0.546875	13.8906	17	0.66929	8	203.2
9/16	0.5625	14.2875	18	0.70866	9	228.6
37/64	0.578125	14.6844	19	0.74803	10	254.0
19/32	0.59375	15.0812	20	0.78740	11	279.4
39/64	0.609375	15.4781	21	0.82677	12	304.8
5/8	0.625	15.8750	22	0.86614	13	330.2
41/64	0.640625	16.2719	23	0.90551	14	355.6
21/32	0.65625	16.6687	24	0.94488	15	381.0
43/64	0.671875	17.0656	25	0.98425	16	406.4
11/16	0.6875	17.4625	26	1.02362	17	431.8
45/64	0.703125	17.8594	27	1.06299	18	457.2
23/32	0.71875	18.2562	28.	1.10236	19	482.6
47/64	0.734375	18.6531	29	1.14173	20	508.0
3/4	0.75	19.0500	30	1.18110	21	533.4
49/64	0.765625	19.4469	31	1.22047	22	558.8
25/32	0.78125	19.8437	32	1.25984	23	584.2
51/64	0.796875	20.2406	33	1.29921	24	609.6
13/16	0.8125	20.6375	34	1.33858	25	635.0
53/64	0.828125	21.0344	35	1.37795	26	660.4
27/32	0.84375	21.4312	36	1.41732	27	685.8
55/64	0.859375	21.8281	37	1.4567	28	711.2
7/8	0.875	22.2250	38	1.4961	29	736.6
57/64	0.890625	22.6219	39	1.5354	30	762.0
29/32	0.90625	23.0187	40	1.5748	31	787.4
59/64	0.921875	23.4156	41	1.6142	32	812.8
15/16	0.9375	23.8125	42	1.6535	33	838.2
61/64	0.953125	24.2094	43	1.6929	34	863.6
31/32	0.96875	24.6062	44	1.7323	35	889.0
63/64	0.984375	25.0031	45	1.7717	36	914.4

Index

Haynes Motorcycle Manuals – The Complete List

Title	Book No.
BMW	
BMW 2-valve Twins (70 - 96)	0249
BMW K100 & 75 2-valve Models (83 - 96)	1373
BMW R850 & R1100 4-valve Twins (93 - 97)	3466
BSA	
BSA Bantam (48 - 71)	0117
BSA Unit Singles (58 - 72)	0127
BSA Pre-unit Singles (54 - 61)	0326
BSA A7 & A10 Twins (47 - 62)	0121
BSA A50 & A65 Twins (62 - 73)	0155
DUCATI	
Ducati 600, 750 & 900 2-valve V-Twins (91 - 96)	3290
HARLEY-DAVIDSON	
Harley-Davidson Sportsters (70 - 99)	0702
Harley-Davidson Big Twins (70 - 99)	0703
HONDA	
Honda NB, ND, NP & NS50 Melody (81 - 85) ◊	0622
Honda NE/NB50 Vision & SA50 Vision Met-in (85 - 95) ◊	1278
Honda MB, MBX, MT & MTX50 (80 - 93)	0731
Honda C50, C70 & C90 (67 - 99)	0324
Honda CR80R & CR125R (86 - 97)	2220
Honda XR80R & XR100R (85 - 96)	2218
Honda XL/XR 80, 100, 125, 185 & 200 2-valve Models (78 - 87)	0566
Honda CB100N & CB125N (78 - 86) ◊	0569
Honda H100 & H100S Singles (80 - 92) ◊	0734
Honda CB/CD125T & CM125C Twins (77 - 88) ◊	0571
Honda CG125 (76 - 99) ◊	0433
Honda NS125 (86 - 93) ◊	3056
Honda MBX/MTX125 & MTX200 (83 - 93) ◊	1132
Honda CD/CM185 200T & CM250C 2-valve Twins (77 - 85)	0572
Honda XL/XR 250 & 500 (78 - 84)	0567
Honda XR250L, XR250R & XR400R (86 - 97)	2219
Honda CB250 & CB400N Super Dreams (78 - 84) ◊	0540
Honda CR250R & CR500R (86 - 97)	2222
Honda Elsinore 250 (73 - 75)	0217
Honda CBR400RR Fours (88 - 99)	3552
Honda VFR400 (NC30) & RVF400 (NC35) V-Fours (89 - 98)	3496
Honda CB400 & CB550 Fours (73 - 77)	0262
Honda CX/GL500 & 650 V-Twins (78 - 86)	0442
Honda CBX550 Four (82 - 86) ◊	0940
Honda XL600R & XR600R (83 - 96)	2183
Honda CBR600F1 & 1000F Fours (87 - 96)	1730
Honda CBR600F2 & F3 Fours (91 - 98)	2070
Honda CB650 sohc Fours (78 - 84)	0665
Honda NTV600 & 650 V-Twins (88 - 96)	3243
Honda Shadow VT600 & 750 (USA) (88 - 99)	2312
Honda CB750 sohc Four (69 - 79)	0131
Honda V45/65 Sabre & Magna (82 - 88)	0820
Honda VFR750 & 700 V-Fours (86 - 97)	2101
Honda VFR800 V-Fours (97 - 00)	3703
Honda CB750 & CB900 dohc Fours (78 - 84)	0535
Honda CBR900RR FireBlade (92 - 99)	2161
Honda ST1100 Pan European V-Fours (90 - 97)	3384
Honda Shadow VT1100 (USA) (85 - 98)	2313
Honda GL1000 Gold Wing (75 - 79)	0309
Honda GL1100 Gold Wing (79 - 81)	0669
Honda Gold Wing 1200 (USA) (84 - 87)	2199
Honda Gold Wing 1500 (USA) (88 - 98)	2225
KAWASAKI	
Kawasaki AE/AR 50 & 80 (81 - 95)	1007

Title	Book No.
Kawasaki KC, KE & KH100 (75 - 99)	1371
Kawasaki KMX125 & 200 (86 - 96) ◊	3046
Kawasaki 250, 350 & 400 Triples (72 - 79)	0134
Kawasaki 400 & 440 Twins (74 - 81)	0281
Kawasaki 400, 500 & 550 Fours (79 - 91)	0910
Kawasaki EN450 & 500 Twins (Ltd/Vulcan) (85 - 93)	2053
Kawasaki EX & ER500 (GPZ500S & ER-5) Twins (87 - 99)	2052
Kawasaki ZX600 (Ninja ZX-6, ZZ-R600) Fours (90 - 97)	2146
Kawasaki ZX-6R Ninja Fours (95 - 98)	3541
Kawasaki ZX600 (GPZ600R, GPX600R, Ninja 600R & RX) & ZX750 (GPX750R, Ninja 750R) Fours (85 - 97)	1780
Kawasaki 650 Four (76 - 78)	0373
Kawasaki 750 Air-cooled Fours (80 - 91)	0574
Kawasaki ZR550 & 750 Zephyr Fours (90 - 97)	3382
Kawasaki ZX750 (Ninja ZX-7 & ZXR750) Fours (89 - 96)	2054
Kawasaki 900 & 1000 Fours (73 - 77)	0222
Kawasaki ZX900, 1000 & 1100 Liquid-cooled Fours (83 - 97)	1681
MOTO GUZZI	
Moto Guzzi 750, 850 & 1000 V-Twins (74 - 78)	0339
MZ	
MZ ETZ Models (81 - 95) ◊	1680
NORTON	
Norton 500, 600, 650 & 750 Twins (57 - 70)	0187
Norton Commando (68 - 77)	0125
PIAGGIO	
Piaggio (Vespa) Scooters (91 - 98)	3492
SUZUKI	
Suzuki GT, ZR & TS50 (77 - 90) ◊	0799
Suzuki TS50X (83 - 99) ◊	1599
Suzuki 100, 125, 185 & 250 Air-cooled Trail bikes (79 - 89)	0797
Suzuki GP100 & 125 Singles (78 - 93) ◊	0576
Suzuki GS, GN, GZ & DR125 Singles (82 - 99) ◊	0888
Suzuki 250 & 350 Twins (68 - 78)	0120
Suzuki GT250X7, GT200X5 & SB200 Twins (78 - 83) ◊	0469
Suzuki GS/GSX250, 400 & 450 Twins (79 - 85)	0736
Suzuki GS500E Twin (89 - 97)	3238
Suzuki GS550 (77 - 82) & GS750 Fours (76 - 79)	0363
Suzuki GS/GSX550 4-valve Fours (83 - 88)	1133
Suzuki GSX-R600 & 750 (96 - 99)	3553
Suzuki GSF600 & 1200 Bandit Fours (95 - 97)	3367
Suzuki GS850 Fours (78 - 88)	0536
Suzuki GS1000 Four (77 - 79)	0484
Suzuki GSX-R750, GSX-R1100 (85 - 92), GSX600F, GSX750F, GSX1100F (Katana) Fours (88 - 96)	2055
Suzuki GS/GSX1000, 1100 & 1150 4-valve Fours (79 - 88)	0737
TRIUMPH	
Triumph Tiger Cub & Terrier (52 - 68)	0414
Triumph 350 & 500 Unit Twins (58 - 73)	0137
Triumph Pre-Unit Twins (47 - 62)	0251
Triumph 650 & 750 2-valve Unit Twins (63 - 83)	0122
Triumph Trident & BSA Rocket 3 (69 - 75)	0136
Triumph Triples & Fours (carburettor engines) (91 - 99)	2162
VESPA	
Vespa P/PX125, 150 & 200 Scooters (78 - 95)	0707
Vespa Scooters (59 - 78)	0126
YAMAHA	
Yamaha DT50 & 80 Trail Bikes (78 - 95) ◊	0800
Yamaha T50 & 80 Townmate (83 - 95) ◊	1247
Yamaha YB100 Singles (73 - 91) ◊	0474

Title	Book No.
Yamaha RS/RXS100 & 125 Singles (74 - 95)	0331
Yamaha RD & DT125LC (82 - 87) ◊	0887
Yamaha TZR125 (87 - 93) & DT125R (88 - 95) ◊	1655
Yamaha TY50, 80, 125 & 175 (74 - 84) ◊	0464
Yamaha XT & SR125 (82 - 96)	1021
Yamaha 250 & 350 Twins (70 - 79)	0040
Yamaha XS250, 360 & 400 sohc Twins (75 - 84)	0378
Yamaha RD250 & 350LC Twins (80 - 82)	0803
Yamaha RD350 YPVS Twins (83 - 95)	1158
Yamaha RD400 Twin (75 - 79)	0333
Yamaha XT, TT & SR500 Singles (75 - 83)	0342
Yamaha XZ550 Vision V-Twins (82 - 85)	0821
Yamaha FJ, FZ, XJ & YX600 Radian (84 - 92)	2100
Yamaha XJ600S (Diversion, Seca II) & XJ600N Fours (92 - 99)	2145
Yamaha YZF600R Thundercat & FZS600 Fazer (96 - 99)	3702
Yamaha 650 Twins (70 - 83)	0341
Yamaha XJ650 & 750 Fours (80 - 84)	0738
Yamaha XS750 & 850 Triples (76 - 85)	0340
Yamaha TDM850, TRX850 & XTZ750 (89 - 99)	3540
Yamaha FZR600, 750 & 1000 Fours (87 - 96)	2056
Yamaha XV V-Twins (81 - 96)	0802
Yamaha XJ900F Fours (83 - 94)	3239
Yamaha FJ1100 & 1200 Fours (84 - 96)	2057
ATVS	
Honda ATC70, 90, 110, 185 & 200 (71 - 85)	0565
Honda TRX300 Shaft Drive ATVs (88 - 95)	2125
Honda TRX300EX & TRX400EX ATVs (93 - 99)	2318
Polaris ATVs (85 to 97)	2302
Yamaha YT, YFM, YTM & YTZ ATVs (80 - 85)	1154
Yamaha YFS200 Blaster ATV (88 - 98)	2317
Yamaha YFB250 Timberwolf ATV (92 - 96)	2217
Yamaha YFM350 Big Bear and ER ATVs (87 - 95)	2126
Yamaha Warrior and Banshee ATVs (87 - 99)	2314
ATV Basics	10450
TECHNICAL TITLES	
Motorcycle Basics Manual	1083
MOTORCYCLE TECHBOOKS	
Motorcycle Electrical TechBook (3rd Edition)	3471
Motorcycle Fuel Systems TechBook	3514
Motorcycle Workshop Practice TechBook (2nd Edition)	3470

◊ = not available in the USA   **Bold type** = *Superbike*

The manuals on this page are available through good motorcycle dealers and accessory shops.
In case of difficulty, contact: **Haynes Publishing**
(UK) +44 1963 440635   (USA) +1 805 4986703
(FR) +33 1 47 78 50 50   (SV) +46 18 124016
(Australia/New Zealand) +61 3 9763 8100

MCL08.09/99